Routledge Revivals

Fluvial Geomorphology

It was not until the 20th century that many geomorphologists began to integrate the work of hydraulic engineers and hydrologists with geomorphic concepts in fluvial systems. Statistical manipulations and concepts aided greatly in the development of ideas on the relation of the magnitude and frequency of hydrologic events on the fluvial system. Also arising primarily as a result of the statistical outlook was the approach where characteristics of the river and its basin were looked upon as random variables in a population. Different approaches, whether old or new, are tools used to aid in better understanding the morphology of the earth's surface features and the processes by which a river works to create and modify the form of the land. The papers in this volume, originally published in 1973, reflect these approaches but more than that they unify and integrate the results of research with basic concepts and principles of fluvial morphology and process.

Fluvial Geomorphology

A Proceedings Volume of the Fourth Annual Geomorphology Symposia Series

Edited by Marie Morisawa

Routledge
Taylor & Francis Group

First published in Great Britain in 1981 by George Allen & Unwin Ltd

This edition first published in 2026 by Routledge
4 Park Square, Milton Park, Abingdon, Oxon, OX14 4RN

and by Routledge
605 Third Avenue, New York, NY 10158

Routledge is an imprint of the Taylor & Francis Group, an informa business

ISBN: 978-1-041-33372-2 (hbk)
ISBN: 978-1-003-78748-8 (ebk)
ISBN: 978-1-041-33375-3 (pbk)

Book DOI 10.4324/9781003787488

FLUVIAL GEOMORPHOLOGY

Edited by: MARIE MORISAWA
State University of New York at Binghamton

A Proceedings Volume of the Fourth Annual Geomorphology Symposia Series held at Binghamton, New York, September 27-28, 1973

London
GEORGE ALLEN & UNWIN
Boston Sydney

This volume is dedicated to Brandy — devoted friend, companion, and assistant for his lifetime of sixteen years.

First published in 1973
First published by George Allen & Unwin in 1981

GEORGE ALLEN & UNWIN LTD.
40 Museum Street, London, WC1A 1LU

CONTENTS

CONTRIBUTORS

Victor R. Baker
Department of Geological Sciences, University of Texas at Austin, Austin, Texas 78712

J.L. Bassett
Department of Geology and Water Resources Research Center, Indiana University, Bloomington, Indiana 47401

W.C. Bradley
Department of Geological Sciences, University of Colorado, Boulder, Colorado 80302

James Brice
Department of Earth Sciences, Washington University, St. Louis, Missouri 63130

James Carl
Department of Geological Sciences, State University of New York at Potsdam, Potsdam, New York, 13676

G.H. Dury
Departments of Geography and Geology, The University of Wisconsin-Madison, Madison, Wisconsin, 53706

R.K. Fahnestock
Department of Geology, State University of New York at Fredonia, Fredonia, New York 14063

John T. Hack
US Geological Survey, Washington, DC 20244

Iaakov Karcz
Department of Geological Sciences, State University of New York, Binghamton, New York, 13901

E.A. Keller
Department of Geological Sciences, The University of California at Santa Barbara, Santa Barbara, California, 93106

W.N. Melhorn
Department of Geosciences, Purdue University, West Lafayette, Indiana 47906

Marie Morisawa
Department of Geological Sciences, State University of New York, Binghamton, New York 13901

Irwin D. Novak
Department of Physical Sciences and Engineering, University of Maine at Portland-Gorham, Portland, Maine 04103

R.V. Ruhe
Department of Geology and Water Resources Research Center, Indiana University, Bloomington, Indiana 47401

S.A. Schumm
Department of Earth Resources, Colorado State University, Fort Collins, Colorado 80523

J.S. Smart
IBM Thomas J. Watson Research Center, Yorktown Heights, New York 10598

D.G. Smith
Department of Geography, University of Calgary, Calgary, Alberta T2N 1N4

John B. Stall
Illinois State Water Survey, P.O. Box 232, Urbana, Illinois 61801

James Street
Department of Geology and Geography, St. Lawrence University, Canton, New York 13617

Chih Ted Yang
Illinois State Water Survey, P.O. Box 232, Urbana, Illinois 61801

PREFACE

These are the proceedings of the Fourth Annual Geomorphology Symposium held in the Department of Geological Sciences at State University of New York at Binghamton, New York on September 27-28, 1973. Its theme was Fluvial Geomorphology and it was the second of the symposia to emphasize a geomorphic process. Previous symposia topics and dates are as follows:

First - 1970, Environmental Geomorphology

Second - 1971, Quantitative Geomorphology

Third - 1972, Coastal Geomorphology

Each symposium has been well attended by participants from industry, government and academia, including professors and students. It has been our pleasure to act as hosts and supply the opportunity for those interested in geomorphology to get together and exchange ideas. Our authors, too, come from a variety of occupations but all are experts in the theme. This volume presents their papers in written form. Part I contains those presentations primarily quantitative in outlook or which use a quantitative approach to a fluvial problem. Part II has three papers that examine the flow in a stream, not simply as hydrology, but in relation to the effects of discharge. Part III includes the papers concerned with channel pattern - meandering and braiding. Part IV draws together papers that are more philosophical dealing with basic concepts in tone, and principles of fluvial process and form.

The appendix includes abstracts of three papers on education, i.e. getting undergraduate students involved in studying fluvial systems. These people were not asked to write up their presentations as they were not concerned with geomorphologic principles per se.

The success of a conference depends on many people and I would like especially to thank those who took part by leading the sessions: Bryce Hand of Syracuse University, Nicolay Timofeeff of State University of New York at Binghamton, and James Bugh of State University of New York -College at Cortland, Cortland, New York. My deep appreciation also to the geomorphology students at Binghamton for all their help in front and behind the scenes and to my colleague Don Coates.

Our next symposium in the fall of 1974 will be on Glacial Geomorphology.

Marie Morisawa
Editor

INTRODUCTION

There is no need to justify the theme of this year's symposium. Both in terms of humanistic needs and scientific inquiry, rivers are important subjects for investigation and communication. The knowledge we have about rivers, their morphology, and how they function is important to mankind.

It is difficult to determine exactly when man began to recognize and acknowledge the efficacy of running water in the denudation of the earth. Certainly Avicenna, the great scientist of the Arab world of the early part of the eleventh century, realized the role of running water in carving out mountains (Adams,1938). In Europe during the "Dark Ages", the Italian genius Leonardo da Vinci was fully aware of the effects of rivers on the landscape as they wore away the hillsides and deposited the material in the valleys. He set down clearly his observations on the building up of deltaic deposits by rivers.

Despite the widely held belief that the surface of the earth was shaped by the Mosaic flood and similar catastrophes, Agricola, in the sixteenth century, explained how the "little brooks" wash away the soil and cut through solid rock (Adams, 1938). He understood that rivers cut deep valleys, lowering and broadening them through time to form fertile floodplains. He saw that rivers not only erode material but carry it away and later deposit it.

Guettard and Desmarest are among the few observers of the eighteenth century who demonstrated that the catastrophists and diluvialists of their day were wrong, simply by describing the work of rivers as they observed them. It was Hutton, in his paper delivered in 1795, who most clearly enunciated the principle that streams were most effective agents in denuding the earth. He also pointed out that as rivers in a watershed increase in size downstream, they decrease in number, thus, in a way forecasting later quantitative laws (Chorley, et al, 1964).

Playfair, in his attempts to explain and prove the ideas of Hutton, indicated that he was extremely knowledgable in the ways of rivers. He saw that they were capable of decreasing the size of material by attrition and that this same process was responsible for the wearing away of channels. He understood the delicacy of the equilibrium in a stream channel and appreciated the way a river adjusted its gradient to its velocity and discharge. He realized also the influence of material already in transport on the ability of a river to carry even more load.

As Chorley, et al (1964) point out, the fluvialists were eventually vindicated by the wealth of observations made while new parts of the world were opened up and explored - Africa, India and the Americas, for example. In the wide-open American west geologists were observing what rivers were doing, and in a region where the geology is most obvious. Not having the biases and preconceived notions of their European counterparts, these scientific explorers were more free to put down in writing what they actually saw - as they saw it. So men like Newberry, Powell, Gilbert and others established without doubt that rivers cut their own valleys and wear away the earth. In so doing they laid down many of the precepts of river denudation which are still viable today and on which whole chronicles of landforms have been built.

Perhaps the most notable of this group of American geomorphologists was Gilbert, whose paper on the transport of material by rivers (1914) is a classic. In trying to solve an environmental problem of aggradation, he set forth the relations of river capacity and competence to characteristics of the channel, the discharge, and the material carried. His observations and concepts were derived from both flume and field studies. Gilbert also realized that rivers were self-adjusting in an effort to maintain an equilibrium. He used the term "grade" as the ideal condition of a stream in balance, where it neither erodes nor deposits material. He spoke in terms of an equilibrium of the energy which gives the ability to transport, and of the load available. He also noted that the smooth profile of a river, which denotes equilibrium, is attained regardless of rock resistance, although he was aware that resistance and gradient were closely related. Moreover, he stated that a river system was an integrated whole, with one part affecting another, Gilbert (1877). Indeed, it seems to me that we are greatly indebted to Gilbert for most of our concepts about fluvial geomorphology in some way or another.

Davis took the work of Gilbert and many others and from them developed a comprehensive and integrated system of landscape evolution by rivers, called the fluvial cycle (Davis, 1899). This Davisian system dominated the thinking of fluvial geomorphologists for many decades and is still very influential today. He established the concept of grade deeply into the consciousness of American geomorphologists, stating it as a balance of erosion and deposition reached by rivers in late maturity or old age. Most emphasis was placed on a smooth, concave, longitudinal profile as an indication of the graded condition. This concept of a graded river in a delicate equilibrium condition was most clearly defined by Mackin (1948).

During this same time advances in understanding the mechanics and hydraulics of rivers were being made by engineers. Surell, Beardmore and others had concerned themselves with problems of river erosion and deposition and understood that rivers controlled their own development by adjustment of interrelated variables such as velocity, slope and discharge. Most of this information had been derived because of attempts to solve problems associated with the fact that man is so dependent upon rivers in many ways. Because of this dependency he had to learn how rivers function in order to cope with them. He had to find out how to prevent both natural and man-made channels from siltation which interferred with both irrigation and navigation. He had to know how to protect his land from excessive erosion and from flood waters which overflowed the river banks. So many of the nineteenth and twentieth century advances in hydraulics, hydrology and fluid mechanics were reached because man came into direct conflict with natural events which in some cases were man-accelerated. Many of these engineering studies supplied empirical relationships among stream parameters, which supplied a basis for latter-day investigators in fluvial systems.

However, despite the quantitative analyses done by engineers and by geologists such as Gilbert, most fluvial geomorphologists tended to devote themselves to historical and descriptive studies of rivers. The many papers of Davis, Johnson, Bryan and their contemporaries attempted to trace the development of drainage basins and describe their effects on the landscape. It was not until the middle of the twentieth century that many geomorphologists began to integrate the work of hydraulic engineers and hydrologists with geomorphic concepts in fluvial systems. The extraordinary papers by

Hjulstrom (1935) and Sundborg (1956) are examples of this. But it was really the work by Horton (1945) that gave an impetus to the application of mathematical and statistical analysis to drainage systems.

As explained by Salisbury (1972) the quantitative emphasis in fluvial geomorphology was less concerned with description than with process. In fact, although Horton's paper was quantitative, it was also descriptive. It did, however, give a basis for methodology by which quantitative approaches could be made to river basins. He among other things, proposed a set of empirically derived equations which could describe relations among the topologic and morphologic characteristics of river systems. Horton's laws were further elucidated and related to physical principles by Strahler and his students (see Melton 1958, Morisawa 1964, and Woldenberg 1966). Investigators from the U.S. Geological Survey dealt mostly with the statistical relationships among characteristics of channel geometry and discharge, which quantified previous concepts of equilibrium between stream flow and river morphology, (for example, see Leopold and Maddock 1953, Leopold and Miller 1956, Langbein, et al. 1947, and Hack 1957). In other words, quantification was a way by which exactness was given to previously enunciated but vaguely defined principles.

The emphasis in many recent studies, starting with Horton, has been to treat a river and its tributaries as a system (Chorley 1962). This of course is not new but refers back to Hutton and Playfair who conceived the branches of a river as a "body" whose parts were intimately inter-related and interdependent on each other. This treatment has resulted in a concept of dynamic equilibrium of energy balance and adjustment in a stream which is time independent. This is distinct from the Davisian idea of grade as historical. This topic of energy equilibrium will be discussed further in Part IV.

Statistical manipulations and concepts aided greatly in the development of ideas on the relation of the magnitude and frequency of hydrologic events on the fluvial system (Brush, 1961 and Dury, 1962). Also arising primarily as a result of the statistical outlook was the approach where characteristics of the river and its basin were looked upon as random variables in a population. This has led to simulation studies of stream systems where both random and non-random models have been used (Howard, 1972). It must be kept in mind that all these approaches, whether new or old, are tools used to aid us to better understand the morphology of the earth's surface features and the processes by which a river works to create and modify the form of the land.

The papers of this symposium reflect the trends that have been mentioned, but more than that they show an effort to unify and integrate the results of recent research with basic concepts and principles of fluvial morphology and process.

REFERENCES CITED

Adams, F. D., 1938, The birth and development of the geological sciences: Dover Publ. Inc., New York, N.Y., 506 pp.

Brush, L. M., 1961, Drainage basins, channels, and flow characteristics of selected streams in central Pennsylvania: U.S.G.S. Prof. Paper 282-F, 180 pp.

Chorley, R.J., 1962, Geometry and general systems theory: U.S.G.S. Prof. Paper 500-B, 10 pp.

Chorley, R. J., Dunn, A. J., and Beckinsale, R. P., 1964, The history of the study of landforms, Vol I geomorphology before Davis: Methuen and Co., Ltd., London, England, 678 pp.

Davis, W. M., 1899, The geographical cycle: Geogr. Jour., XIV, pp. 481-504.

Dury, G. H., 1962, Bankful discharge: an example of its statistical relationships: Intern. Assoc. Sci. Hydrol., Bull. 6, pp. 48-55.

Gilbert, G. K., 1877, Report on the geology of the Henry Mountains: U. S. Geol. and Goeg. Survey of the Rocky Mountain Region, Washington, D. C., 170 pp.

Hack, J. T., 1957, Studies of longitudinal stream profiles in Virginia and Maryland: U. S. G. S. Prof. Paper 294-B, 97 pp.

Hjulstrom, F., 1935, Studies of the morphological activity of rivers as illustrated by the River Fyris: Univ. of Upsala Geol. Inst. Bull. 25, pp. 221-527.

Horton, R. E., 1945, Erosional development of streams and their drainage basins: hydrophysical approach to quantitative morphometry: Geol. Soc. Amer., Bull. v. 56, pp. 275-370.

Howard, A. D., 1972, Problems in interpretation of simulation models of geologic processes: in Quantitative Geomorphology::some aspects and application, edited by M. Morisawa, Publ. in Geomorphology, Binghamton, N. Y., pp. 6-182.

Langbein, W. B., et al., 1947, Topographic characteristics of drainage basins: U. S. G. S. Water Supply Paper 986-C, pp. 125-155.

Leopold, L. B., and Maddock, T., Jr., 1953, The hydraulic geometry of stream channels and some physiographic implications: U. S. G. S. Prof. Paper 252, 56 pp.

Leopold, L. B. and Miller, J. P., 1956, Ephemeral streams - hydraulic factors and their relation to the drainage net: U. S. G. S. Prof. Paper 282-A, 36 pp.

Mackin, J. H., 1948, Concept of the graded river: Geol. Soc. Amer. Bull., v. 59, pp. 463-512.

Melton, M.A., 1958, Correlation of morphimetric properties drainage systems and their controlling agents: Jour. Geol., v. 66, pp. 442-460.

Morisawa, M., 1964, Development of drainage systems on an upraised lake floor: Amer. Jour. Sci., v. 262, pp. 340-354.

Salisbury, N., 1972, Threads of inquiry on Quantitative Geormorphology: in Quantitative Geormorphology, some aspects and applications, M. Morisawa, editor, Public. in Geormorph., Binghamton, N. Y., pp. 9-60.

Sundborg, A., 1956, The River Klaralven - a study of fluvial processes: Geogr. Ann., v. 38, p. 127-316.

Woldenberg, N.J., 1966, Horton's laws justified in terms of allometric growth and steady state systems: GSA Bull. v. 77, p. 431-434.

Marie Morisawa
Department Geological Sciences
State University of New York
Binghamton, New York 13901

PART I

The first group of papers are concerned with quantitative aspects of drainage basins and inferences that can be drawn from such factors. They demonstrate a variety of ways in which quantification can be used in studies of fluvial systems.

The first paper by Novak is an examination and clarification of the well-known curve of erosion and deposition made by Hjulstrom who investigated the erosion velocity required to move sand grains. From his experiments Hjulstrom drew a curve representing velocity versus particle size. Novak is interested in determining the validity of extending the Hjulstrom curve by extrapolation into the coarser grain sizes. Experimental flume data as well as measurements from a variety of field situations such as catastrophic Pleistocene as well as recent floods, proglacial streams and engineering works are used to define the trend of the erosion velosity for larger particles. The curve so described indicates a difference between the velocity actually required for sediment transport and that predicted by extension of the Hjulstrom curve. A number of reasons can be advanced explaining this difference. This paper exemplifies one type of investigation where the observer is interested in stream competence and movement of the bed-load.

Since he propounded them in 1945, Horton's laws of drainage composition have been further developed, discussed and tested by many investigators. In the second paper Smart deals with the concept of randomness in basin topology and tests Horton's laws against a drainage net developed by a random growth simulation. He discusses the refinement and clarification of ideas on drainage development that has resulted from the pioneering paper of Horton and contrasts the idea of orderly growth implied by Horton's laws with the idea of randomness. He reviews the Horton model and its descriptive and predictive properties. By comparison, the random model is such that geomorphic properties are assumed to be sets of random variables from a population. Specifically Smart finds that the random model predicts actual numbers of streams and stream lengths in the basins chosen for study better than does the Horton model. In addition, the random model seems to explain many other quantitative relationships in the drainage basin. However, this in no way negates the concept of orderly growth in a watershed, since during such orderly growth the distribution of fluvial characteristics in a dendritic basin should be normal.

In the final paper of this section, Hack attempts to use a quantitative measure to determine the degree of adjustment of rivers in the Appalachian region and thus, to shed some light on the geomorphic history of the area. Over a period of time the gradient of a river should depend on the discharge and rock type over which it flows. Gilbert expressed this as an equilibrium reached when the ratio of the erosive action is equal to the ratio of rock resistance (Gilbert, 1877). Ordinarily, a river would have a steeper gradient as it moves over resistant rocks and a gentler gradient over non-resistant material. Lack of this kind of relation indicates disequilibrium. So Hack examines the rivers using a gradient index, the SL ratio, which is the slope of a given reach times the total length of river from its head through the reach. His study is a fine example of the use of morphometric tools to determine the geomorphic history of the rivers and, thus, of a region.

1

PREDICTING COARSE SEDIMENT TRANSPORT: THE HJULSTROM CURVE REVISITED

Irwin D. Novak

ABSTRACT

Many geologists utilize the Hjulstrom curve (or modifications of this curve) to predict erosion, transportation, and deposition in terms of velocity and particle size. The curve offers a simple, graphical representation of the possible behavior of a sedimentary particle as the velocity regime changes. An empirical curve derived and extrapolated from data in the sand to fine gravel size range, the Hjulstrom curve has been misused because it has not been fully verified for all size ranges under natural stream conditions. Field and laboratory observations led to modification of the curve in the clay and silt size range to allow for cohesion between grains and the attendant variation in erosion velocities. The Hjulstrom curve has also provided the geologist with a theoretical framework from which the velocities necessary to entrain coarse sediment could be estimated.

Field measurements and estimates of mean water velocities range from 1.0 to 26.0 feet per second for particles 0.09 to 20.0 feet in diameter in transport studies of streams, Pleistocene and recent floods, engineering works, dam failures, and flumes. Analysis of these data shows that a curve of mean velocity for overturning and a curve of bottom velocity for sliding defines a zone that predicts coarse sediment transport better than the Hjulstrom curve or its subsequent modifications. In general, the field measurements and estimates of mean velocity versus particle size agree closely with engineering curves for rock stability.

Revision of the Hjulstrom-type curves in the gravel to boulder size range is strongly recommended for application to the fluvial environment.

INTRODUCTION

Research in fluvial hydraulics has been valuable in increasing the geologist's understanding of the forces acting upon sedimentary particles. Many mathematical equations and theoretical formulas have been utilized to predict bedload transport (Scheidegger, 1970; Einstein, 1950). While this approach has provided reasonably good agreement with laboratory experiments, the application of theory to field interpretation and natural streams has not proved to be as successful. Alternatively, many geologists employ the Hjulstrom (1935) curve or modifications of the Hjulstrom curve in order to predict entrainment, transportation, and deposition in terms of velocity and particle size. The curves offer a simple, graphic representation of the possible behavior of a sedimentary particle as the velocity regime changes.

The Hjulstrom curve (Figure 1) is an empirical curve based on earlier published experiments. Velocity versus particle size data derived for the sand to fine gravel size range were combined with a consideration of the influence of cohesion on clay and silt and the importance of mass in the cobble to boulder size range. In addition, Hjulstrom assumed average velocities, in

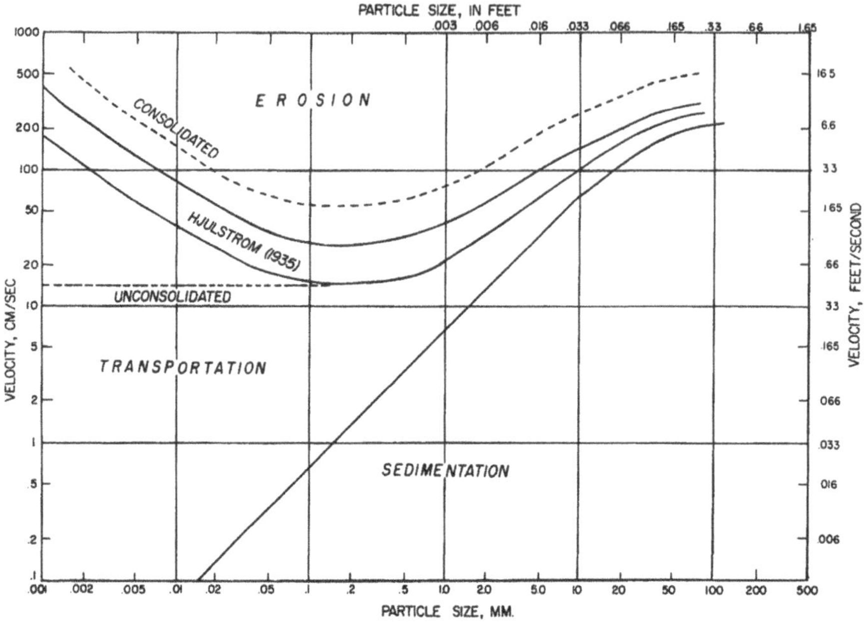

Figure 1 Hjulstrom's (1935) curve (solid lines) including modification by Sundborg (1956) (dashed lines).

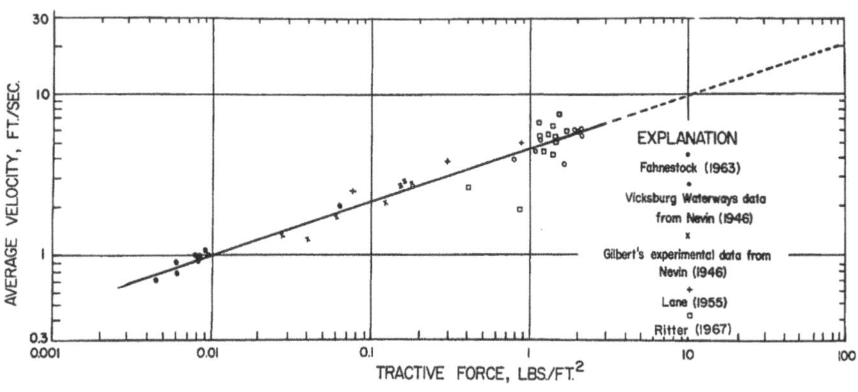

Figure 2 Relationship between tractive force and average velocity.

water at least one meter deep, acting upon grains initially at rest on a bed of equal size grains. The resultant curve combines observational data with an extrapolation into the fine and coarse ends of the sediment spectrum.

The validity of the Hjulstrom curve has not been fully verified under natural stream conditions or by laboratory studies for the clay, silt, fine gravel, cobble, and boulder size ranges. Sundborg (1956) introduced modifications to allow for the cohesion of clay, silt, and sand, and also to allow for a channel slope that may not be small (Scheidegger, 1970; see dashed line Figure 1 this paper).

In order to realize the benefits of a Hjulstrom-type curve, data must be assembled from natural streams or laboratory models in which non-uniformly sorted sediment is considered in addition to more precise velocity measurements. Completion of a fully utilitarian curve might best be accomplished by investigating individual size ranges within the sediment spectrum. The following investigation of coarse sediment transport is offered as a step in this direction.

VELOCITY MEASUREMENTS IN STREAMS AND ENGINEERING WORKS

Hjulstrom, in deriving his curve, used "the average velocity across a transverse profile of a river...in water at least one meter deep" as a variable affecting sediment transport. He added, prophetically, that "the choice of average velocity as an independent variable must be regarded as a temporary substitute until more data are available" (Hjulstrom, 1939, p. 9). Since the first publication of Hjulstrom's curve (1935) considerable confusion concerning transport *versus* particle size has arisen in the literature because authors have related particle size variously to tractive force, velocity, critical erosion velocity, mean velocity, bed velocity, and other similar properties.

Nevin (1946), in a widely quoted paper, discussed or mentioned critical tractive force, critical-traction velocity, critical bed velocity, and effective bed velocity in an effort to clarify the terms and make them more meaningful to geologists. To be sure, all of the terms discussed by Nevin (1946) are related to each other because they are all properties of a moving current of water measured or calculated at different levels in the flow. For example, the critical-tractive (or just tractive) force is a measure of the frictional drag of a current upon the stream bed; units are in force per unit area.

The critical traction velocity of Nevin (1946, p. 665) was the "average current velocity taken near, but not directly on, the stream bed"; units are in length per unit time. The critical-traction velocity would be greater than the actual bed velocity. In order to rectify the differences caused by measuring critical-traction velocity somewhat above the bed, Nevin (1946) suggested "penalizing" or reducing the critical-traction velocity. He therefore, adopted Rubey's (1938) method of estimating the bed velocity, i.e. the velocity that is most nearly the equivalent of the actual velocity near the bed. The critical bed velocity, which Nevin (1946, p. 669) used inter-changeably with effective bed velocity, is synonymous with Rubey's (1938) "bed" velocity. The quotation marks were used by Rubey to signify that the velocity was not in reality measured on the bottom, but an estimate of the average velocity at the boundary between the laminar and turbulent zones near the bottom (Rubey, 1938, p. 129). The mathematical relationships between the velocity measures described above are discussed in Inman (1949).

Recent geologic literature utilizes velocity, while

TABLE I

PARTICLE SIZE, AVERAGE VELOCITY, TRACTIVE FORCE FROM THE FLUVIAL ENVIRONMENT

LARGEST PARTICLE MOVED, FT.	AVERAGE VELOCITY IN FT./SEC.	TRACTIVE FORCE FT./LB.
HELLEY (1969, TABLE I)		
1.30	7.90	
1.10	6.78	
1.00	7.78	
.85	5.32	
1.45	8.00	
1.25	8.74	
.90	5.53	
.90	5.45	
.65	4.83	
1.10	6.45	
1.35	7.62	
1.00	9.63	
.96	7.40	
1.50	8.09	
1.10	7.95	
.90	5.84	
1.70	8.44	
.95	6.38	
1.40	9.30	
1.10	7.16	
1.00	5.80	
.80	6.17	
1.05	7.61	
.75	5.92	
.75	6.22	
.70	5.98	
.85	6.60	
.50	3.86	
.55	4.65	
.95	6.20	
.60	4.61	
.70	5.25	
.75	5.07	
.70	5.61	
BIRKELAND (1968, FIGURE 2)		
3.5		2.0
5.0		5.0
10.0		5.0
25.0		5.0
5.5		27.0
10.0		27.0
20.0		27.0
WOLMAN & EILER (1958, TABLE I)		
5.0	9	
2.7	11	
2.5	7	
1.0	7	
SCOTT & GRAVLEE (1968, TABLE 4)		
10.80		101.0
7.80		115.7
4.74		71.6
6.00		64.5
1.63		13.7
2.95		11.9
1.57		7.3
1.50		6.8
2.05		5.5
MALDE (1968, TABLE 6)		
10	24	
12	26	
2	13	
1	10	
4	17	
3	15	
.5	8	
RITTER (1967)		
.03	1.90	0.99
.19	4.14	1.27
.10	4.73	1.51
.30	5.53	1.37
.11	5.84	1.77
.27	7.49	1.60
.30	6.46	1.41
.40	6.73	1.22
.08	5.45	1.22
.15	6.17	1.37
.21	4.80	1.41
.14	2.54	0.42
FAHNESTOCK (1963, TABLE 8)		
.32	3.47	1.53
.37	5.8	2.04
.25	3.76	.81
.7	5.6	2.04
.7	5.72	1.91
.55	5.95	1.85
.75	5.55	1.24
.5	4.66	1.07
WILCOCK (1971, FIGURES 16 & 17)		
50 CUSECS		
.26	1.82	
.12	2.30	
.16	2.7	
.14	1.5	
.10	1.4	
.12	1.4	
.08	1.0	
.11	1.2	
100 CUSECS		
.29	2.34	
.13	3.47	
.18	3.6	
.15	2.2	
.11	2.3	
.13	2.0	
.09	1.4	
.12	1.6	
200 CUSECS		
.14	5.3	
.20	5.0	
.17	3.2	
.12	3.8	
.15	2.9	
.10	1.9	
.13	1.9	
400 CUSECS		
.18	4.6	
.16	4.0	
.11	3.3	
.14	3.2	
800 CUSECS		
.20	6.8	
.18	5.5	
.12	5.1	
.16	4.4	

engineering literature uses tractive force, in discussions concerning particle transport. The difference between tractive force and velocity is clearly expressed by Morisawa (1968, p. 47ff):

> "The force required to entrain a given grain is called the critical tractive force, and the velocity at which this force operates on a given slope is the erosion velocity...The critical erosion velocity (or competent velocity) is the lowest velocity at which grains of a given size, loose on the bed of a channel, will move... The value of erosion velocity varies, depending upon the characteristics of the water and of the grain to be moved. Grains of the same size, but with different densities, lying on different slopes and with different distributions on the bottom require different velocities to be moved. Moreover, the erosion velocities vary with the depth and density of the water."

In order to compare the information about transport and coarse particle size, it would be beneficial to have data from different sources in the same units. Fortunately, Nevin (1946) reported both the tractive force and average velocity necessary for the movement of the same size sediment. Utilizing Nevin's (1946, Tables 2, 3, 5, and 6) data it is possible to express graphically, a relationship between tractive force and velocity (Figure 2). The largest particle moved in the experiments reported by Nevin (1946) was 0.02 feet (6.08 mm.). At first glance, it would appear that the tractive force and velocity relationships expressed for fine grained sediment could not be extended to the cobble and boulder size range because of differing flow conditions, bed roughnesses, and slopes encountered where these sediments are found. However, when the tractive force and velocity data reported by Lane (1955, Table 7), Fahnestock (1963, Table 8) and Ritter (1967) for pebbles, cobbles and boulders, are added to Figure 2, the exponential relationship is reinforced.

The particle size <u>versus</u> average velocity data from a variety of sources (Table I) are plotted on Figure 3 along with the upper part of Hjulstrom's curve. When the data were reported in tractive force <u>versus</u> particle size, it was converted to average velocity using the regression equation derived from Figure 2. When the data were reported as average velocity <u>versus</u> particle size, this information was recorded directly on Figure 3. Figure 3 will be referred to frequently in the following evaluation of the various data sources.

Pleistocene catastrophic floods

Pleistocene catastrophic floods provide a source of data on large particle movement, but because of the absence of direct velocity measurements, estimates have to be made. The velocities necessary to move sediment deposited by the Late Pleistocene Bonneville flood were calculated by Malde (1968) by the formula $V_t = 10 \times d^{1/2.6}$, where V_t is the traction velocity in feet per second, and the diameter, d, is in feet. This formula reflects Nevin's (1946) idea that the particle size varies as the 2.6 power of the velocity.

A similar approach was used by Birkeland (1968) for the sediments deposited by the Pleistocene floods along the Truckee River. Birkeland used the Manning formula $V = 1.5/n \times R^{2/3}S^{1/2}$ (where V is the mean velocity in feet per second, <u>n</u> is the coefficient of roughness, R the hydraulic radius in feet, and S the slope) to "give a rough estimate of the mean velocity of the flood". Birkeland calculated two mean velocities for different points along the flood path by using different values of <u>n</u> (Birkeland, 1968,

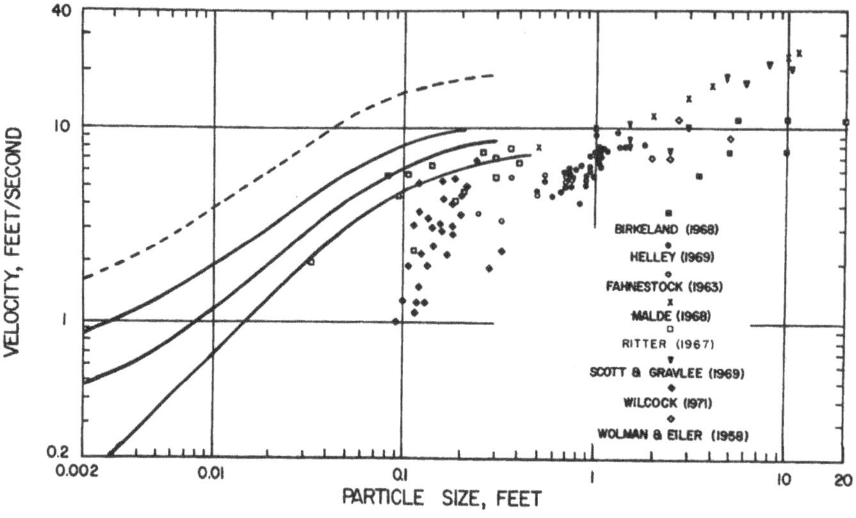

Figure 3 Velocity *versus* particle size data from streams, Pleistocene catastrophic floods, recent floods, and engineering works. Hjulstrom's (1935) curve shown at left.

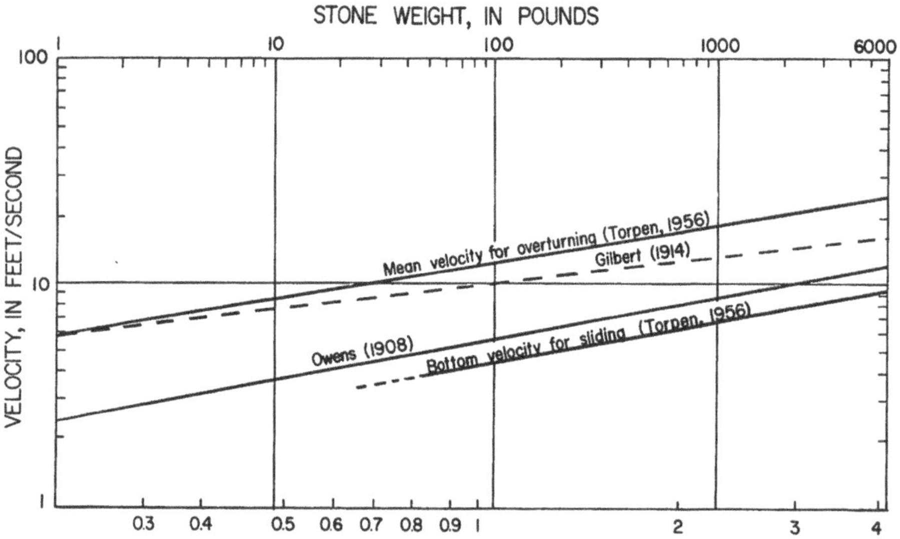

Figure 4 Velocity *versus* particle size curves from Owens (1908), Gilbert (1914), and Torpen (1956).

Table 2). In addition, he estimated the tractive force required to initiate boulder movement on the Truckee Meadows during the Pleistocene flood. By using Figure 2 and Birkeland's tractive force data, the resulting average velocities agree closely with the minimum mean velocities calculated by him with the lower n value. This supports the validity of using Figure 2 to convert data to a common unit for comparative purposes.

Recent floods

Velocity versus particle size data for the August 1955 flood in Connecticut were reported by Wolman and Eiler (1958, Table 1). They estimated mean velocity by dividing the flood discharge by the cross-sectional area of the stream. This method is a common one employed by stream hydrologists to determine velocity. Discharge information is obtained from the records of stream gaging stations and the cross-sectional area can be measured in the field. Wolman and Eiler's (1958) data were reported as mean velocity versus particle size and can be used (Figure 3) without conversion.

Streams

The grain size of bedload sediments and the velocities required to transport them were measured by Fahnestock (1963) in the shallow water of a proglacial stream. Bedload sediment already in motion was trapped by screens or caught by hand, while velocities were measured with a current meter or by timing floats travelling on the surface of the stream. Some size measurements were made on boulders deposited on bars and the velocity immediately upstream from the bar recorded. Fahnestock (1963, p. A30) concluded:

> "The relations of particle size to velocity (see his Figure 30, and Figure 3 of this report) indicates that the boulders from White River were moving at lower stream velocities than would be predicted from other data cited (Hjulstrom's curve)."

The apparent departure from Hjulstrom's curve was attributed to the fact that the measured velocities were the transporting and not necessarily the eroding velocities (Fahnestock, 1963, p. A30). Fahnestock's study was the first attempt to substantiate the water velocities necessary for large particle transport in streams as theorized by Hjulstrom (1935).

Another source of data on large particle transport is Helley (1969). Thirty-six rocks of various sizes, shapes, and lithology were selected, tagged, and placed on the bed of a stream with a small drainage basin that had known discharge characteristics. The particles were placed in the stream for the purpose of studying the initiation of bed-particle motion under natural conditions. A modified Pygmy current meter was used to measure velocities.

The velocities were compared with those predicted by the theory that at the initiation of motion, turning moments of the fluid forces of drag and lift are balanced by the resisting moment of the submerged particle weight. Helley's data were also compared to the Hjulstrom curve, to a curve published by Mavis and Laushey (1949), and to Fahnestock's (1963) measurements.

Helley (1969) concluded that his field measurements agreed closely with the range of velocities predicted by the balanced moment theory in particular, and the previously published curves, in general. The major difference in relations between velocity for the initiation of motion and size result from differences in particle shape, density, and orientation (Helley, 1969, p. G12). His conclusions are supported to some degree by bed-load experiments made by Keller (1970). Keller concluded that 69 percent of the variability in the distance a bed-load particle will move "can best be explained by the variability of the particle parameters and effective bottom velocity."

For experiments designed to measure transport rates and competence limits on the River Hodder, Yorkshire, United Kingdom, Wilcock (1971) placed painted particles ranging in size from -4.0 to -7.5 phi on the stream bed. Velocity measurements were obtained by an acoustic current meter and floats. Wilcock (personal communication, 1972) kindly provided, in tabular form, the data concerning velocity and particle size relationships for the Hodder River basin. These data are included in Table I and Figure 3.

Twelve 1-foot square stations were established by Ritter (1967) beneath a gaging station cableway across the Middle Fork Eel River, California, in order to determine the velocities and the tractive forces necessary to move bed material. Each station was photographed so that the size of transported material could be determined by noting which particles were subsequently removed from the squares. Average velocity for each station was determined from a discharge measurement and the tractive force calculated by using the DuBoys equation. These data are included in Table I and the velocity data shown in Figure 3.

Fahnestock (1963) was concerned with particles already in motion and Ritter (1967), Helley (1969), and Wilcock (1971) with those initially at rest. Considering this, and the difficulties under which their field data were gathered, differences between the authors work are to be expected. However, their close agreement with each other and with predicted values indicate that their data are within the limits of sampling error.

Engineering works and flume studies

Engineering projects using large stones for construction can supply additional information regarding velocity and particle transport. The failure of the partly completed rock-fill, Hell Hole Dam, on the Rubicon River in California, provided a point source of boulders which, combined with the computed discharge characteristics of the flood surge, allowed for an approximation of the flow competence (Scott and Gravlee, 1968). For comparison with data from other sources, the tractive forces reported by Scott and Gravlee have been converted to average velocities using Figure 2 and are included in Figure 3.

The type of material selected by hydraulic engineers for the construction of cofferdams, river diversions, closures, and dams is partly influenced by the anticipated current velocities (Torpen, 1956). In the design of channel reventments, the armor stone along the channel slope should be capable of withstanding anticipated current velocities without being displaced (U.S. Army Coastal Engineering Research Center, 1966). Semi-empirical equations and curves based on laboratory and field data have been developed to aid in the determination of the stability of stone under different velocity regimes (Campbell, 1966, pl. 5; U. S. Army Coastal Engineering Research Center, 1966, Figure 4-27).

A graph relating water velocity to the weight of stones at 165 pounds per cubic foot (equivalent to a specific gravity in air of 2.65) was presented by Torpen (1956) from a variety of sources. Stone weight partly avoids consideration of the influence of shape on stability. However, the equivalent stone diameter (the diameter of a sphere having the same weight) permits minimum stability conditions to be estimated because spheres would be the first shape to move under a given velocity regime, all other factors being equal. Torpen's upper curve represents the mean velocity for over-turning and the lower curve represents the bottom velocity for sliding (Figure 4). The two curves are drawn from the equations $d = 0.08\ V^2$ and $d = 0.06\ V^2$, where d is the particle diameter in inches, and V the water velocity in feet per second.

A similar equation was derived from experiments conducted in a channel between tide-pools (Owens, 1908). Measuring velocities by means of floats, Owens tested the ability of currents to move pebbles 0.5 to 6 inches in diameter, over a channel bed of sand. From his experiments, Owens concluded that the diameter, D, of a particle measured in feet, is related to the velocity, V, in feet per second, by the equation:

$$D = \frac{0.059\ V^2}{G-1} \qquad (1)$$

where G is the density of the sediment. This equation may be rewritten for a rock density of 2.65 as:

$$D = 0.036\ V^2 \qquad (2)$$

As one aspect of his classic flume studies, Gilbert (1914) presented the following equation relating competent mean velocity to particle transport:

$$D = \frac{0.042\ V^{2.7}}{G-1} \qquad (3)$$

with units the same as in equation one. Equation three may also be rewritten for a rock with a density of 2.65 as:

$$D = 0.0025\ V^{2.7} \qquad (4)$$

The curves drawn from Gilbert's (1914) and Owen's (1908) equations are presented in Figure 4. The difference between the coefficients in equations two and four reflect the bed conditions over which particles in each experiment moved; Owens' experiments were conducted with pebbles on a sand bed, while the particles used in Gilbert's experiments rested among similar sized sediment and were less readily dislodged (Gilbert, 1914. p. 163).

Torpen's (1956) graph is actually an envelope within which the engineering curves of rock stability and the curves of Owens (1908) and Gilbert (1914) fit. As such, Torpen's graph summarizes a number of theoretical and semi-empirical equations (see also: Neill, 1967) relating particle size and transport velocity (Figure 4).

DISCUSSION

The information shown in Figure 3 represents a compilation of recent, first-hand observations or estimates of stream velocities that have transported coarse sediment. Early laboratory experiments and the theoretical approaches summarized in Figure 4 provide additional sources of information on the transport of sediment by flowing water. The data of Fahnestock (1963), Helley (1969), Ritter (1967), and Wilcock (1971) are especially important because they represent experiments carried out under natural conditions using currently accepted measuring techniques for streams. The other sources are useful because they show that coarse particles respond to transport by water in the same way under a variety of flow conditions.

The most outstanding feature of Figure 3 is that different velocities for coarse sediment transport are required than are predicted by Hjulstrom's curve. Possible reasons why the data do not follow Hjustrom's prediction include: 1) a velocity, other than the average velocity, was measured or estimated; 2) bed roughness becomes significant in comparing Hjulstrom's curve with other studies; 3) the effective density increase due to sand and silt in suspension lowers the velocity at which cobbles and boulders are transported; or 4) a combination of the first three variables.

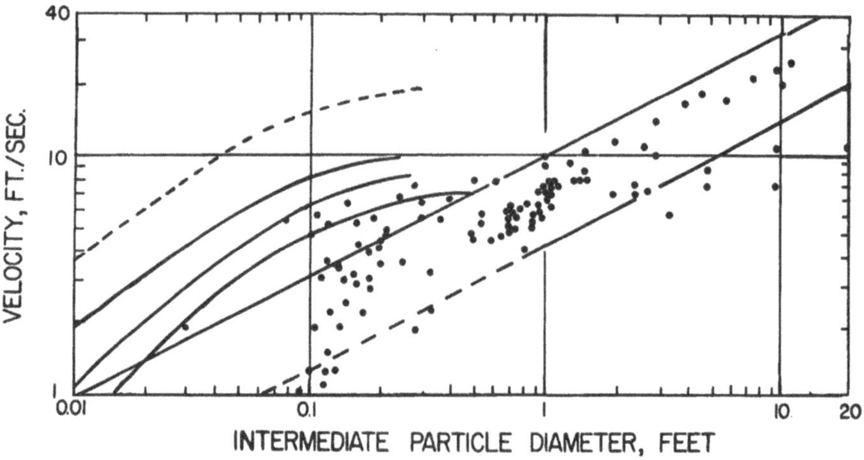

Figure 5 Data from streams, Pleistocene catastrophic floods, recent floods, and engineering works with Hjulstrom's (1935) curve and the curves of stone stability.

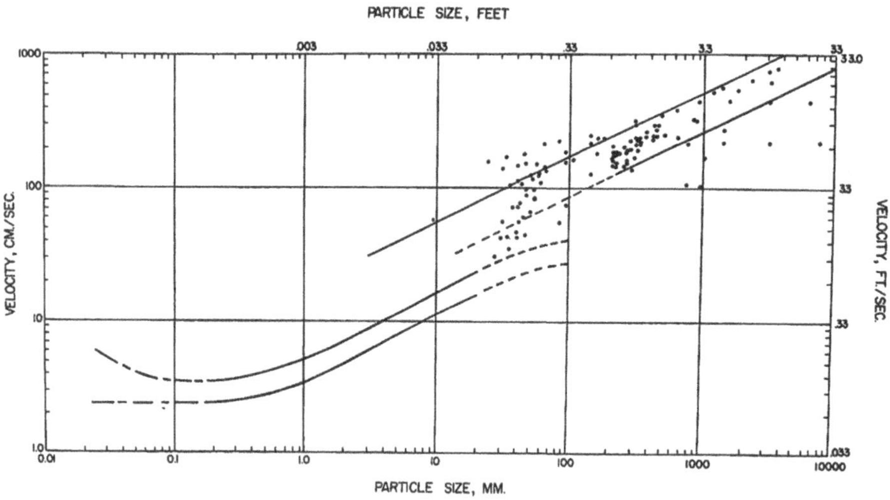

Figure 6 Data from streams, Pleistocene catastrophic floods, recent floods, and engineering works with Sundborg's (1967) curve and the curves of stone stability. (Short dashed lines added by present author).

The variety of velocities reported for the same size particles no doubt reflects differences in the velocities actually measured or estimated. Ostensibly, all of the velocities shown in Figure 3 are mean velocities, but the assumptions used in the absence of direct observation (especially for the Pleistocene and recent flood data) makes it likely that the velocities vary somewhat from the mean. Despite these possible variations the seven data sources utilized show a well-defined velocity increase with increasing size.

The size of the bed material influences the type of flow (laminar or turbulent) and the velocity profiles (Briggs and Middleton, 1965). The stream beds in the studies by Fahnestock (1963), Helley (1969), Wilcock (1971), Wolman and Eiler (1958), Ritter (1967), and Scott and Gravlee (1968), were composed of a natural range of particle sizes. Hjulstrom (1935) assumed uniform size bed materials in the work that led to his curve. The importance of bed roughness is difficult to assess especially for the historical flood data of Malde (1968) and Birkeland (1968). Briggs and Middleton (1965) have suggested that even for beds with constant bottom roughness average velocities and erosion velocities are not directly proportional. This relationship may be of even greater significance when comparing stream beds with differing bottom roughnesses.

The data derived from catastrophic Pleistocene floods (Malde, 1968; Birkeland, 1968) fall within the same range as data reported for much smaller streams. The magnitude and turbulence of Pleistocene floods were such that huge amounts of sand and silt were most certainly carried by the water. The effective density increase of the fluid was not then taken into account in the velocity calculations, but Malde (1968, p.47) conceded that increased density probably decreased the velocity at which large blocks were transported. Because the fluid density was probably greater than that of clear water, the velocity estimates of Malde (1968) and Birkeland (1968) should be taken as minimums for the particle sizes that were transported. To a lesser extent, the same objections regarding fluid density may be raised against the recent flood data reported by Wolman and Eiler (1958). The magnitude of the density increase would by no means be as great as that which occured during Pleistocene catastrophic floods.

CONCLUSIONS

Field measurements and estimates of mean velocities necessary to move coarse sediment in the fluvial environment agree closely with the velocities predicted by the engineering curves for rock stability. When the data shown in Figure 3 are plotted with Torpen's (1956) graph, most of the data fall within the envelope (Figure 5). Torpen's curve is preferred over Hjulstrom's curve for predicting coarse particle transport in streams because it is based on the theory of, and experiments with, coarse sediment movement. Hjulstrom's curve is an extrapolation of data determined for sand size sediment. A more recently published curve by Sundborg (1967) also fails to adequately predict coarse sediment competency (Figure 6).

This analysis of recent studies on gravel sediment transport by flowing water in streams, floods, and engineering works shows that a curve of mean velocity for overturning and a curve of bottom velocity for sliding defines a zone that predicts coarse sediment transport better than the Hjulstrom-type curves. It seems apparent that an adjustment of Hjulstrom's curve is required for predicting stream transport velocities in the cobble and boulder size range. The nature of the adjustment is suggested by Figure 5 but could best be determined by future flume and field measurements on the transport velocities necessary to move coarse sand and pebbles. However, the usefulness of the Hjulstrom curve as a theoretical predictor of coarse sediment transport by swash on beaches (Novak, 1972) and for the smaller size ranges in streams should not be minimized.

REFERENCES

Birkeland, P. W., 1968, Mean velocities and boulder transport during Tahoe-age floods of the Truckee River, Calif. - Nevada: Geol. Soc. Amer. Bull., V. 79, p. 137-192.

Briggs, L. I., and Middleton, G. V., 1965, Hydromechanical principles of sediment structure formation. In: Middleton, G. V. (Editor), Primary sedimentary structures and their hydrodynamic interpretation: Soc. Econ. Paleon. and Mineralogists, Spec. Publ. 12, p. 5-16.

Campbell, F. B., 1966, Hydraulic design of rock rip-rap: U. S. Army Engin. Waterways Exper. Stat. M. P., No. 2-777, 24 p.

Einstein, H. A., 1950, The bed-load function for sediment transportation in open channel flows: U. S. Dept. of Agric., Soil Conser. Serv. Tech. Bull. No. 1026, 71 p.

Fahnestock, R. K., 1963, Morphology and hydrology of a glacial stream, White River, Mt. Rainier, Wash.: U. S. Geol. Surv. Prof. Paper, 422-A, 70 p.

Gilbert, G. K., 1914, The transportation of debris by running water: U. S. Geol. Surv. Prof. Paper, 86, 261 p.

Helley, E. J., 1969, Field measurements of the initiation of large bed particle movement in Blue Creek near Klamath, Calif.: U. S. Geol. Surv. Prof. Paper, 562-G, 19 p.

Hjulstrom, F., 1935, Studies of the morphological activity of rivers as illustrated by the River Fyris: Geol. Instit. Univ. Upsala Bull., V. 25, p. 221-527.

Hjulstrom, F., 1939, Transportation of detritus by moving water. In: Trask, P. O., (Editor), Recent Marine Sediments: Amer. Assoc. Petrol. Geol., Tulsa, p. 5-31.

Inman, D. L., 1949, Sorting of sediments in the light of fluid mechanics: Jour. Sedim. Petrol., V. 19, p. 51-70.

Keller, E. A., 1970, Bed-load movements experiments: Dry Creek, California: Jour. Sedim. Petrol., V. 40, p. 1339-1344.

Lane, E. W., 1955, Design of stable channels: Amer. Soc. Civ. Engin. Trans., V. 120, p. 1234-1260.

Malde, H., 1968, The catastrophic Late Pleistocene Bonneville flood in the Snake River plain, Idaho: U. S. Geol. Surv. Prof. Paper, 596, 52 p.

Mavis, E. T., and Laushey, L. M., 1949, Formula for velocity of beginning of bedload movement reappraised: Civ. Engin. V. 19, p. 38-39.

Morisawa, M., 1968, Streams: their dynamics and morphology. McGraw-Hill, New York, 175 p.

Neill, C. R., 1967, Mean-velocity criterion for scour of coarse uniform bed material: Internat. Assoc. Hydraulic Research Congr., 12th, Proc., V. 13, p. 6.1-6.9.

Nevin, C., 1946, Competency of moving water to transport debris: Geol. Soc. Amer. Bull., V. 57, p. 651-674.

Novak, I. D., 1972, Swash-zone competency of gravel-size sediment: Marine Geol., V. 13, p. 335-345.

Owens, J. S., 1908, Experiments on the transporting power of sea currents: Geogr. Jour., V. 31, p. 415-420.

Ritter, J. R., 1967, Bed-material movement, Middle Fork Eel River, California: U. S. Geol. Surv. Prof. Paper, 575-C, p. C219-C221.

Rubey, W. W., 1938, The force required to move particles on a stream bed: U. S. Geol. Surv. Prof. Paper, 189-E, p. 121-141.

Scheidegger, A. E., 1970, Theoretical geomorphology, 2nd Edition: Springer-Verlog, New York, 435 p.

Schiffman, A., 1965, Energy measurements in the swash-surf zone: Limnology and Oceanography, V. 10, p. 225-260.

Scott, K. M., and Gravlee, G. C., 1968, Flood surge on the Rubicon River, Calif. - hydrology, hydraulics, and boulder transport: U. S. Geol. Surv. Prof. Paper, 422-M, 40 p.

Sundborg, A., 1956, The River Klaralven: a study of fluvial processes: Geograf. Ann., V. 38, p. 125-316.

Sundborg, A., 1967, Some aspects on fluvial sediments and fluvial morphology, I. General views and graphic methods: Geograf. Ann., V. 49A, p. 333-343.

Torpen, B. E., 1956, Large rocks in river control works: Civ. Engin., V. 26, p. 56-61.

U. S. Army Coastal Engineering Research Center, 1966, Shore Protection, Planning and Design: Tech. Report No. 4, 3rd Edition, 401 p.

Wilcock, D. N., 1971, Investigation into the relations between bedload transport and channel shape: Geol. Soc. Amer. Bull., V. 82, p. 2159-2176.

Wolman, M. G., and Eiler, J. P., 1958, Reconnaissance study of erosion and deposition produced by the flood of August, 1955 in Connecticut: Amer. Geophys. Union Trans., V. 39, p. 1-14.

2

THE RANDOM MODEL IN FLUVIAL GEOMORPHOLOGY*

J. S. Smart

ABSTRACT

This paper describes a model of drainage basin composition that is primarily based on two postulates suggested by investigations of the role of randomness in drainage network development. The postulates are (1)the topology of channel networks is random, (2)link lengths are independent random variables drawn from a common distribution. A third assumption, introduced to help explain areal properties, is that the dimensionless constant $K=\mathcal{L}^2/(2n-1)\mathcal{A}$ is approximately unity (n is magnitude; $\mathcal{L}$ total stream length; $\mathcal{A}$, basin area). Data on 95 natural networks were used to compare the random model and Horton's laws with respect to estimation of bifurcation and stream length ratios. The random model proved better in both cases and was generally easier to use.

As examples of other possible applications, the random model was employed to explain such diverse observations as apparent regularities in successive individual bifurcation ratios for stream networks in Malaysia and the correlation of first and third order stream lengths for networks in Uganda. By defining new topologic parameters that are analogous to geometric ones, such geomorphic "laws" as the 0.6 power relation between mainstream length and basin area can be predicted.

More detailed description of channel network structure is possible by using dimensionless parameters derived from the interior and exterior link lengths and their associated drainage areas. Such description permits the characterization of small but easily identifiable differences in dendritic network structure due to environmental factors and degree of maturity.

*Supported in part by the Geography Programs, Office of Naval Research.

INTRODUCTION

The quantitative characterization of the physiographic properties of a drainage basin is an extremely complex task. Strahler (1964) lists about 30 different quantitative parameters that have been used frequently in geomorphic research, and a similar set is presented by Doornkamp and King (1971) in their text on numerical analysis in geomorphology. It is well recognized, however, that the commonly-used morphometric parameters are not all independent of each other, but rather are interrelated in a complicated and subtle fashion. Consequently, it is not necessary to measure all possible geomorphic variables in order to obtain an adequate description of a drainage basin; instead, the measurements can be confined to some appropriately chosen subset and the remaining variables estimated with reasonable accuracy. In this paper, description is used in the quantitative sense and refers to the use of measurements on a drainage basin, or of results derived from the measurements, to characterize that basin. As as example, the list of link lengths for a channel network provides some descriptive information about the network. Estimation here means the use of measured quantities to predict unmeasured ones, either for the same basin or for other basins. Thus the data on link lengths might be used to estimate the area of the watershed, or to estimate the distribution functions for link lengths of other drainage basins in the vicinity.

Statistical analysis is a standard procedure for sorting large quantities of data on intercorrelated variables, for deciding which variables should be measured, and for providing quantitative estimates of the others. Doornkamp and King (1971, Part 1) give a clear exposition of the application of various statistical techniques (regression analysis, factor analysis, discriminant analysis, etc.) to morphometric data on drainage basins. An alternative (and, to the writer, more satisfactory) approach is the development of a conceptual model of drainage basin composition. A conceptual model of a system consists of certain primary assumptions or postulates from which properties of the system can be deduced. Thus a satisfactory model of drainage basin composition, in addition to providing efficient and accurate methods of description and estimation, should explain previously observed relationships and predict new ones. There is of course no reason why conceptual models and statistical methods cannot be combined in the analysis of geomorphic data. In fact, Doornkamp and King in their discussion of statistical methods employed a simple model (Horton's stream laws) to help decide which parameters were to be included in the original data collection.

Recent empirical and theoretical studies of drainage basin composition have strongly suggested that many properties can be explained by assuming that drainage basins develop through random processes. This work was initiated by important papers by Leopold and Langbein (1962) and Shreve (1966) and has been carried out by a number of different authors. Readers are referred to a recent paper by Smart (1972a) for a detailed review of the subject.

The principal results obtained from these investigations of randomness can be abstracted into postulates to form a model of drainage basin composition. This paper discusses the effectiveness of the random model in explanation, description, and estimation of drainage basin properties. (The term "random model" is admittedly an oversimplification, since there are a great many possible stochastic models. This particular stochastic model has been deliberately chosen to be in accord with natural networks and at the same time to be as simple as possible.) The discussion here is confined to planimetric properties, but it seems likely that the approach could be extended to three-dimensional ones as well.

HORTON'S STREAM LAWS

Before introducing the random model, we briefly review Horton's (1945) laws of drainage composition, which constitute the oldest and best known model of drainage basin properties. Clearly, any new model must stand comparison with Horton's laws.

The law of stream numbers states that

$$N_\omega \approx R_B^{(\Omega-\omega)} \qquad 1 \leq \omega \leq \Omega \tag{1}$$

where N_ω is the number of streams of order ω in a basin of order Ω, and R_B is the bifurcation ratio. In this and further discussion, stream order is defined by Strahler's (1952) modification of Horton's original ordering scheme.

The law of stream lengths is

$$\bar{L}_\omega \approx R_L^{(\omega-1)} \bar{L}_1 \tag{2}$$

where $\bar{L}_\omega$ is the mean length of streams of order ω and R_L is the stream length ratio.

Horton also suggested that there should be an analogous relationship for areas, and one was later stated explicitly by Schumm (1956).

$$\bar{A}_\omega \approx R_A^{(\omega-1)} \bar{A}_1 \tag{3}$$

where $\bar{A}_\omega$ is the mean area drained by streams of order ω (including their tributaries of lower order) and R_A is the basin area ratio.

Data on stream numbers, stream lengths, and basin areas are often presented graphically in so-called Horton diagrams by plottinng $\ell n N_\omega$, $\ell n \bar{L}_\omega$, and $\ell n \bar{A}_\omega$ against ω. If Horton's laws were satisfied exactly, these plots would be straight lines. According to Horton, the bifurcation ratio R_B is determined by finding the straight line that passes through $(\Omega,0)$ and gives the best fit to the stream number

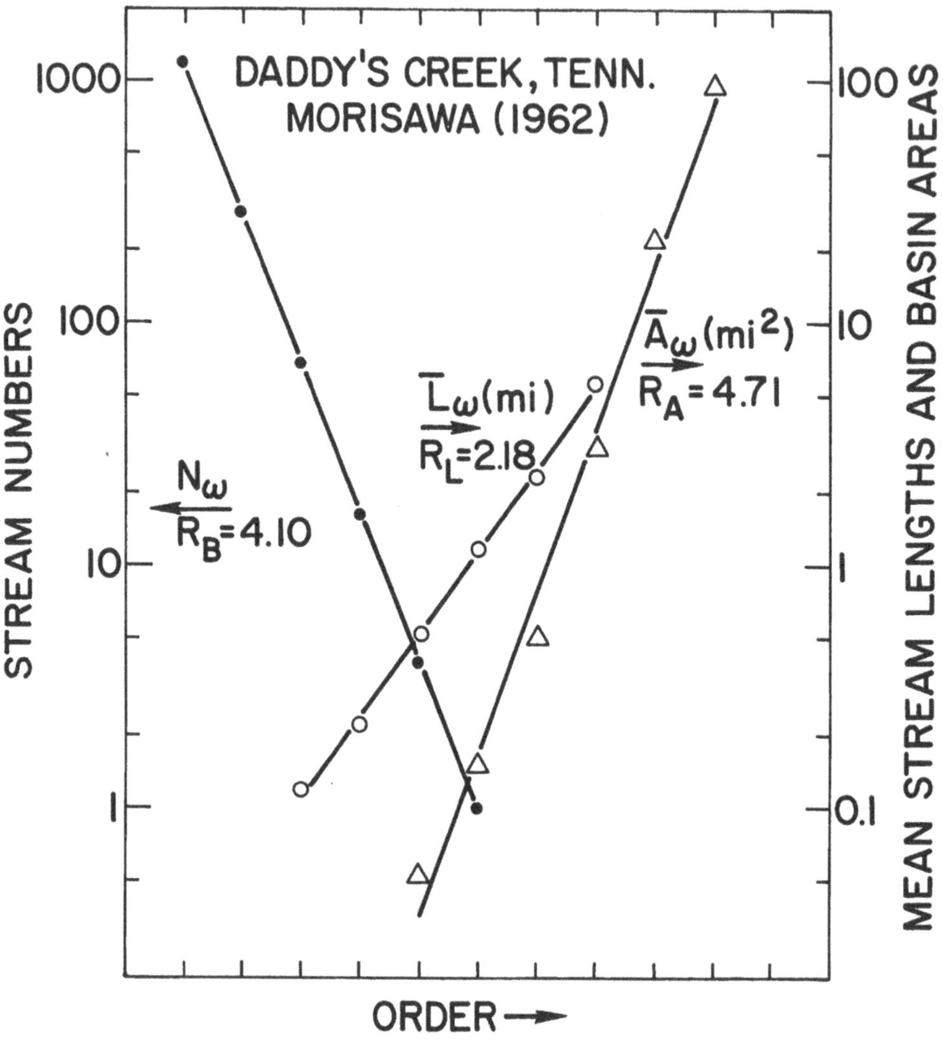

Figure 1 Horton diagrams for Daddy's Creek, Tenn. Data from Morisawa (1962).

points; the slope of this line is $-\ln R_B$. (It should be noted that several other kinds of bifurcation ratios have been defined and used in the literature; see Smart (1972a, p. 311-314) for a review). The Strahler stream length and basin area ratios are found by analogous procedures except that the best-fit line is not required to pass through any particular point. Figure 1 shows the Horton diagrams for Daddy's Creek, Ohio, prepared from data taken by Morisawa (1962).

Horton's laws have mainly been used for descriptive purposes, that is, the numbers R_B, R_L, and R_A which characterize a drainage basin are derived from data taken on that basin. Horton, however, apparently felt that the three parameters are functions of the local geology and climate, and that values determined for one basin could be used to estimate the properties of other basins in comparable regions. Specifically, he (Horton, 1945, p. 290) claimed that the bifurcation ratio "is generally higher for hilly, well-dissected drainage basins than for rolling basins". The published data on which this conclusion was based involves a sample of 11 basins, and later, more extensive studies by Strahler and his students have not supported this claim. In a review article, Strahler (1957, p. 914) observed that the bifurcation ratio "is highly stable and shows a small range of variation from region to region or environment to environment, except where powerful geologic controls dominate".

A quantitative evaluation of the accuracy of description and estimation by Horton's laws is presented in a later section.

THE RANDOM MODEL

The basis for a random model of drainage basin composition is lucidly described by Krumbein and Shreve (1970, p. 40). The evolution of geomorphic features such as channel networks and hillslopes involves many complex processes, e.g., rock weathering, soil creep, fluid flow, sediment transport. Many variables are required to specify even a small system, and initial states are generally unknown. Thus, although exact predictions of the development of a drainage basin may be possible in principle (the question is to some extent a philosophical one), they are clearly not possible in practice.

Instead of trying to predict geomorphic properties exactly, we should regard them as random variables drawn from specified populations. A qualitative argument will perhaps make the point clearer. The processes involved in drainage basin development are influenced by both geologic and climatic factors; for example, the rate of rock weathering depends both on the chemical composition of the rock and on precipitation and temperature. The combination of geologic and climatic influences will be called, for convenience, the environment; it seems reasonable to assume then that drainage basins developed under the same environment and selected according to some specific attribute such as area or order will have similar, but not identical, geomorphic properties. The observed values of some particular variable, such as drainage density, will vary randomly from basin to basin but will belong to a well-defined distribution determined by the environmental conditions.

The transition from these general ideas to quantitative analysis of drainage basin properties is achieved by introducing two basic postulates of the random model for mature drainage basins:

(1) In the absence of geologic controls, channel networks are topologically random (Shreve, 1966).

(2) For drainage basins developed under comparable environmental conditions, the exterior and interior link lengths are independent random variables with a single common distribution for each type (Smart, 1968).

Although these postulates are generally supported by observations on natural networks, it should be emphasized that they are not presented as being infallible. Both have been shown to be incorrect for certain specific areas or in certain specific details (e.g., Smart, 1969, p. 1768-1772; James and Krumbein, 1969; Howard, 1971, p. 1371-1374). The justification for this model is purely pragmatic. Starting with these two postulates, many other features of drainage basin composition relating to topology and channel lengths can be explained quantitatively and with minimum recourse to adjustable parameters; also, the associated description and estimation procedures are a considerable improvement over those previously used.

These two assumptions are analogous to the first two Horton laws in the sense that the first deals with purely topologic properties and the second deals with length properties. In order to extend the scope of the model, a third postulate relating to areal properties is needed. One possible and obvious procedure would be to make some statement about probability distributions for drainage areas, thus placing the third assumption in the framework of the random model. Some attempts along these lines have been made (Shreve, 1969; Smart, 1973).

For the purposes of this paper, however, we have found another approach to be more fruitful, and will introduce a length-area relationship through the dimensionless network parameter K (Melton, 1960; Shreve, 1967; Smart, 1972b)

$$K = \mathcal{L}^2/(2n-1)\mathcal{A} \qquad (4a)$$

where $\mathcal{L}$ is the total channel length, $\mathcal{A}$ is the total drainage area, and $n(=N_1)$ is the magnitude of the network. K can also be written in a number of other interesting and suggestive ways:

$$K = D^2/F \qquad (4b)$$

$$= D\bar{\ell} \qquad (4c)$$

$$= D^2\bar{a} \qquad (4e)$$

$$= \bar{\ell}^2/\bar{a} \qquad (4f)$$

where $D=\mathcal{L}/\mathcal{A}$ is the drainage density, $F=(2n-1)/\mathcal{A}$ is the link density (Shreve, 1967, p. 185), $\bar{\ell}$ is the mean link length (including both interior and exterior links) and $\bar{a}$ is the mean link drainage area. Since a channel network with n exterior links has $(n-1)$ interior links, the quantities $\bar{\ell}$ and $\bar{a}$ for networks of magnitude $n>>1$ are well approximated by

$$\bar{\ell} = \frac{1}{2}(\bar{\ell}_e + \bar{\ell}_i) \qquad (5a)$$

$$\bar{a} = \frac{1}{2}(\bar{a}_e + \bar{a}_i) \qquad (5b)$$

where the subscripts $\underline{e}$ and $\underline{i}$ refer to exterior and interior links respectively.

The third basic postulate of our model of mature drainage basins is

(3) The parameter K is unity (Melton, 1960, and private communication; Shreve, 1967).

Like the first two postulates, this one is abstracted from observations and is not expected to be literally satisfied under all conditions, even in the restricted sense that the mean value is unity. There is some evidence, for example, (Smart, 1972b), that K depends somewhat on the environment, i.e., its mean may be greater than 1 in some areas and less than 1 in others.

For convenience, in the remainder of this paper, we shall consider the random model as including all three postulates, although the third has no obvious relation to randomness.

A partial list of drainage basin properties that are explained, at least semi-quantitatively, by the random model includes:

(1) The distribution of bifurcation ratios (Shreve, 1966).

(2) The tendency of plots of $\ln N_\omega$ versus ω to be concave upward (Shreve, 1966).

(3) The distribution of stream length ratios (Shreve, 1967; Smart, 1968; Shreve, 1969).

(4) Melton's (1959) observation that R_B and R_L are correlated (Smart, 1968).

(5) Melton's (1958) discovery of the approximate proportionality of Strahler stream frequency, F_S, and D^2 (Shreve, 1967, p. 184-185).

(6) The 0.6 power relation between mainstream length and basin area (Shreve, 1970; Werner and Smart, 1973).

(7) The relation between the Langbein (1947) area-distance distribution parameter and basin area (Werner and Smart, 1973).

(8) The relation between mainstream length and mean length of travel (Werner and Smart, 1973).

It should be noted that these results are achieved without adjustable parameters and fitting procedures. The successful explanation of such varied features of drainage basin composition by the application of three simple premises encourages more widespread use of the model. In the next section, the effectiveness of the random model in description and estimation of drainage basin properties is compared with that of Horton's laws.

COMPARISON OF HORTON'S LAWS AND RANDOM MODEL

Horton (1945, p. 296-300) noted that the determination of stream numbers and stream lengths for large basins could be quite laborious. He proposed as an alternative that the properties of the lower order streams could be accurately estimated from measurements made on small scale maps on which only the higher order streams appear. The validity of this suggestion, however, depends on Horton's belief that the bifurcation and stream length ratios are characteristic of the region, or at least the network. As mentioned previously, investigations by Strahler and his colleagues have thrown considerable doubt on this idea. Despite these results, applications of Horton's suggestion still appear occasionally in the literature and it seems appropriate to conduct a quantitative test of the matter.

We first consider the estimation of stream numbers. Let N_1, N_2, ..., $N_{\Omega-1}$, 1 be the stream numbers determined from a large scale map of a basin of order Ω. Now suppose that the same network is outlined on a map whose scale is just enough smaller so that none of the original first order streams are depicted but all of the second and higher order streams remain. The small scale network will be of order Ω-1 and will have N_2 first order streams, N_3 second order streams, and so forth. Then we can determine the Strahler bifurcation ratio for this truncated set of stream numbers, extrapolate the regression line back to "zeroth" order (small scale) and thus obtain an estimate of the number of first order streams on the large scale map. Since the exact value of N_1 is known, we can then determine the accuracy of the estimate. As an example of how the procedure is applied, consider Morisawa's (1962) data on Daddy's Creek. The original set of stream numbers is 1181, 288, 68, 16, 4, 1 corresponding to a Strahler bifurcation ratio of 4.101. The bifurcation ratio for the set 288, 68, 16, 4, 1 is 4.088 (thus in this instance supporting Horton's claim that the bifurcation ratio for subnetworks should be about the same as that for the complete network). Plotting the best fit straight line and extrapolating back one order gives an estimate of 1141 for N_1, to be compared with the actual value of 1181.

The random model can also be used to estimate the numbers of lower order streams. According to Shreve (1967, p. 184), the individual bifurcation ratios $N_\omega/N_{\omega+1}$ approach 4 as the number of streams increase without limit. Thus in the case of Daddy's Creek, the random model predicts that N_1 is approximately 4 x 288=1152. The estimation procedure could in principle be improved by detailed calculations that give the exact expectation value of $N_\omega/N_{\omega+1}$ as a function of $N_{\omega+1}$ but the corrections are small and do not seem worth the extra effort.

Stream number data on 38 networks of fifth or higher order were used to test and compare the accuracy of the Horton and random model methods of estimating lower order stream numbers. This data was taken from networks selected on the basis of dendritic patterns and uniform drainage density by Smart (1969, 1972b) for other studies. The comparison was restricted to $\Omega \geq 5$ because neither method could be expected to work very well for lower orders, the random model because the number of streams must be large and Horton's method because it would have three or fewer points from which to establish a bifurcation ratio.

The results of the comparison are shown in Table 1. The random model gave the better estimate for 32 out of the 38 cases. A more quantitative statistic is the root mean square reduced error ε.

$$\varepsilon = \left(\sum(N_1 - \hat{N}_1)^2/NN_1^2\right)^{1/2} \qquad (6)$$

where N_1 is the correct value, $\hat{N}_1$ the estimate, and N the sample size (38 in this case). The reduced error was chosen because of the wide range of N_1-values (a factor of 6 or more).

Table 1. Estimates of N_1

Method of Estimation	Number of Times Better	ε	Number of Times Too Low
Horton	6	0.309	34
Random Model	32	0.123	20

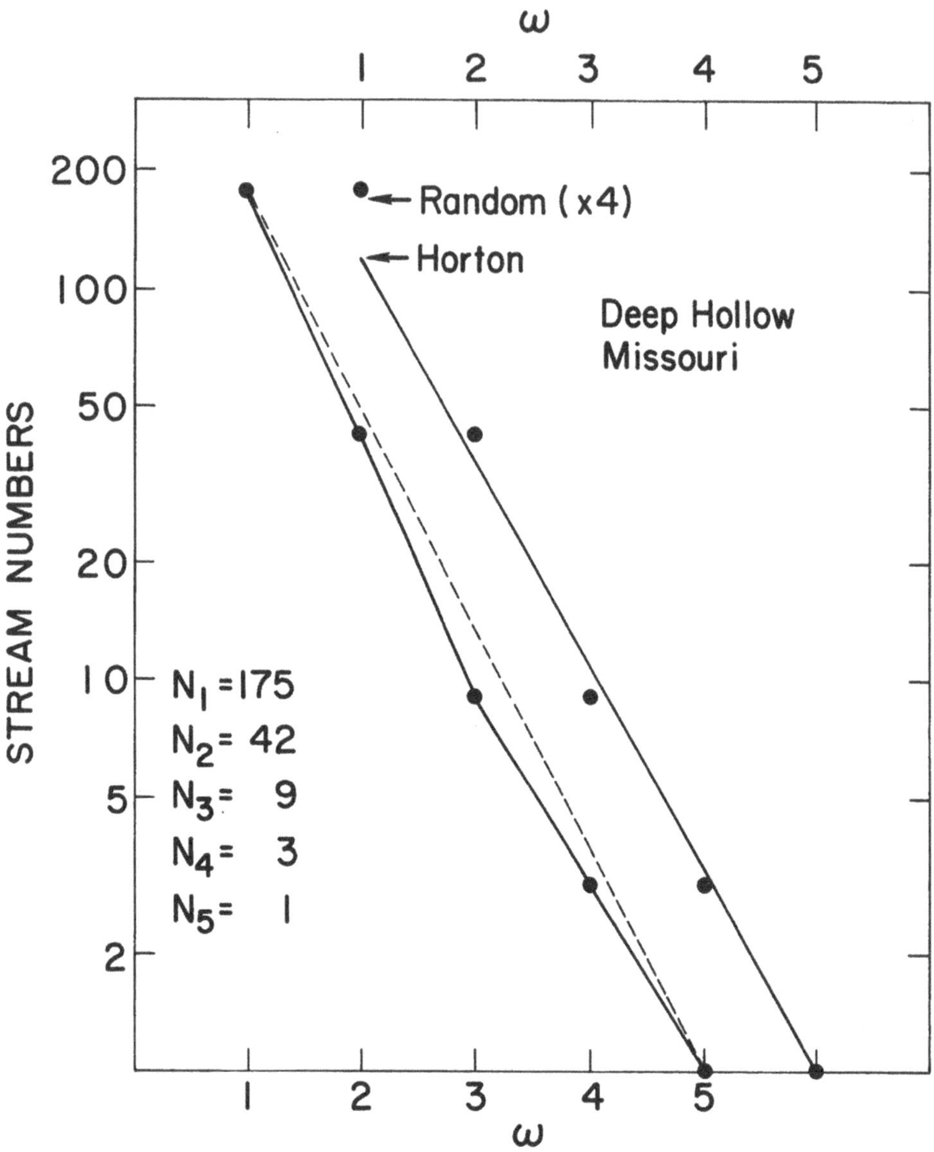

Figure 2 Two plots of log N_ω versus ω for Deep Hollow, Missouri. Left side shows that the plot is concave upward. Right side illustrates why Horton's extrapolation method gives estimates for N_1 that are too low in such cases.

ε for the Horton method is more than twice that for the random model. Finally, as indicated by the last column of Table 1, the Horton estimates are badly biased; 34 out of 38 estimates are too low. The random model, on the other hand, is too low in 20 cases and too high in 18 cases. It should also be noted that the random model estimates, in addition to being more accurate, are also much easier to obtain; multiplying by 4 is definitely less effort than the Horton regression-extrapolation procedure.

The reason why the Horton method is so inaccurate is shown in Figure 2. Here the stream numbers for Deep Hollow, Missouri are plotted twice. The left hand curve shows that the actual plot is concave upwards in the sense that the straight line connecting the two extreme points lies above the intermediate points. Schumm (1956, p. 603) and Maxwell (1960, p. 12) have noted that stream number plots tend to be concave upward more frequently than they are concave downwards or mixed (intermediate points on both sides of the straight line). Their observations were based on small samples but later Shreve (1966, p. 24-26) came to the same conclusion in an analysis of stream numbers for 246 networks. He also showed that this behavior is a straightforward consequence of the assumption of topological randomness.

The curve on the right side of Fig. 2 gives an example of how a stream number plot that is concave upward leads to too low an estimate for N_1. In this particular case, the estimated value is 119, compared with the actual value 175. The random model estimate is 168.

Yang and Stall (1971) have proposed a refinement and extension of Horton's method in which data taken on a small scale map is used to estimate results for a large scale map by knowing the difference in order δ (in general, non-integral) between the two map systems. The results reported above, however, refute their claim for the special case δ=1, and it is hard to see why the method should work for any other value of δ. Also, the single example provided by Yang and Stall does not seem to support their arguments. They report that stream numbers for the upper Rogue River, Oregon are 111, 26, 6, 1 on a 1:250,000 scale map and 177, 50, 12, 4, 1 on a 1:24,000 scale map, and conclude that "the two sets of data follow the same Horton-Strahler straight lines". Actually, the bifurcation ratios are 4.76 and 3.63 for the small and large scale maps, respectively. (The authors do not state specifically how the straight lines should be determined. The Maxwell (1960, p. 12) bifurcation ratios are quoted here because, judging from Yang and Stall's Figure 1, that seems to have been what they intended; in any event, the difference between the two Strahler bifurcation ratios is even greater.) If only the small scale data were available, as would be the case when the method is put into practice, the estimated number of second order streams on the large scale map would be 314, whereas the actual number is 177.

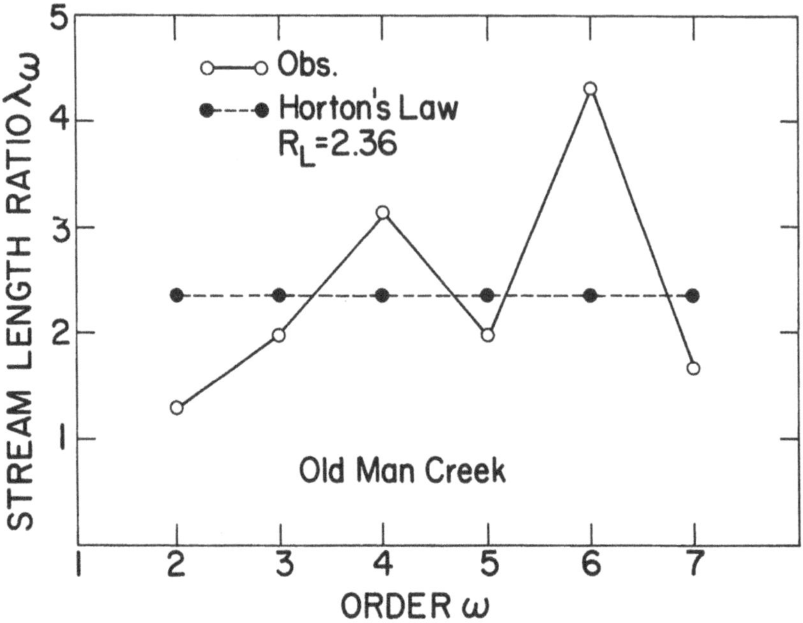

Figure 3 Observed stream length ratios $\bar{L}_\omega/\bar{L}_{\omega-1}$ for Old Man Creek, Iowa and Horton's law approximation.

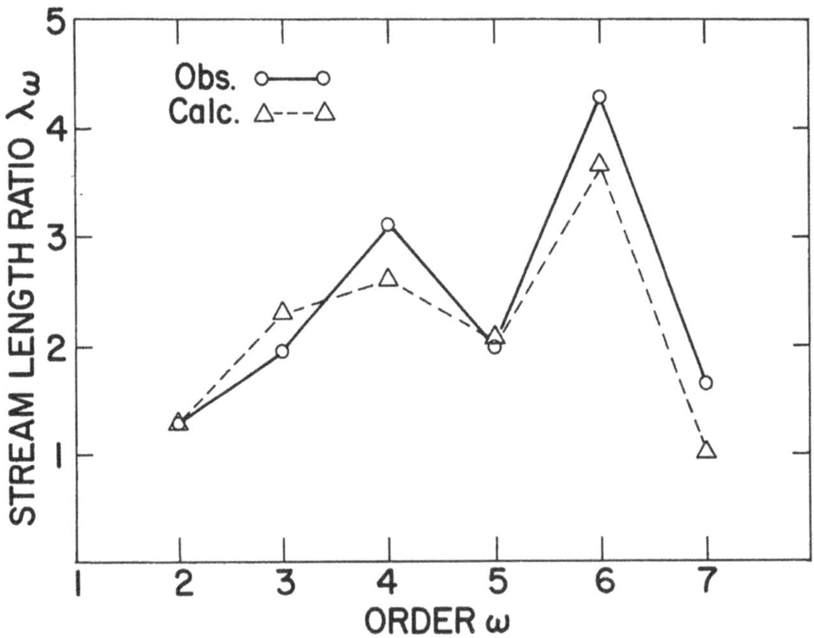

Figure 4 Observed and calculated (eqs. 7a and 7b) stream length ratios $\bar{L}_\omega/\bar{L}_{\omega-1}$ for Old Man Creek, Iowa. (Reprinted from Smart (1972a, p. 327) by permission of Academic Press. Copyright 1972).

The effectiveness of Horton's method in estimating low order stream lengths can of course be tested in the same way. However, since bifurcation ratios and stream length ratios are known to be rather highly correlated (Melton, 1958; Smart, 1968), it seems unlikely that the method would produce good results in one case and not in the other. Also, and more important, Horton's law of stream lengths is not even very good for describing the length properties of the network on which the measurements were made. This point has been noted frequently in the literature (e.g., Broscoe, 1959, p. 5; Maxwell, 1960, p. 23; Bowden and Wallis, 1964, p. 769), but one additional example is presented here. Figure 3 shows the individual stream length ratios $\overline{L}_\omega/\overline{L}_{\omega-1}$ versus ω for Old Man Creek, Iowa; the data are derived from Table 1 of Milling and Tuttle (1964). The observed values display an irregular variation which is not very well represented by the constant R_L value of 2.36.

Smart (1968, 1972a) has used the random model to derive expressions for the mean values of individual stream length ratios. These means, taken over a topologically random population of channel networks with a specific set of stream numbers, can be used as estimates.

$$\overline{L}_2/\overline{L}_1 \approx (N_1-1)/(2N_2-1)\lambda \qquad (7a)$$

where $\lambda=\overline{\ell}_e/\overline{\ell}_i$ is the ratio of mean exterior link length to mean interior link length.

$$\overline{L}_\omega/\overline{L}_{\omega-1} \approx (N_\omega-1)/(2N_{\omega-1}-1) \qquad (7b)$$

$$\omega=3,4,\ldots\Omega$$

Since the random model does not provide any information about the parameter λ, this factor, which appears in the expression for $\overline{L}_2/\overline{L}_1$, must be determined empirically; it is not, however, adjustable. The estimates of the higher order ratios do not require any length data at all; only the stream numbers are needed. Figure 4 compares the values obtained from equations 7a and 7b with the observations on Old Man Creek. Without any quantitative test, it is apparent that the random model gives a much better representation of the observations than does Horton's law of stream lengths. In effect, for the higher order ratios, the random model estimates based on stream numbers alone are superior to the Horton's law description based on the actual stream lengths.

In one sense, the example chosen was favorable to the random model because the network is large (N_1=5156) and estimates based on mean values would be expected to be better for large networks than for small ones. Comparison of the two methods for 95 natural networks indicated that the superiority of the random model decreases with decreasing magnitude until about magnitude 25, where the two methods appear to be equally bad. Again it should be noted that the random model is much easier to use, since determining λ and evaluating a few fractions requires much less labor than determining all stream lengths, computing their averages, and performing a regression analysis.

APPLICATIONS OF THE RANDOM MODEL

One of the most attractive features of the random model is its ability to help interpret all kinds of empirical results, ranging from detailed observations on a single small basin to the general relationships often referred to as geomorphic "laws". Some examples of both kinds are given in this section.

The first example has been discussed previously by Smart (1972a, p. 342). Eyles (1968) reported that 8 of 19 fourth order basins in West Malaysia had individual bifurcation ratios such that $N_1/N_2 > N_2/N_3 > N_3/N_4$. He stated that the expected number of such occurrences was 19/6 (apparently on the assumption that the three ratios were independent and identically distributed), and proposed that the difference was due to rejuvenation processes that generated new first-order streams. It seems unlikely that the networks were drawn from a topologically random population of fourth order basins, because the mean magnitude for the sample is 28.5, compared with the expected value of 43 (Shreve, 1969, p. 181). This discrepancy is probably at least in part due to the selection procedure in which the basins were taken from cleared areas that were "limited in extent and did not conform to natural landscape units". (This sampling procedure was of course quite suitable for Eyles' purposes.) If the random model is applicable to this region, however, each of the 19 networks chosen by Eyles should have been drawn from a topologically random population of appropriate magnitude. Calculations based on this assumption indicate that for the specific set of magnitudes observed, the expected number of occurrences of the inequalities $N_1/N_2 > N_2/N_3 > N_3/N_4$ is 7.67 and the most probable number is 8. Thus there appears to be no need to go outside the random model for an explanation of Eyles' results.

In a study of fourth-order drainage basins in southern Uganda, Doornkamp and King (1971, p. 71) noted a significant correlation between the total first order stream length $\mathcal{L}_1$ and the total third order stream length $\mathcal{L}_3$. They commented that "there is no obvious reason why the total length of first-order streams should have any influence on the total length of third-order streams..."

By definition

$$\mathcal{L}_1 = N_1 \bar{\ell}_e \quad (8a)$$

$$\mathcal{L}_3 = N_3 \bar{L}_3 \quad (8b)$$

The value of $\mathcal{L}_3$ can be estimated from the random model (Smart, 1968, p. 1006-7) as

$$\mathcal{L}_3 \approx (N_1 - 1) \frac{N_2 - 1}{2N_2 - 1} \frac{N_3}{2N_3 - 1} \bar{\ell}_i \quad (8c)$$

This result, incidentally, is applicable to basins of any order. For fourth and higher order basins, $N_1 >> 1$, so that we can replace $(N_1 - 1)$ by N_1 without much loss in accuracy. The factor that depends on N_2 is slightly less than 1/2 and the factor that depends on N_3 is slightly greater than 1/2, so that their product is close to 1/4. Finally, substituting $N_1 = \mathcal{L}_1 / \bar{\ell}_e$ from equation 8a, we have

$$\mathcal{L}_3 \approx \mathcal{L}_1 / 4\lambda \qquad (9)$$

Note that the random model not only explains the correlation but predicts that the quantities are directly proportional and gives the proportionality factor in terms of the empirical parameter λ.

This analysis can easily be extended to produce general expressions of the form

$$\mathcal{L}_\omega \approx \mathcal{L}_1 / 2^{\omega-1}\lambda \qquad 2 \leq \omega \leq \Omega-1 \qquad (10a)$$

$$\mathcal{L}_\Omega \approx \mathcal{L}_1 / 2^{\Omega-2}\lambda \qquad (10b)$$

As an example of general relationships that can be explained by the random model, we first consider Melton's (1958) discovery that the Strahler stream frequency F_S (number of Strahler streams per unit area) and the square of the drainage density are approximately proportional. A regression analysis of 156 watersheds covering a wide range of both F_S and D gave the result

$$F_S / D^2 \approx 0.69 \qquad (11)$$

Equation 11 is evidently rather closely related to equation 4b

$$F/D^2 = 1/K \qquad (4b)$$

and since

$$F_S = N_S / a = N_S F/(2n-1) \qquad (12)$$

where N_S is the total number of Strahler streams in a network, we can write

$$F_S / D^2 = N_S / (2n-1)K \qquad (13)$$

This result is exact since it merely involves combining the definitions of the various quantities.

The random model can be used to estimate the right hand side of equation 13 (Shreve, 1967, p. 185). For networks of magnitude $n >> 1$, the number of Strahler streams is approximately $n + n/4 + n/16 + \ldots = 4n/3$. Then, setting K=1 and replacing (2n-1) by 2n, we have

$$F_S / D^2 \approx 2/3 \qquad (14)$$

in excellent agreement with Melton's observations.

Although it is clear at this point that the random model can be effectively applied to geomorphic variables derived from Horton-Strahler ordering, it should also be emphasized that the model is independent of ordering techniques. In fact, the explanation and prediction of relations between Horton-Strahler variables is not very rewarding because these quantities are byproducts of the ordering procedure and their relation to more fundamental network properties is often unclear. For example, Shreve (1967, p. 185) has noted that the use of equation 11 instead of equation 4b introduces additional uncertainty due to variations in network topology and deflects attention from the conceptually simple parameter K (or 1/K).

One of the best known empirical relationships in fluvial geomorphology is that between mainstream length L and drainage area. Hack (1957, p. 63-64) reported that his measurements on 90 drainage basins in Virginia and Maryland and measurements by Langbein et al. (1947) on about 400 basins in the northeastern United States could be well represented by the relation.

$$L = 1.4\mathcal{A}^{0.6} \qquad (15)$$

where L is measured in miles and $\mathcal{A}$ in square miles. Later work by Gray (1961) gave essentially the same results.

To apply the random model to this result, we must first introduce a new topologic parameter, the network <u>diameter d</u> (Werner and Smart, 1973). If we start at the outlet and count off the number of links to each source in turn, we obtain the set of <u>exterior path lengths</u> for the network. The largest exterior path length is called the diameter; clearly the diameter is closely related to the mainstream length. In particular, for reasonably large networks

$$L \approx d\bar{\ell}_i \qquad (16a)$$

and similarly

$$\mathcal{A} = (2n-1)\bar{a} \qquad (16b)$$

The last equation is exact if $\bar{a}$ is the mean link drainage area for the network and approximate if $\bar{a}$ is the mean link drainage area for a region including the network.

Thus the random model predicts that the mainstream length-area relationship should be largely determined by the way in which network diameter depends on the magnitude. Although d is a conceptually simple topologic parameter, it has not been possible to derive any simple relations for the statistical properties of d as a function of n for a topologically random population. Monte Carlo simulation is an obvious recourse in such cases. It should be noted, however, that the coefficient 1.4 in equation 15 is not dimensionless and that, regardless of whether theoretical or simulation methods are used to develop relations between d and n, a quantitative comparison of random model predictions with observation will require some way of transforming the d-n relations to the appropriate set of units.

Shreve (1970) used a digital computer and Monte Carlo techniques to generate a sample of topologically random channel networks with magnitudes ranging up to 10^6. Link lengths and link drainage areas were randomly assigned from actual data for a region in eastern Kentucky. For networks with drainage areas up to about 10^3 mi^2 ($n \approx 20{,}000$), the location, trend, and scatter of the plots of the simulated mainstream lengths and basin areas agreed very well with the Hack-Langbein observations. For larger basins, there was some unexplained deviation, with the simulated mainstream lengths falling below the actual ones.

Werner and Smart (1973) also simulated topologically random networks, but with magnitudes randomly chosen in the restricted range 20-200. The best-fit power law relation, established by regression of ℓn d on ℓn (2n-1) was found to be

$$d = 1.4\ (2n-1)^{0.58} \tag{17}$$

Quantitative relations between the two coefficients and the two exponents of equations 15 and 17 can be derived with the aid of equations 16a and 16b. We find that the two exponents should be equal and that the coefficients should satisfy the relation

$$C_o/C_s = 2K^{1-b}D^{2b-1}/(1+\lambda) \tag{18}$$

where C_o and C_s are the coefficients for the observed and simulated data, respectively, and b is the common exponent. An evaluation of the right hand side of equation 18 for 20 drainage basins in the eastern United States gave values ranging between 0.95 and 1.24 with a mean of 1.12. Thus the predicted relations between both exponents and coefficients are quite well satisfied. (The expression for the coefficient ratio C_o/C_s is an improvement over the one originally used by Werner and Smart, which essentially assumed $\lambda=1$). In evaluating the coefficient ratio, we did not set K=1, because the calculations involved the empirical parameter λ and there is some evidence that K and λ are correlated.

The results of Shreve and Werner and Smart strongly suggest that the empirical "law", equation 15, is in great part a consequence of the topological and geometrical randomness of natural drainage networks.

DETAILED DESCRIPTION OF DENDRITIC CHANNEL NETWORKS

The usual qualitative classification of drainage patterns as dendritic, trellis, radial, etc. is based on the relative orientations of channels and networks in the system and is carried out by visual inspection alone. Dendritic drainage patterns are the most common type and are generally considered to be associated with the absence

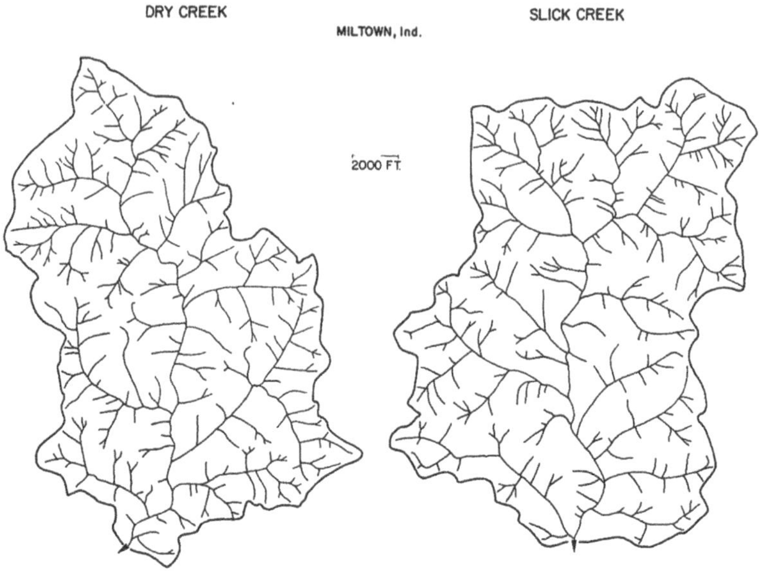

Figure 5 Dry Creek and Slick Creek channel networks, Miltown quadrangle, Indiana (Reprinted from Smart (1972b, p. 1488)) by permission of the American Geophysical Union. Copyright 1972).

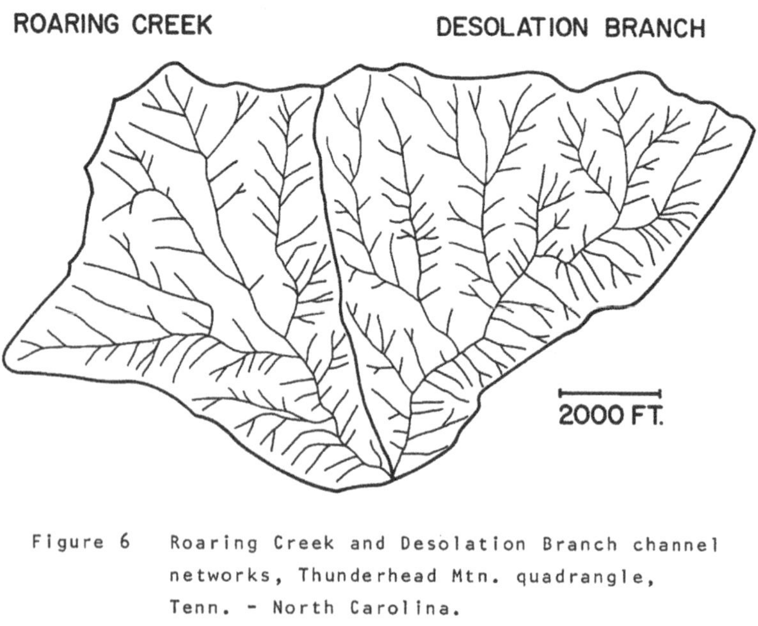

Figure 6 Roaring Creek and Desolation Branch channel networks, Thunderhead Mtn. quadrangle, Tenn. - North Carolina.

of significant geologic controls; thus dendritic channel networks may be expected to be topologically random. However, dendritic networks developed in one environment are usually quite easily distinguished from those developed in another environment. The distinguishing factors are again identified by eye, and are in general related to non-topologic features such as junction angles, relative lengths of exterior and interior links, etc.

As an example, Figure 5 shows two networks from southern Indiana and Figure 6 shows two networks from the Great Smoky Mountains, North Carolina. Although all four networks would be classified as dendritic, the eye readily identifies members of a pair as being more like each other than like members of the other pair.

Smart (1972) has proposed a method for quantitative characterization of these small but detectable differences in dendritic networks. Each network is described by four dimensionless variables derived from the elementary link lengths and drainage areas.

$$\lambda = \bar{\ell}_e/\bar{\ell}_i \tag{19a}$$

$$\alpha = \bar{a}_e/\bar{a}_i \tag{19b}$$

$$K_e = \bar{\ell}_e^{\,2}/\bar{a}_e \tag{19c}$$

$$K_i = \bar{\ell}_i^{\,2}/\bar{a}_i \tag{19d}$$

Actually only three of these parameters are independent since they are related by

$$\lambda^2/\alpha = K_e/K_i \tag{20}$$

Specification of all four is, however, helpful in interpretation.

Figure 7 shows a profile plot of λ, $\alpha^{\frac{1}{2}}$, $K_e^{\frac{1}{2}}$, and $K_i^{\frac{1}{2}}$ for the networks depicted in Figures 5 and 6. It is clear that members of the same pair of profiles tend to be more like each other than like the members of the other pair, just as for the corresponding channel networks.

Additional data and a detailed discussion of the method are given by Smart (1973). It seems likely that these procedures could also be used to characterize non-dendritic drainage patterns.

ACKNOWLEDGEMENT

I am indebted to R. L. Shreve and C. Werner for several helpful comments and suggestions.

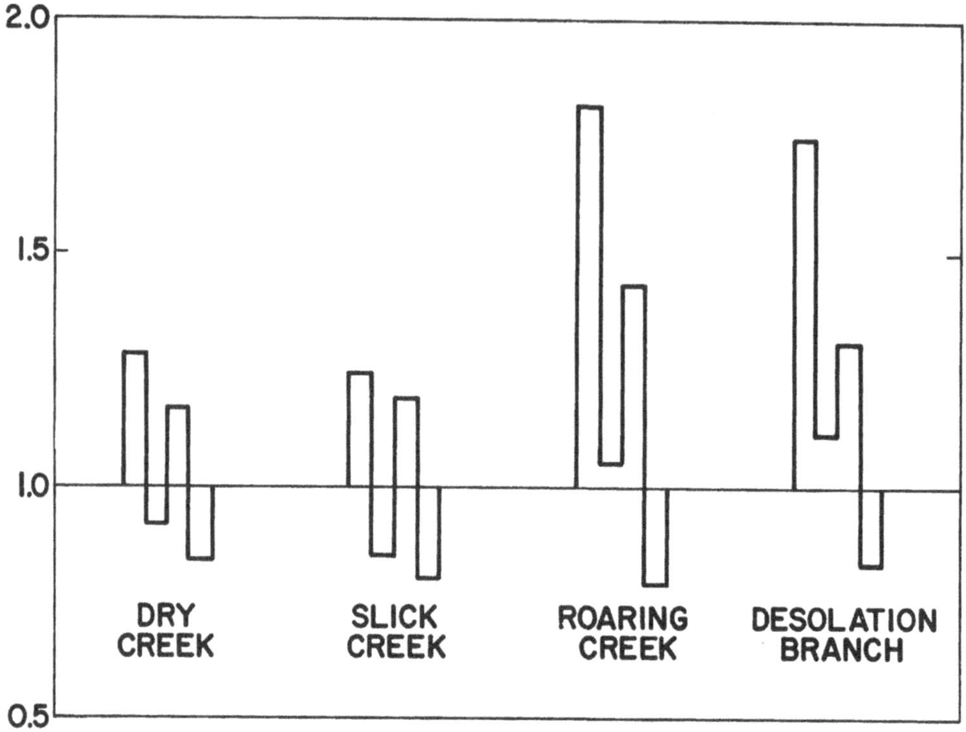

Figure 7 Profile plots of dimensionless variables for the channel networks of Figures 5 and 6. In each profile, values of λ, $\alpha^{\frac{1}{2}}$, $K_e^{\frac{1}{2}}$ and $K_i^{\frac{1}{2}}$ are plotted from left to right.

REFERENCES

Bowden, K. L., and Wallis, J. R., 1964, Effect of stream-ordering technique on Horton's laws of drainage composition: Geol. Soc. Amer. Bull., v. 75, p. 767-774.

Broscoe, A. J., 1959, Quantitative analysis of longitudinal stream profiles of small watersheds: Dept. Geology, Columbia University, Project NR389-042, Tech. Rept. no. 18, 73p.

Doornkamp, J. C., and King, C. A. M., 1971, Numerical analysis in geomorphology: London, Edward Arnold, 372p.

Eyles, R. J., 1968, Stream net ratios in West Malaysia: Geol. Soc. Amer. Bull., v. 79, p. 701-712.

Gray, D. M., 1961, Interrelationship of watershed characteristics: Jour., Geophys. Research, v. 66, p. 1215-1223.

Hack, J. T., 1957, Studies of longitudinal stream profiles in Virginia and Maryland: U. S. Geol. Survey Prof. Paper 294-B, p. 45-97.

__________, 1973, Stream-profile analysis and stream gradient index: Jour. Research U. S. Geol. Survey, v. 1, p. 421-429.

Horton, R. E., 1945, Erosional development of streams and their drainage basins: hydrophysical approach to quantitative morphology; Geol. Soc. Amer. Bull., v. 56, p. 275-370.

Howard, A. D., 1971, Simulation model of stream capture: Geol. Soc. Amer. Bull., v. 82, p. 1355-1376.

James, W. R., and Krumbein, W. C., 1969, Frequency distributions of stream link lengths: Jour. Geol., v. 77, p. 544-565.

Krumbein, W. C., and Shreve, R. L., 1970, Some statistical properties of dendritic channel networks: Dept. Geol. Sci., ONR Task No. 389-150, Tech. Rept. no. 13; also Dept. Geol., UCLA, Spec. Proj. Rept. NSF Grant 6A-1137, 117p.

Langbein, W. B., et al., 1947, Topographic characteristics of drainage basins: U. S. Geol. Survey Water-Supply Paper 947-C, p. 125-155.

Leopold, L. B., and Langbein, W. B., 1962, The concept of entropy in landscape evolution: U. S. Geol. Survey Prof. Paper 500-A, A1-A20.

Maxwell, J. C., 1960, Quantitative geomorphology of the San Dimas National Forest, California: Dept. Geology, Columbia University, Project NR389-042, Tech. Rept. no. 19, 95p.

Melton, M. A., 1958, Geometric properties of mature drainage systems and their representation in an E_4 phase space: Jour. Geol., v. 66, p. 35-56.

Melton, M. A., 1960, Concept of state of a drainage system: Geol. Soc. Amer. Bull., v. 71, p. 1928.

Milling, M. E., and Tuttle, S. D., 1964, Morphometric study of two drainage basins near Iowa City, Iowa: Proc. Iowa Acad. Sci., v. 71, p. 304-319.

Morisawa, M. E., 1962, Quantitative geomorphology of some watersheds in the Appalachian Plateau: Geol. Soc. Amer. Bull., v. 73, p. 1025-1046.

Schuum, S. A., 1956, Evolution of drainage systems and slopes in badlands at Perth Amboy, New Jersey: Geol. Soc. Amer. Bull., v. 67, p. 597-646.

Shreve, R. L., 1966, Statistical law of stream numbers: Jour. Geol., v. 74, p. 17-37.

__________, 1967, Infinite topologically random channel networks: Jour. Geol., v. 75, p. 179-186.

__________, 1969, Stream lengths and basin areas in topologically random channel networks: Jour. Geol., v. 77, p. 397-414.

__________, 1970, Mainstream length vs. basin area in topologically random channel networks: Geol. Soc. Amer., Abstracts with Programs, v. 2, p. 682.

Smart, J. S., 1968, Statistical properties of stream lengths: Water Resources Research, v. 4, p. 1001-1013.

__________, 1969, Topological properties of channel networks: Geol. Soc. Amer. Bull., v. 80, p. 1757-1774.

__________, 1972a, Channel networks: from Advances in Hydroscience, v. 8, New York, Academic Press, V. T. Chow, Ed., p. 305-346.

__________, 1972b, Quantitative characterization of channel network structure: Water Resources Research, v. 8, p. 1487-1496.

__________, 1973, Joint distribution functions for link lengths and drainage areas: Proc. Internat. Assoc. Math. Geologists Symp. on Random Processes in Geology, Montreal, 1972, in press.

Strahler, A. N., 1952, Hypsometric (area-altitude) analysis of erosional topography: Geol. Soc. Amer. Bull., v. 63, p. 1117-1142.

__________, 1957, Quantitative analysis of watershed geomorphology: Trans. Amer. Geophys. Union, v. 38, p. 913-920.

__________, 1964, Quantitative geomorphology of drainage basins and channel networks: in Handbook of Applied Hydrology, New York, McGraw Hill: V. T. Chow, Ed., p. 4-39 to 4-76.

Werner, C., and Smart, J. S., 1973, Some new methods of topologic classification of channel networks: Geog. Analysis (in press).

Yang, C. T., and Stall, J. B., Note on the map scale effect in the study of stream morphology: Water Resources Research, v. 7, p. 709-712.

3

DRAINAGE ADJUSTMENT IN THE APPALACHIANS

John T. Hack

ABSTRACT

The degree of adjustment of the topography can be analyzed by study of longitudinal stream profiles. Streams of the same size whose profiles are mutually adjusted tend to have equal gradients on rocks of equal resistance to erosion. If their gradients are different, adjustment has not been achieved. Stearns' concept of the subenvelope is used as a means of generalizing the vertical dimensions of the stream network and at the same time generalizing the topography. A contour map of the central and southern Appalachians drawn on the stream channels but eliminating all streams less than 20 miles (32 km) in length shows that the general height of land is just west of the Atlantic-Mississippi drainage divide. The drainage divide and general height of land do not, however, parallel the structural trend of the rocks but trend more to the north. The divide crosses the Appalachian Valley south of Roanoke, Va., at the place where the structural pattern changes from one of imbricate thrust sheets to more open folds.

Selected stream profiles on both sides of the divide are analyzed by means of Hack's gradient index concept. This analysis and examination of regional lithologies suggest that the position of the highlands is determined by a partial adjustment of the profiles of the major streams to the regional rock pattern. The most resistant rock units are the sandstones of Early Pennsylvanian age in the Appalachian Plateau and various rocks of late Precambrian age along the northwest margins of the Blue Ridge highland which include the Chilhowee and Great Smoky Groups and volcanic rocks of low metamorphic grade. The least resistant rocks are shales of Silurian and Devonian age, shales and carbonates of Cambrian and Ordovician age, mostly in the Valley and Ridge, and crystalline metamorphic rocks of the Blue Ridge and Piedmont.

Some areas of the Appalachians are not in adjustment. In particular the southeastern border of the Blue Ridge Highland south of Roanoke is not in equilibrium as shown by over-steepened profiles of headwater streams. This is a zone of shifting divides probably related to a zone of uplift or faulting that may have occurred in Triassic time. Other areas of disequilibrium on a smaller scale are probably common in the Appalachians and could be related either to tectonism or to adjustments required by changes in the pattern of the rocks as the erosional topographic surface is lowered by weathering and erosion.

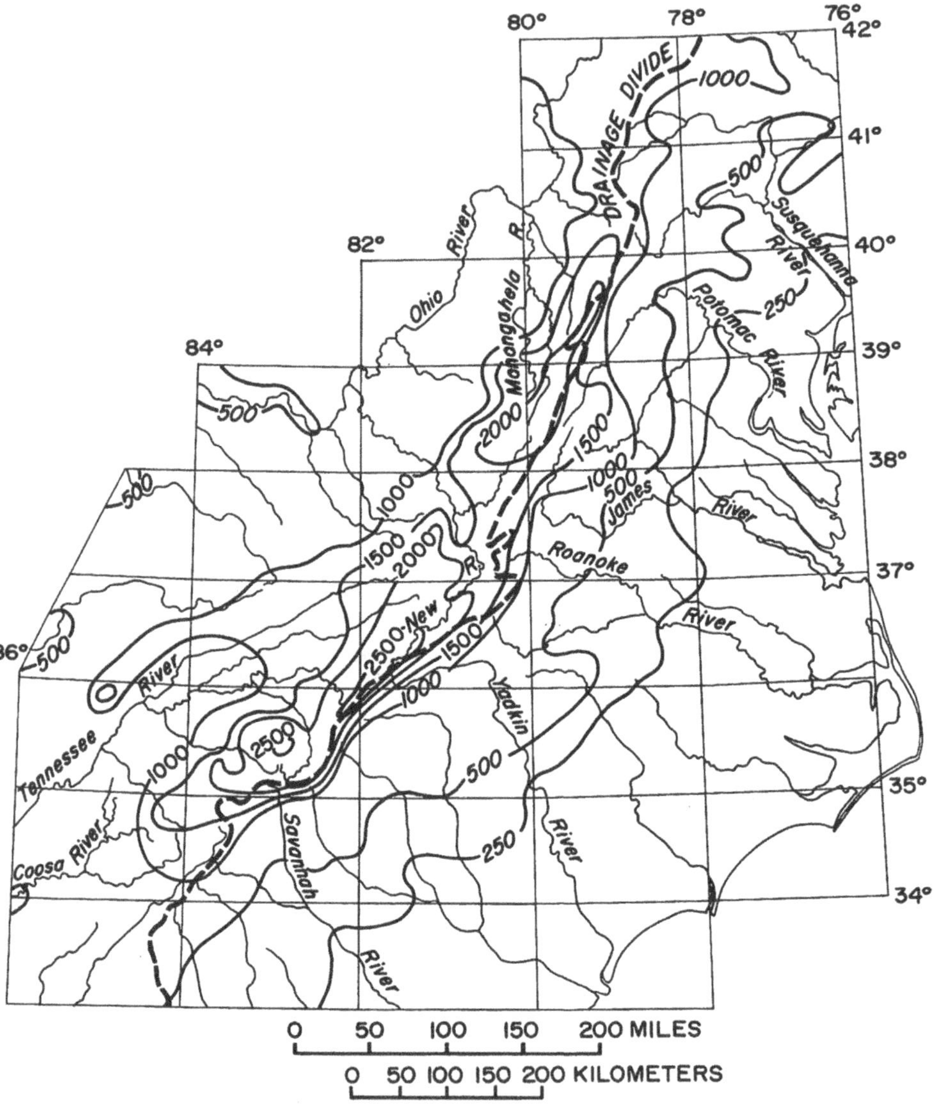

Figure 1 Subenvelope map of the southern and central Appalachians showing elevations on stream channels over 20 miles (32 km) in length. Atlantic-Gulf drainage divide is shown by heavy dashed line. Elevation data are from U. S. Geological Survey 1:250,000-scale series of topographic maps.

INTRODUCTION

The evolution of Appalachian drainage has fascinated many geologists and dozens of references on the subject could be cited. Many thought that the original height of land lay to the southeast of the present Atlantic-Gulf divide and southeast of the Appalachian Valley. This was a reasonable idea because the oldest rocks lay in that region (Davis, 1889). It was assumed, however, that the area had been reduced to a peneplain by erosion before the drainage reversal which occurred during a later period of dissection. The idea of an extensive divide migration, however, during the dissection of a peneplain poses serious problems and aroused much discussion among geomorphologists (reviewed by Thornbury, 1965, p. 82). The concept of drainage reversal is more reasonable if peneplanation is not involved and if it occurred as part of the process of long continued erosion.

Whatever the past history of the drainage network, it is now for the most part well adjusted to the geology. When observed in plan the adjustment is remarkable in the Ridge and Valley province where the contrast between hard and soft rocks is extreme. Most of the larger rivers avoid the resistant rocks. For example, the Tennessee passes out of the valley and ridge and crosses the Appalachian Plateau without crossing the outcrop of the massive sandstones of Early Pennsylvanian age, except for a short reach at Walden Ridge where it has cut through them into carbonate rocks of Mississippian age. The main stem of the Potomac stays almost entirely in the nonresistant rocks and crosses the belt of Tuscarora Sandstone where the sandstone is faulted out for a short distance.

In the weathered crystalline rocks of the Blue Ridge and the Piedmont, as well as in the flat-lying sedimentary rocks of the plateau, the criteria for determining adjustment of the stream pattern are not so evident. This is probably because these rocks do not have strongly contrasting structural patterns at the surface which impose a constraint on erosion. The New River is unique among the larger streams of the Appalachians in that it is the only one that flows all the way across the Appalachian highlands; its course shows no adjustment to structure in the Appalachian Valley, which it crosses at its highest segment.

The vertical dimension of the topography and its adjustment to geology can also be analyzed by study of stream profiles and stream channel gradients. As will be shown, the Appalachian drainage network is adjusted to the rocks in the vertical dimension as well as the horizontal. The adjustment is not complete, however, and many anomalous areas probably exist in which the erosional energy of the streams may be greater or less than the norm. The Blue Ridge highland is believed to be an outstanding example which can be analyzed now that the details of its geology are emerging as a result of recent work.

GENERAL CONFIGURATION OF THE APPALACHIAN DRAINAGE SYSTEM

Study of a drainage system as large and complex as that of the Appalachians is time-consuming simply because of the number of streams involved. This study of so large a system was made feasible by a detailed manuscript map of the drainage at 1:1,000,000 scale, courtesy of the late F. E. Matthes. The map depicts a drainage density of approximately one-half mile of stream length per square mile, although first order and many second order streams are not shown. The use of the map at a reduced scale simplified location of drainage divides and selection of data points.

The general elevation of the drainage network in the area south of the New York state line is shown on Figure 1. This map is a general contour map of the Appalachians similar to the subenvelope map of R. G. Stearns (1967, p. 1114). As the contours are drawn on the channels of the main streams, it shows the configuration of the bottom of the topography. It was constructed as follows: All streams having a length greater than 20 miles (32 km) were selected using the manuscript drainage map referred to above. Each of these streams was then located on sheets of the 1:250,000-scale series of topographic maps and the location of channel elevations at 250, 500, 1,000, 1,500, 2,000 and 2,500 feet were plotted on the drainage map at reduced scale. The rule

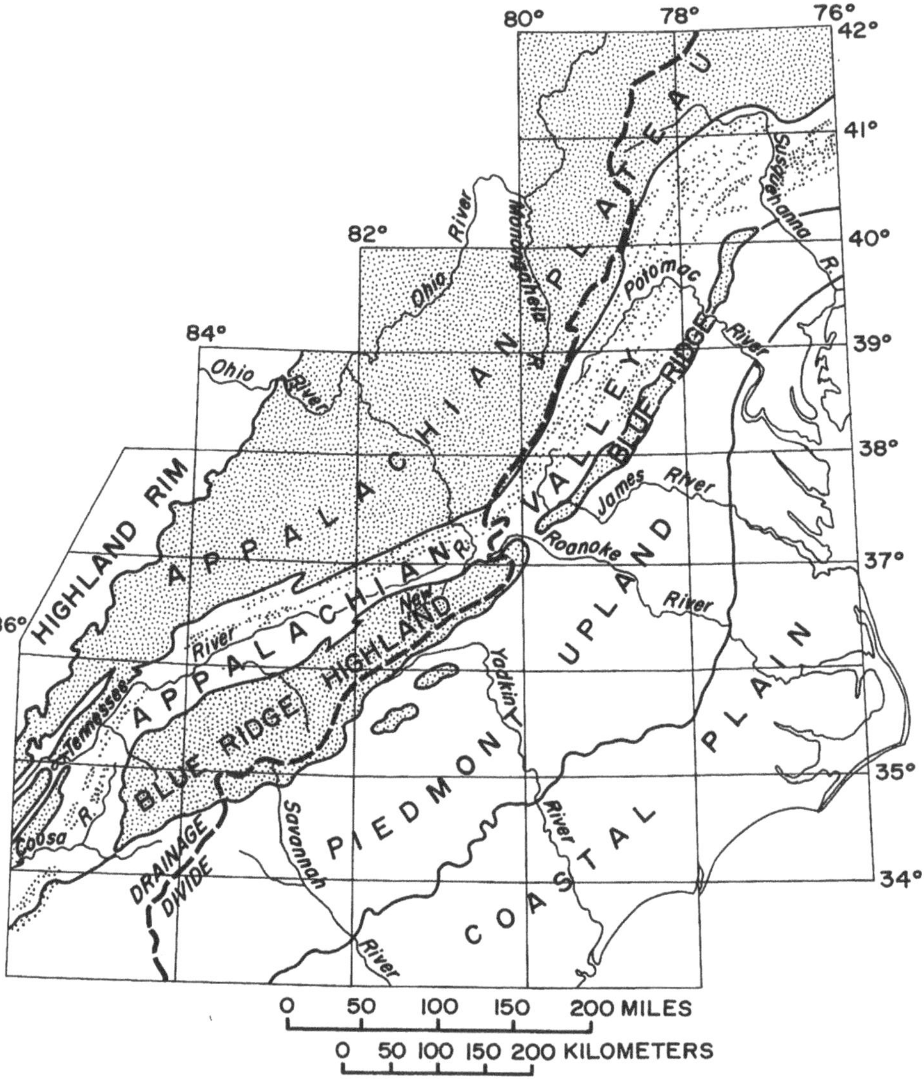

Figure 2 Physiographic map of the southern and central Appalachians. Stippled pattern represents mountain and highland areas. Compilation is based on tectonic maps of southern and central Appalachians (Fisher, 1970; U. S. Geological Survey and Bureau Mines, 1968, Plate 2; and Fenneman, 1938, Plates I-III).

followed was to omit all elevation points less than 20 miles (32 km) from the head of a stream. The map thus shows a highly generalized surface on the valley bottoms. Local mountain topography, of course, is not shown because elevation data along all short streams are omitted. The final product (Figure 1) is based on approximately 360 data points.

The general height of land in the Appalachians is consistently west of the drainage divide (Figure 1). The profiles on either side of the divide are asymmetric, as the streams flowing to the southeast drop off more steeply from the divide than those on the northwest. Comparison with Figure 2 shows that the drainage divide is not parallel to the structural trend of the Appalachians but crosses it at a sharp angle in a more northerly direction. Note also that the Potomac and Susquehanna Rivers cross the folded belt of the Appalachian Valley at very low elevations. All the streams that cross the rocks of Early Pennsylvanian age on the Appalachian Plateau, however, maintain a high elevation for some distance from the drainage divide. Other features shown on Figure 1 are the high elevation of the New River where it crosses the Appalachian Valley and the high elevation of the southern Blue Ridge belt between the Atlantic slope on one side and the Tennessee and Coosa River systems on the other.

RELATION OF DRAINAGE CONFIGURATION TO GEOLOGY

The rational basis for this analysis of Appalachian drainage is the concept that streams in any reach tend to adjust their gradients both to discharge and to the character of the rocks or other material making up the channel. For a given discharge streams in rocks that are resistant to both chemical and physical attack will have steeper gradients than streams in softer rocks. When resistant material becomes part of the bed load, the material affects roughness and channel cross section for some distance downstream from its source. As the drainage system is essentially the skeletal frame work of the topography, the interaction between the streams and the rocks determines the general form of the landscape.

Studies in the Potomac River basin (Hack, 1957, p. 74) indicate that rock materials such as sandstone, quartzite, vein quartz and chert persist in streams for long distances below their source, whereas carbonate rock, shale and metamorphic rock, like mica schist, break up within or close to the reach in which they crop out. Unpublished studies on the Blue Ridge escarpment of North Carolina indicate that quartz-rich, coarse-grained and massive igneous and metamorphic rocks support steeper channel slopes than sheared, fissile or highly micaceous rocks. For example, in the Grandfather Mountain area (Bryant and Reed, 1970) the steep gorges in the area are in massive granites (Beech Granite, Brown Mountain Granite) and especially in arkosic quartzite (Grandfather Mountain Formation).

The structural attitude of the rock and its length of outcrop in or near the stream also affect gradient. A stream flowing through a massive sandstone that is horizontal or gently dipping is more affected by the resistant material than a stream crossing a steeply dipping bed in a short reach. This is because a nearly horizontal sandstone layer has a long outcrop area and contributes to greater roughness even though it may be high above the channel, for large blocks can enter the stream by creep, land falls and small tributaries.

METHOD OF STUDY

Study of stream gradients is complicated by the fact that discharge as well as rock materials affect slope. Generally the greater the discharge the gentler the slope. In streams of great length, that is, several hundred miles or more, the rate of increase of discharge with distance is small so that sharp changes in geology along the channel are apt to be the primary cause of sharp changes in slope. The effect of increasing discharge, however, even in small streams, can be masked by plotting the stream profile on a semilogarithmic graph in which the distance from the source is plotted on a logarithmic scale. This foreshortens the profile in proportion to the increased discharge, so that the gradient of an upstream reach can be compared directly with that of a downstream reach.

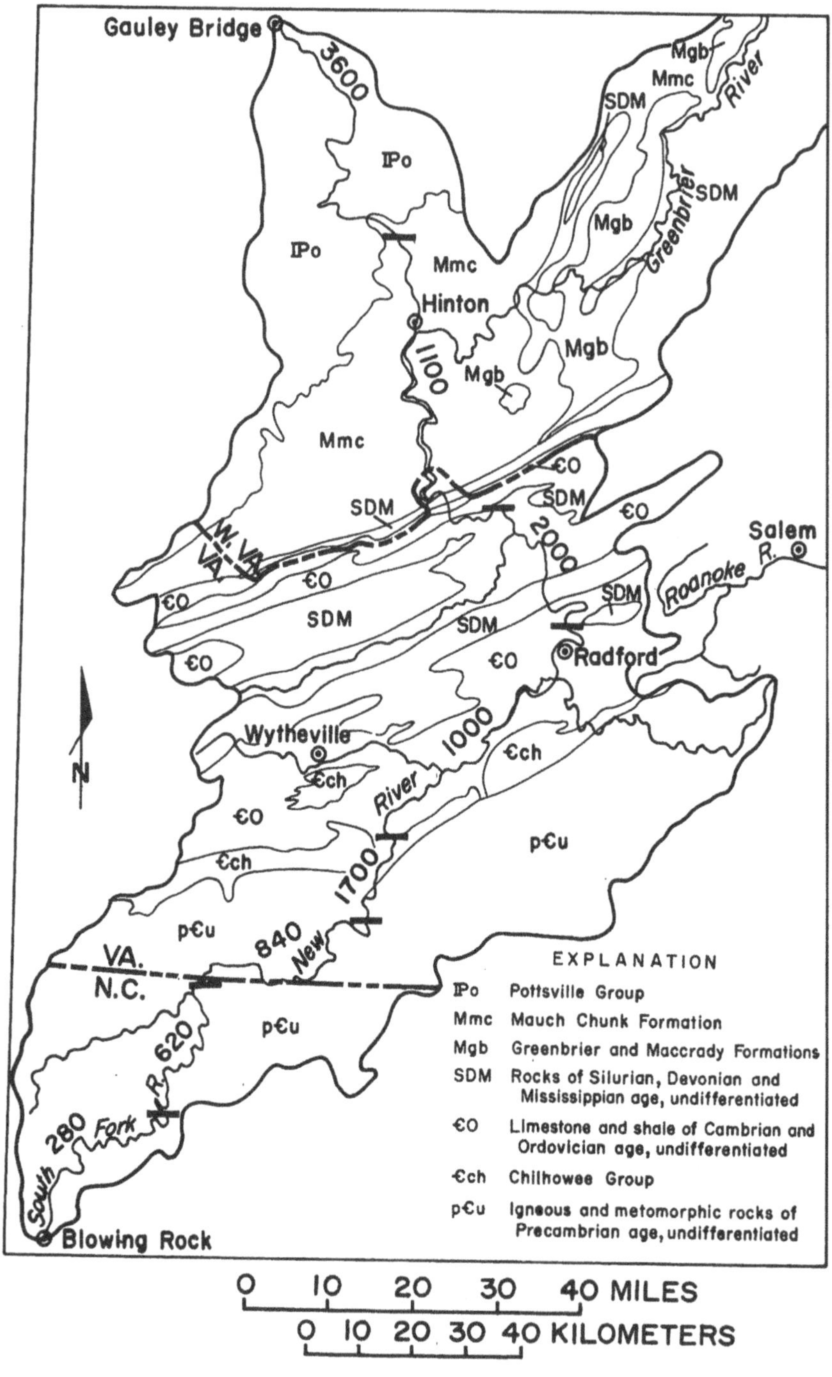

Figure 3 Generalized geologic map of New River basin showing SL values of reaches (separated here by rectangular bars) having different bedrock lithology. Geology from state geologic maps of Virginia, West Virginia and North Carolina.

It has been found that the steepness of the logarithmic curve is in a very rough way proportional to a function of the size of bed material and can be used as an indication of the competence of a stream and to compare channel slopes of streams of different sizes (Hack, 1957, p. 88-89, and 1973). The slope of the curve serves as an index called the stream gradient index (or SL value) and can be calculated by two methods:

1. Gradient index = $$\frac{H_1 - H_2}{\log_e L_2 - \log_e L_1}$$

where H_1 and H_2 are the altitudes of each end of a given reach and L_1 and L_2 are the distances from each end of the reach to the source of the stream measured along the course of the longest tributary.

2. Gradient index = SL

where L is the distance from the midpoint of the given reach to the source in miles and S is the channel slope of the reach in feet per mile. Equation 2 generally will not yield quite the same value as equation 1 but is useful when a large number of stream reaches are to be compared quickly. The method of study used in this paper is illustrated by the profile of the New River.

New River

The New River rises on the Blue Ridge. Its longest headwater, the South Fork, begins at the town of Blowing Rock, North Carolina, at the edge of the Blue Ridge escarpment. It flows northward to cross the Appalachian Valley at Radford and at length 343 miles (550 km), as measured on topographic maps, joins the Gauley River at Gauley Bridge to form the Kanawha. The geology of the basin is shown in Figure 3. The numbers along the river are average SL values for the reaches between the rectangular bars.

In the upper course the river flows mostly in Precambrian rocks of high metamorphic grade belonging to the Ashe Formation (Rankin, 1970, p. 232). This is a fine-grained layered mica gneiss and schist cut by numerous thin quartz veins. The quartz erodes in cobble- and boulder-size fragments and becomes more and more prevalent in the stream bed downstream and probably accounts in large part for the rise in SL values from 280 to 840 (Figure 3). In the Chilhowee Group of Cambrian age the SL values rise to an average of 1700 and the profile actually steepens. The rocks are highly siliceous and contain beds of brittle quartzite that underlie a low mountain range south of the Appalachian Valley. In the reach above Radford, underlain by limestone and shale, the gradient index drops to an average of 1000. It increases again below Radford as the river crosses ridges of resistant Silurian and Devonian quartzite and sandstone. In the shaly rocks of the Mauch Chunk Formation SL values again decline. In the course below Hinton the river enters the Pottsville Group of Early Pennsylvanian age which contains many gently dipping thick strata of hard sandstone. In this section the profile steepens abruptly and has SL values over 3600.

The long profile of the New River is shown on both a logarithmic graph (Figure 4A) and an arithmetic graph (Figure 4B). Altitudes are shown by dots, the numbers are SL values for the segments shown by dashed lines. In profile (A) the dashed line connecting the two ends of the stream has a slope or SL value of 560, which represents the average for the entire river. In profile (B) the New River is shown on an arithmetic scale with a vertical exageration of 2640 times. The smooth dashed curve is the same simple logarithmic profile as the dashed straight line in profile (A). This probably approximates the profile the river would have if it maintained the same competence throughout its length. It is interesting to note that the actual profile falls only 700 feet in the first 70 miles of the river course. At the lower end, where a gentle gradient would be expected because of the high discharge, it falls the same amount in only 60 miles. In conclusion it appears that adjustment of the river to increasingly resistant materials downstream gives it a slightly convex upward profile and accounts for the height of the Blue Ridge upland in that part of the New River basin south of the Appalachian Valley.

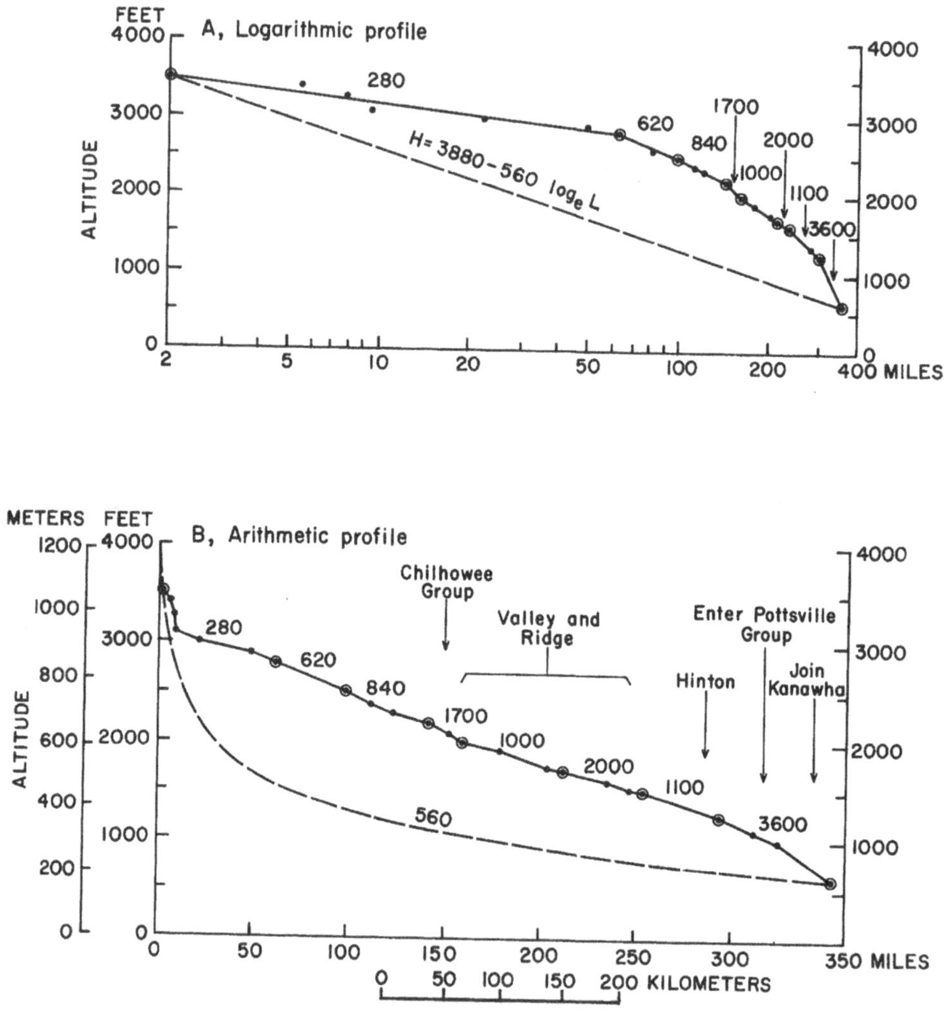

Figure 4 Longitudinal profiles of New River on logarithmic graph (A) and on arithmetic graph (B). Data points are shown by dots. Numbers between open circles represent SL values.

In the analysis that follows the stream profiles will be shown on arithmetic scales as in profile (B). The use of SL values permits comparison of different streams and different reaches.

STREAMS OF THE SOUTHERN BLUE RIDGE

The southern Appalachians are here considered to comprise the Blue Ridge highland and Appalachian Valley south of the Roanoke River and the Appalachian Plateau south of the New River (Figure 2). The Blue Ridge highland, unlike the Blue Ridge northeast of Roanoke, is a broad upland with complex geology (Figure 5). Like the Piedmont it is underlain by crystalline metamorphic rocks, but they are mostly of Precambrian age rather than Paleozoic age. Its northwestern margin is an escarpment in many places. It essentially follows contacts on which the resistant rocks of the Chilhowee Group and Ocoee Supergroup are faulted against nonresistant carbonate rock and shale of lower Paleozoic age. The resistant rocks on the Blue Ridge side of this contact are rock sequences of low metamorphic grade containing massive quartzites, volcanics and gneisses (Colton, 1970, Hadley, 1970, Morgan, 1972, and Rankin, 1970).

The southeastern margin of the Blue Ridge highland is also a steep escarpment but is more ragged in outline, having indentations and several outlying ranges disconnected from the main highland mass. In addition, on this side of the Blue Ridge the rocks are mostly medium and high grade metamorphic rocks like those of the adjacent Piedmont (Morgan, 1972). The top of the escarpment on the southeast corresponds to the Atlantic-Gulf drainage divide, and it is this divide rather than the differences in rock resistance that defines the boundary between the Blue Ridge and the Piedmont.

Selected streams draining the Blue Ridge were studied in order to determine the relation of the stream profiles to the geology. The streams that cross the low grade metamorphic rocks and flow northwestward into the Appalachian Valley are first considered (Figure 5; Figure 6, profiles A-D). The streams are similar in that each profile contains a large hump in the lower part of the course before leaving the Blue Ridge highland to enter the Appalachian Valley. SL values vary in the upper courses but increase downstream. These streams are like the New River in that a high proportion of their fall is in the lower course. As soon as they enter the Appalachian Valley their gradients flatten in adjustment to nonresistant carbonate rocks and shale, and the coarser fractions of their load are dropped in fans or broad terraces.

The French Broad River (Figure 5; Figure 6A) is shown in the geologic map of the Knoxville 1:250,000 scale quadrangle (Hadley and Nelson, 1971). The steep reach at the head of the French Broad (SL-750) is in folded biotite gneiss of Precambrian age containing concordant intrusives of granitic gneiss and amphibolite of Paleozoic age. The profile flattens abruptly as the river crosses the Brevard zone and follows the strike in a long belt of sheared biotite gneiss of the Henderson Gneiss (SL-110). In the lower part of this reach the river turns northward recrossing the Brevard zone to enter an area underlain by the Great Smoky Group of Late Precambrian age. Here the Great Smoky Group contains finer grained rocks, more pelitic with thinner bedded sandstones than to north and west, and the metamorphic grade is higher (Hadley and Nelson, 1971). High SL values occur in basement rocks of Middle Precambrian age composed of layered gneiss. SL values exceed 2300 in the Max Patch Granite and Cranberry Gneiss which are more massive in structure. On crossing the rocks of Late Precambrian and Early Cambrian age (SL values 1200), the gradient flattens. These rocks are mainly shale, siltstone and fledspathic sandstone.

The Tuckaseegee-Little Tennessee River system (Figure 5; Figure 6B) has a profile similar to that of the French Broad. The upper profile is across a complex group of metamorphic and igneous rocks of medium to high metamorphic grade. The river steepens below Bryson City in the Great Smoky Group, which here contains coarse clastic rocks of low metamorphic grade. The steepest segment (SL-1800) is in the quartzitic rocks of the Early Cambrian Chilhowee Group.

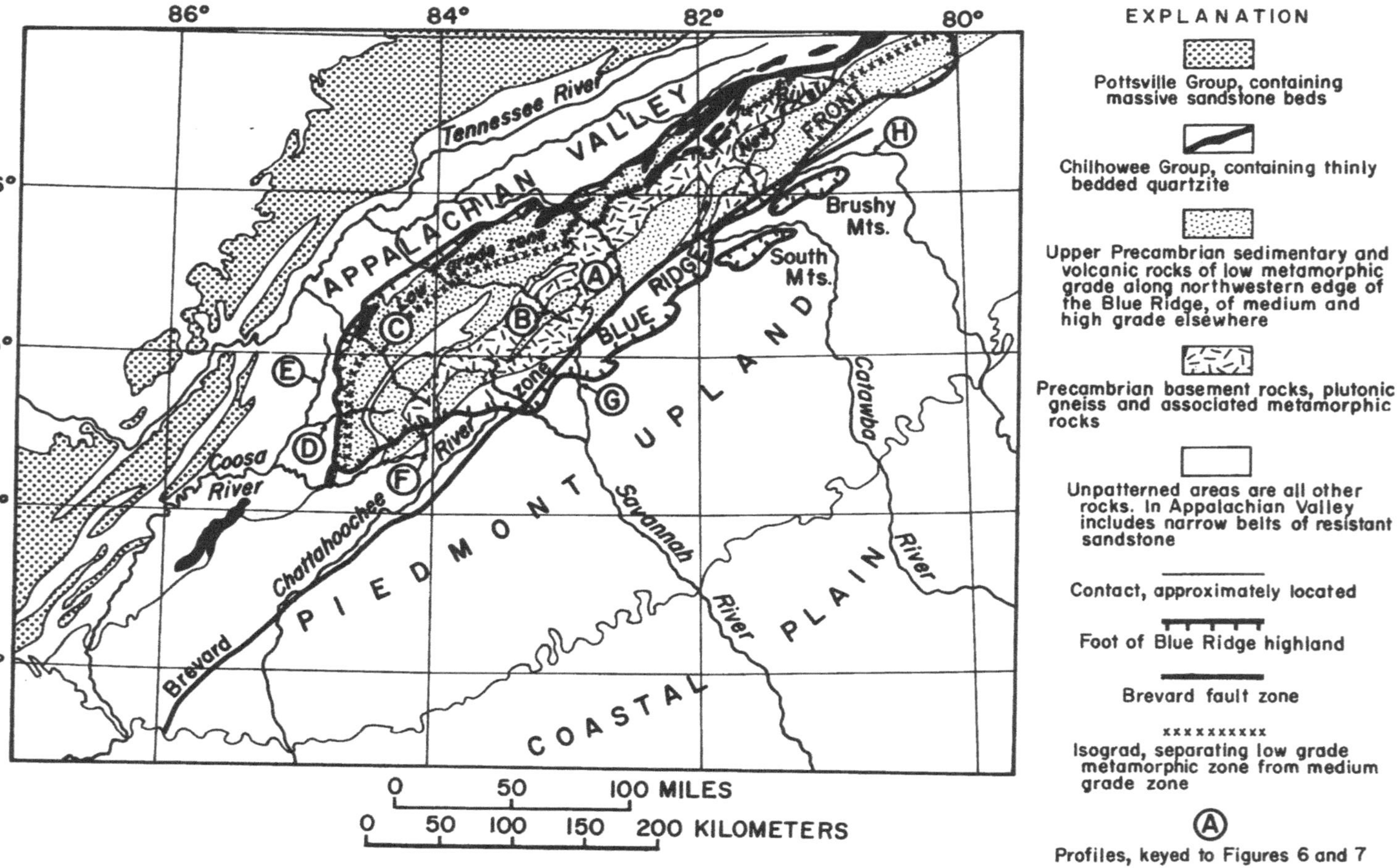

Figure 5 Simplified geologic map of southern Appalachians showing location of streams in Figures 6 and 7. Modified from Fisher (1970) and Morgan (1972).

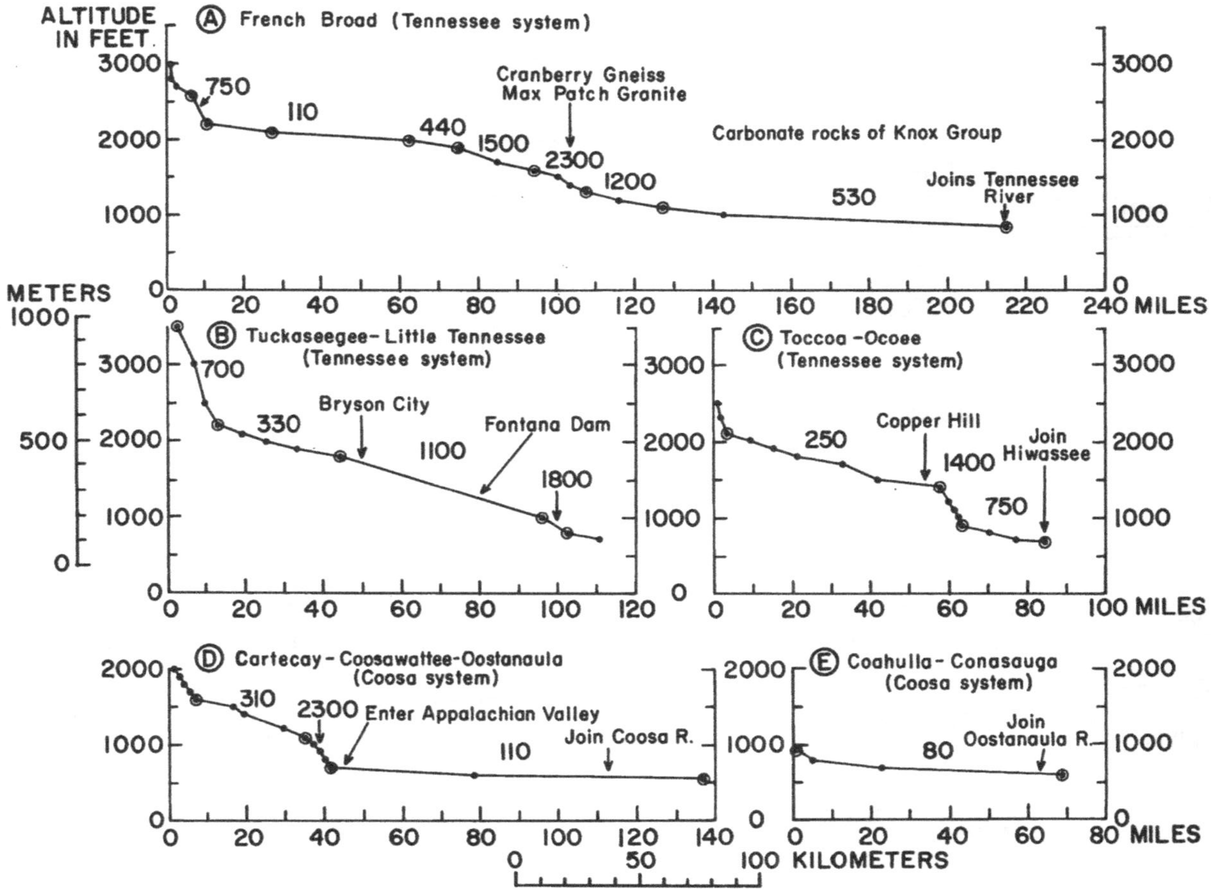

Figure 6 Profiles of selected streams flowing northwest and west across the Blue Ridge into the Appalachian Valley. Data points are shown by dots. Numbers between open circles represent SL values.

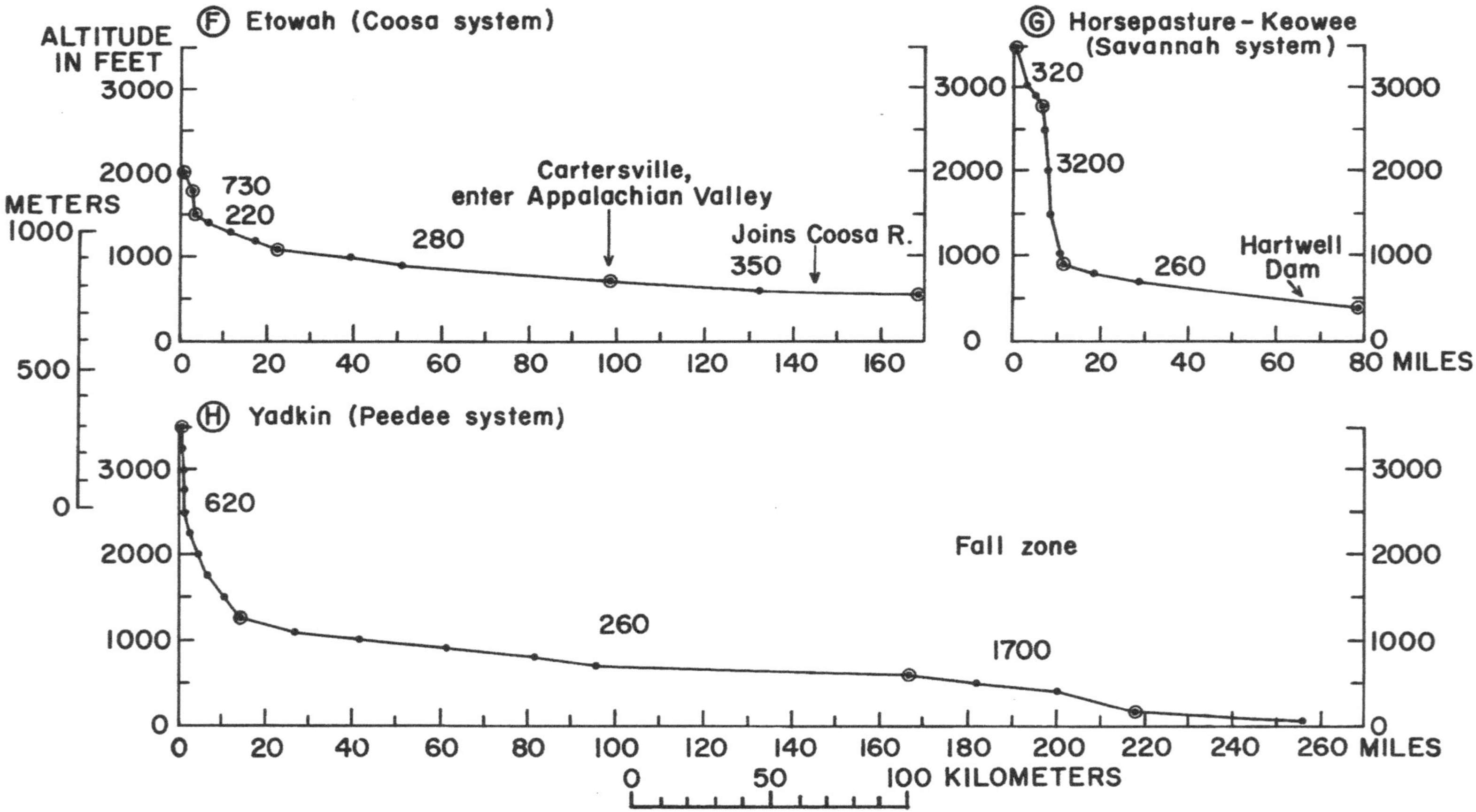

Figure 7 Profiles of selected streams flowing southeastward down the Blue Ridge escarpment into the Piedmont. Data points are shown by dots. Numbers between open circles represent SL values.

The Toccoa-Ocoee and Cartecay-Coosawattee-Oostanaula-Coosa systems (Figure 5; Figure 6C and D) both have steep "humps" in their profiles above their entrance in the Appalachian Valley. In both cases the steepest reaches are in low grade metamorphic rocks of Late Precambrian age equivalent to the Great Smoky Group. Note the very gentle profile (SL-110) of stream D in the Ordovician and Cambrian rocks of the Appalachian Valley.

The Coahulla-Conasauga River system (Figure 5; Figure 6E) is a headwater of the Coosa River. It heads on the floor of the Appalachian Valley in carbonate rocks, and its entire course is in limestone and shale of Early and Middle Cambrian age. The average SL value is only 80, and the profile form approximates a logarithmic curve. The SL value is probably low because the stream is not fed by any large tributaries that cross resistant rocks, and its bed material is comparatively fine-grained.

Three river profiles that typify the escarpment on the southeast margin of the Blue Ridge highland are shown in Figure 7. These profiles are concave and differ markedly from those on the northwest margin. The Etowah River profile (Figure 5; Figure 7F) approximates a logarithmic curve. Its course above Cartersville is entirely in medium to high grade metamorphic rocks and the river enters the Appalachian Valley on a low gradient. At Cartersville it crosses a narrow belt of rocks of the Chilhowee Group containing beds of quartzite, and the SL values increase from 280 to 350. This is only a slight increase, but it may be related to a small input of quartzite in the load.

The Horsepasture-Keowee Rivers, headwaters of the Savannah system (Figure 5; Figure 7G), are in the Atlantic drainage. They are entirely in metamorphic rocks. The profile is concave and oversteepened in the headwaters. It falls 2500 feet in the first 10 miles. The steepest part of the profile is in unsheared Henderson Gneiss, a very massive rock which may be responsible for the steep profile in this section. However, oversteepening and high SL values are characteristic of all the headwater profiles on the Atlantic side of the Blue Ridge highland.

The Yadkin-Peedee River system (Figure 5; Figure 7H) heads just over the hill from the South Fork of the New River near the town limits of Blowing Rock, North Carolina. It flows entirely in schistose and gneissic rocks of the Piedmont for a distance of 180 miles where it enters the fall zone and is lowered to the Coastal Plain. The river accomplishes most of its fall in the first 18 miles of its course, on the edge of the escarpment. At the foot of the escarpment the river turns northeast along the strike, and the gradient abruptly declines. It increases abruptly to high values before entering the Coastal Plain.

The steep headwaters are in gneisses and schists of low to medium metamorphic grade (Bryand and Reed, 1970) that do not appear to be especially resistant. However, the stream contains bed material of cobble and boulder size consistent with high SL values. As will be discussed below in connection with other evidence, the high SL values on the escarpment are probably evidence of more rapid erosion of the escarpment than is taking place on the Piedmont or on the Blue Ridge itself. The high SL values at the fall zone are also probably related to factors other than rock resistance and can be attributed to rapid erosion of the Coastal Plain sediments and Piedmont rocks beneath them following the Late Tertiary lowering of sea level.

ORIGIN OF THE BLUE RIDGE HIGHLAND

Study of stream profiles has shown that the Blue Ridge Province south of the Roanoke River is essentially a highland marginal to the Atlantic-Gulf divide (Figure 2). It is supported by a belt of resistant sedimentary and low grade metamorphic and volcanic rocks that parallel the divide 30 to 40 miles (48 to 64 km) on the northwest. As large areas on the highland are underlain by nonresistant metamorphic rocks of high to medium grade, their relief is gentler than on the upland margin. The northern part of the highland is essentially the upper basin of the New River. This river crosses both the Blue Ridge and the Appalachian Valley at high elevations, because its profile is supported in its lower course by the sandstones of the Pottsville Group.

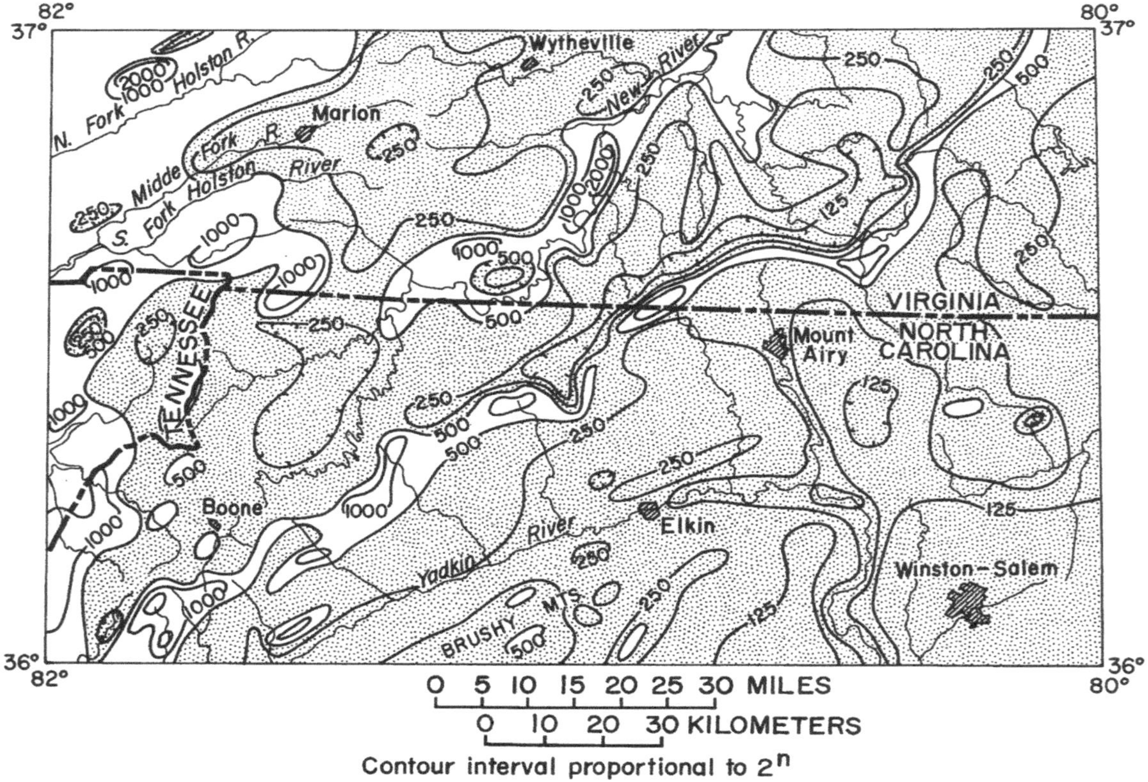

Figure 8 Map of Winston-Salem quadrangle (U. S. Geol. Survey 1:250,000-scale series of topographic maps) showing SL values on streams by contours. Stippled pattern covers areas having SL values less than 500. Note the zone of high values trending northeast along the drainage divide near the crest of the Blue Ridge escarpment.

The Blue Ridge escarpment on the southeast has a different origin, as it is unrelated to geologic boundaries except in limited small areas. The lack of conformity to geology is shown in a general way on Figure 5. The most strikingly discordant feature is probably the Brevard zone, a narrow linear belt of strongly sheared and foliated rocks, traceable from central Alabama through North Carolina. The escarpment clearly has a direct relation to the drainage divide, as was known to W. M. Davis (1903). Davis ascribed the escarpment to headward erosion by streams draining to the Atlantic. These streams have a shorter distance to base level than the streams flowing to the Gulf by way of the Coosa and Tennessee. The data available on modern maps, however, show that the streams on both sides of the divide fall to altitudes as low as 600 to 700 feet within a short distance on either side of the Blue Ridge highland. As shown in this study the flatter profiles of the northwestward-flowing streams on the highland surface are related to the asymmetry of the geology. On the other hand the data support Davis' idea that a disequilibrium exists between the two stream systems.

In order to test this idea the writer made a study of the stream profiles in the Winston-Salem 1:250,000 quadrangle. SL values were obtained on most of the streams in the quadrangle and plotted on a map at a density of about one for every 10 square miles. The map was then contoured to show areas where stream profiles are steep with relation to discharge. The resulting map (Figure 8) shows an anomaly of high SL values along the drainage divide near the heads of the streams draining toward the southeast. The differences in SL values are large, ranging from values of 125 above and below the escarpment to as high as 2000 on the face of the escarpment. Scattered high values and scattered low values also occur in the Blue Ridge highland itself. Most of these values are related to various areas of resistant and nonresistant rocks. For example, the 2000 SL value on the New River southeast of Wytheville is related to the river's crossing of a wide outcrop belt of quartzites of the Chilhowee Group.

An anomalously high belt of SL values is essentially a zone of high stream energy. It could correspond to (1) a belt of resistant rocks, (2) a zone of uplift along a fault, or (3) erosional disequilibrium between two drainage systems. Although some areas of resistant rocks do occur locally along the escarpment, they are scattered and not continuous.

White (1950) postulated a tectonic origin for the escarpment, but recent geologic mapping of the Winston-Salem quadrangle by Rankin and Espenshade[1/] shows no evidence of a fault or fault zone along the face of the escarpment. This is an indication that the escarpment itself is not a fault scarp, but does not rule out the possibility of a tectonic cause related either to uplift or to faulting at some time in the past that has resulted in a disequilibrium between the two stream systems. One hypothesis is that the uplift is quite recent and involves warping close to the escarpment itself. The erosional response is to shift the position of the divide to the northwest. Such an upwarp might leave no evidence of faulting on the escarpment.

Another hypothesis, preferred by the writer, is that the disequilibrium is inherited from Triassic time when the erosional surface was many thousands of feet above the present topographic surface. At this time the Atlantic-Gulf divide may have been some distance southeast of its present position. Because of the resistant rocks on the northwest, the rate of erosion was slower on the Gulf side than the rate on the Atlantic side. Adjustment of the two stream systems has not been completely achieved and a zone of more rapid erosion exists in the headwaters of the Atlantic streams. A very slow northwest migration of the divide continues as the erosional topographic surface continues to be lowered. Presumeably the migration will continue until the divide reaches the belt of resistant rocks. The cartoons in Figure 9 show (A) a northwest-southeast cross section across the Blue Ridge as it is today and (B) the profile as it may have been in Mesozoic or Early Tertiary time.

1/ Espenshade, personal communication

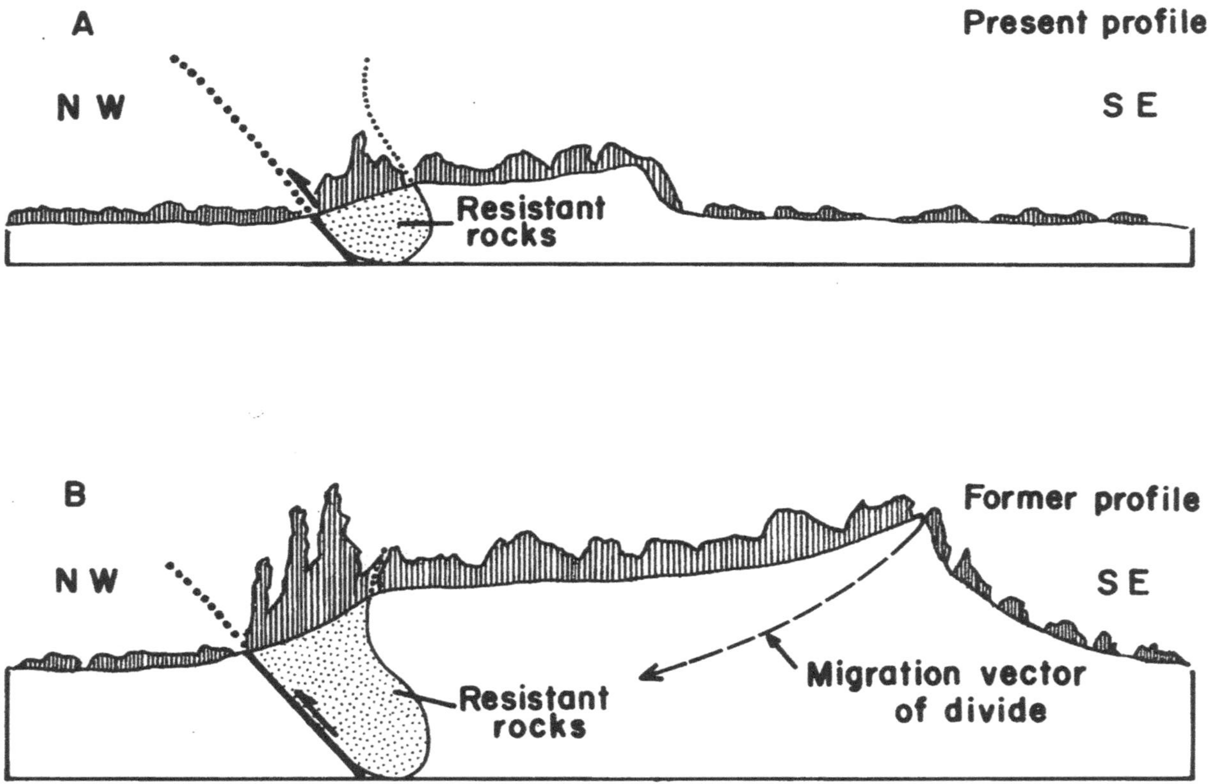

Figure 9 Cartoons of the Blue Ridge showing hypothetical shift of drainage divide to the northwest during Cenozoic and possibly part of Mesozoic time. Shaded pattern represents mountainous topography above stream grade.

One might wonder about the profile at a still earlier time and how the drainage became superimposed across the resistant rocks along the northwest border of the Blue Ridge. During tectonic activity in Triassic time the rate of uplift of the metamorphic rocks in what are now the Blue Ridge and Piedmont may have raised the surface more rapidly than it could be eroded. If this happened the drainage was in effect superimposed on the more resistant rock masses to the northwest.

Divide migration began as soon as the asymmetry of the upland was established. As the relief presumeably was higher, the rate of migration was probably higher. On the other hand the horizontal component of divide shift relative to the vertical component probably became larger as the relief was lowered (Hack, 1969, p. 7).

APPALACHIANS NORTH OF THE ROANOKE RIVER

The Atlantic-Gulf drainage divide forms the eastern edge of the New River basin as it crosses the Appalachian Valley (Figure 2). The escarpment that marks the southeast edge of the Blue Ridge crosses the valley with it, and to anyone traveling down the valley it is quite evident by the steep grade and rough topography, unusual for the limestone country. The divide becomes the eastern edge of the Appalachian Plateau about 60 miles (96 km) north of the Blue Ridge and follows the plateau into Pennsylvania where it gradually shifts still further across the strike, becomes the divide between the Great Lakes-St. Lawrence and Gulf drainage systems and turns westward into Ohio.

The divide remains asymmetric at least as far north as Pennsylvania, as typified by the profiles of the Monongahela and Potomac River systems (Figure 10). No zone of disequilibrium is apparent, however, for the steep profile segments on both sides of the divide are related to resistant rocks. The higher SL values of the Monongahela system are related primarily to the extensive outcrop of Lower Pennsylvanian sandstones. Except in its headwaters, the Potomac main channel avoids the resistant rocks. Locally high SL values are probably related to coarse material brought in by tributaries (Hack, 1973, p. 425.

The Blue Ridge north of the Roanoke River (Figure 2), unlike the Blue Ridge to the south, is in a state of adjustment. The ridge is narrow and its crest at various places is held up by one or more of three resistant rock formations that together form a long narrow belt of Early Cambrian and Late Precambrian age (Hack, 1965, p. 10). No disequilibrium on a grand scale is presently evident, though at some time in the past this mountain ridge may have formed the northwest margin of a broader highland, now marked on its southeast side by scattered outliers and a ragged margin. It may thus have gone through an evolutionary development not yet completed in the southern Blue Ridge.

The sedimentary sequence northwest of the Blue Ridge and southeast of the Plateau contains several ridge-making sandstones, some of which underlie ridges that exceed 4,000 feet (1210 km) in elevation. They rise high above the main rivers which, as shown by Figure 1, flow at low elevations across the softer rocks.

CONCLUSION

The profile data show that the vertical dimension of the drainage network in the Appalachians is in a general way adjusted to the resistance of the rocks in which the streams flow. No two streams have exactly similar profiles for they flow toward the sea with different courses and in so doing traverse different rocks. As the erosion surface is lowered the rock pattern changes and a stream may encounter more resistant or less resistant rocks than it did at a higher erosion level; its profile is adjusted accordingly. Such adjustments between two or more streams produce disequilibrium which may be short- or long-lived depending on the scale of the features involved.

The southern Blue Ridge is an example of large-scale disequilibrium between two stream systems that is inherited from the distant past. Smaller adjustments must be occurring at many places.

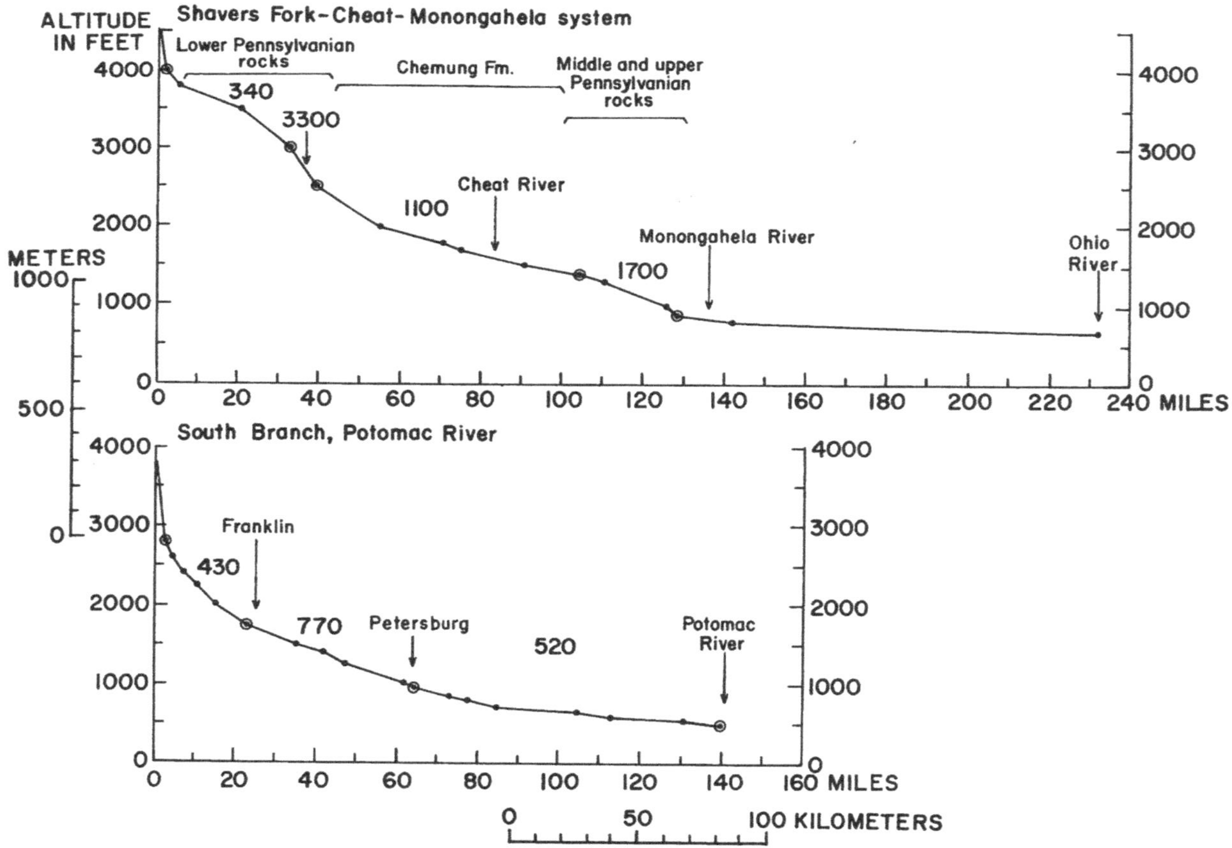

Figure 10 Profiles of Tributaries of the Monongahela and Potomac Rivers. Data points are shown by dots. Numbers between open circles represent SL values.

REFERENCES

Bryant, Bruce, and Reed, John C., Jr., 1970, Geology of the Grandfather Mountain window and vicinity, North Carolina and Tennessee: U. S. Geol. Survey Prof. Paper 615, 190 p.

Colton, G. W., 1970, The Appalachian Basin--its depositional sequences and their geologic relationships, in Fisher, G. W., et al, eds., Studies in Appalachian geology--central and southern: New York, Interscience Publishers, p. 5-47.

Davis, W. M., 1889, The rivers and valleys of Pennsylvania: Nat. Geog. Mag., Vol. 1, p. 183-253.

Davis, W. M., 1903, The stream contest along the Blue Ridge: Geog. Soc. Philadelphia Bull., Vol. 3, p. 213-244.

Fenneman, N. M., 1938, Physiography of the eastern United States: New York, McGraw-Hill Book Co., Inc., 714 p.

Fisher, G. W., et al, eds., 1970, Studies of Appalachian geology--central and southern: New York, Interscience Publishers, 460 p.

Hack, J. T., 1957, Studies of longitudinal stream profiles in Virginia and Maryland: U. S. Geol. Survey Prof. Paper 294-B, p. 45-97.

Hack, J. T., 1965, Geomorphology of the Shenandoah Valley, Virginia and West Virginia, and origin of the residual ore deposits: U. S. Geol. Survey Prof. Paper 484, 84 p.

__________ 1969, The area, its geology--Cenozoic development of the southern Appalachians, in The distributional history of the biota of the southern Appalachians, Part I, Invertebrates (Symposium, Blacksburg, Virginia): Virginia Polytechnic Inst., Research Div. Mon., p. 1-17.

__________ 1973, Stream-profile analysis and stream-gradient index: U. S. Geol. Survey Jour. Research, Vol. 1, No. 4, p. 421-429.

Hadley, J. B., 1970, The Ocoee series and its possible correlatives, in Fisher, G. W., et al, eds., Studies in Appalachian geology--central and southern: New York, Interscience Publishers, p. 247-259.

Hadley, J. B., and Nelson, A. E., 1971, Geologic map of the Knoxville quadrangle, North Carolina, Tennessee and South Carolina: U. S. Geol. Survey Misc. Geol. Inv. Map I-654.

Morgan, B. A., 1972, Metamorphic map of the Appalachians: U. S. Geol. Survey Misc. Geol. Inv. Map I-724.

Rankin, D. W., 1970, Stratigraphy and structure of Precambrian rocks in northwestern North Carolina, in Fisher, G. W., et al, eds., Studies in Appalachian Geology--central and southern: New York, Interscience Publishers, p. 227-245.

Stearns, R. G., 1967, Warping of the Western Highland Rim peneplain in Tennessee by ground-water sapping: Geol. Soc. America Bull. Vol. 78, No. 9, p. 1111-1124.

U. S. Geol. Survey and U. S. Bur. Mines, 1968, Mineral resources of the Appalachian region: U. S. Geol. Survey Prof. Paper 580, 492 p.

White, W. A., 1950, Blue Ridge front--a fault scarp: Geol. Soc. America Bull. Vol. 61, No. 12, pt. 1, p. 1309-1346.

PART II

Since the time of Perrault, who first proved that rivers receive their discharge from rainfall, we have learned a great deal about the place of streams in the hydrologic cycle. We have developed sophisticated techniques for handling hydrologic data and have formulated relationships to predict flow of various magnitude. Hydraulic geometry relations and graphs relating discharge of differing magnitudes to width, depth, velocity and slope of river channels have given us insight into the adjustment of river morphology to discharge. But there are still many unanswered questions and much research to be done on the effects of hydrology on fluvial geomorphology.

The next group of papers are concerned primarily with water in the river channel and with the relation of hydrology and hydraulics to the stream system, denudation, channel morphology and landforms.

In the first paper, Bassett and Ruhe are concerned with outlining the divide of the surface drainage in a watershed which has subterranean passageways. Such a divide is difficult to determine in regions of karst terrain such as in the White River where underground drainage is significant. They accomplish definition of the watershed by dye-tracing which shows that the subsurface drainage divide is not indicated by the surface morphology. They also determine the hydrologic character of the basin, illustrating both surface and subsurface flow. And, by measuring the chemical content of the river waters, they show that corrosion of the watershed is significant in the denudational process. They thus confirm what has been a current emphasis is recent discussions on denudation.

Dury is dealing with problems that are most intriguing to fluvial geomorphologists - the formative discharge of channels, i.e., the magnitude of flow which creates and maintains the river morphology. Another question is the effects of extremely high and low discharges on river form. Is it bankful or mean annual discharge that governs configuration of the channel and floodplain, or is there some other frequency of flow which forms the channel?

It seems reasonable to assume that extremely high discharge and fast velocities should so increase the capacity and competence of a river that channel erosion occurs during flood times or at least during bankful flow. Most texts explain how the stream load is dropped on the floodplain as the river overflows its banks. Yet recent observations during floods of great magnitude, Agnes over Pennsylvania in 1972, for example, show that stretches of river channel may not be greatly eroded under flood conditions, no is silt or clay necessarily deposited on floodplains during overbank flow (J. Moss, personal communication). Lack of deposition may be due to the relation of the fineness of the wash load to flood capacity and competence.

Dury examines channel response to two kinds of alteration of discharge, one an unusually high 1000 year flood and another a major reduction of flow. The response of the river to the high magnitude, low frequency event was very slight to no geomorphic changes. This may indicate that the geomorphic effective discharge is dependent upon the individual river environment which determines a threshold of response (see Schumm, this volume). The reaction of a river to discharge reduced over a period of time was to adjust the channel morphology, mainly width. This result may be applied to the theroy of underfit streams where rivers have had to adjust to a climatic change of less rainfall, hence less discharge.

In the next paper, Baker treats an event which is far off the low end of the frequency scale, the catastrophic Lake Missoula floods which occurred during the Pleistocene. These events of truly great magnitude, far surpassing the threshold of response of any basin, had significant geomorphic results of regional extent. Baker reconstructs the paleohydrology and from that the paleohydraulics showing the mechanism of erosion that took place. The extraordinary discharges with great depths of flow and high velocities must have created great vortices with plucking and cavitation effects. Similar events, but on a much smaller scale even though large, are described by Fahnestock (this volume) with spring bursts of discharge in glacial regions. The mechanism of erosion was much the same as on any river bottom and valley, but in this case it created river bed topography on a giant scale resulting in the morphology of the channeled scablands and adjacent regions. But even with such an unusually large discharge, erosion by flood waters was controlled by lithology and jointing of the rocks, tending to localize effects.

While Baker was dealing with an event on a giant scale, Karcz takes us into the micro-world of hydraulics. River flow is turbulent with movement of the water masses in the stream varying in magnitude and direction. The flow is characterized by various types of phenomena. One type is surges or velocity pulsations, the most extreme of which are flood waves. Another group of flow disturbances result from separation of flow around obstacles where free shear layers of one fluid moving against another enter the main body of flow. The separation of fluid streams and formation of a free shear layer causes a variety of turbulence-forms such as (a) water rollers which are rotary movements about a vertical or horizontal axis, (b) helical flow, or a continuous circular motion in a downstream direction with its axis parallel to the flow, and (c) unstable vortices of local extent.

The Reynolds number is a quantitative expression for the intensity of turbulence in terms of forces which govern the fluid motion, i.e. the relation of inertia forces to viscous forces:

$$Re = \frac{\rho VR}{\mu}$$

where Re is Reynolds number, ρ is density of the fluid, V is velocity of flow, R is depth, and μ is kinematic viscosity.

The Froude number denotes the flow regime which determines the bedforms on a moveable channel bottom.

$$Fr = \quad V= \quad gD$$

where Fr is Froude number, g is gravitational force, D is water depth, and V is velocity of flow. When $Fr<<1$ (sub-critical flow), the channel bed is rippled and, as Fr increases but is still <1, the bottom changes to dune forms. There is a transitional plane bed form as Fr increases above 1. Then as $Fr>1$ (supercritical flow), antidunes form on the bed. These bedforms and associated Froude numbers are important as they cause separation of flow and eddy phenomena, and are themselves morphologic features of the channel.

These are some of the concepts to which Karcz refers in his explanation of longitudinal scour marks and features formed on the bed in his flume experiments.

4

FLUVIAL GEOMORPHOLOGY IN KARST TERRAIN

J.L. Basset and R.V. Ruhe

ABSTRACT

Karst terrain presents a specialized and complex case in fluvial geomorphology. Surface drainage channels and nets discharge waters to subsurface conduits and in turn emerge elsewhere to flow again in surface patterns. Emergence may be within or beyond the topographic limits of the watershed. The combined surface-subsurface drainage system must be defined and can be done by dye tracing which identifies underground flow paths and measures travel times. Conduit flow is analogous to flow through the pipe of a storm-sewer system. Discharge from the underground conduits to a surface drain generally approximates the sum of discharges at surface inlets to the conduit. This balance verifies definition of the surface-subsurface drainage net. Correlation of dissolved load with discharge permits quantitative estimates of rates and amounts of corrosion in this fluvial-geomorphic system. Lost River Watershed in karst terrain in southern Indiana is the model for this study (which is funded in part by the U.S. Department of Interior, Office of Water Resources Research, Grant No. 14-31-0001-3689, Project C-3122).

INTRODUCTION

In a normal drainage basin, lower order streams join higher order and ultimately waters discharge from the basin through the mouth of the highest order stream. Divides are fixed topographically, and runoff descends hillslopes providing waters that discharge through the ordered drainage net. This arrayed system permits ready study and measurement so that the hydraulics of the stream and overland flow can be related to channel and basin morphology and the dynamics involved. All of this is familiar in textbooks on fluvial geomorphology (Leopold, Wolman, and Miller, 1964; Morisawa, 1968; Strahler, 1968).

Karst terrain presents a different and more complex problem, and investigation of karst drainage systems is closely related to the ground water hydrology.

A surface drainage net may partially delineate a karst drainage basin, but a subsurface drainage net also may be part of the system which is more difficult to define. Surface drainage channels may discharge into swallow holes and on the surface in karst springs or rises. The nature of the subsurface flow must often be inferred from indirect observation. Karst springs are often perennial while surface flow is not. In addition, spring sites may be outside the topographic drainage boundary of the watershed. Large portions of karst terrain may lack integrated drainage. Systematic geomorphic and hydrologic study of karst has been limited by proper understanding of the morphology of the drainage system. One objective herein is to present techniques for definition of the drainage system in karst terrain.

Focus here is on the karst of Lost River Watershed in southern Indiana which has had wide textbook exposure from the classic geomorphology approach (Thornbury, 1954, 1965, 1969) and more recently from the point of view of morphometric analysis (McConnell and Horn, 1972). The upper part of the basin contains three distinct physiographic divisions: (1) The eastern part has a parallel, integrated surface drainage net whose streams are oriented along northwest and southwest azimuths and whose stream junctures are at acute and oblique angles (Fig. 1). This angular pattern parallels large scale lineaments shown on ERTS and SKYLAB photographs of southwestern Indiana.

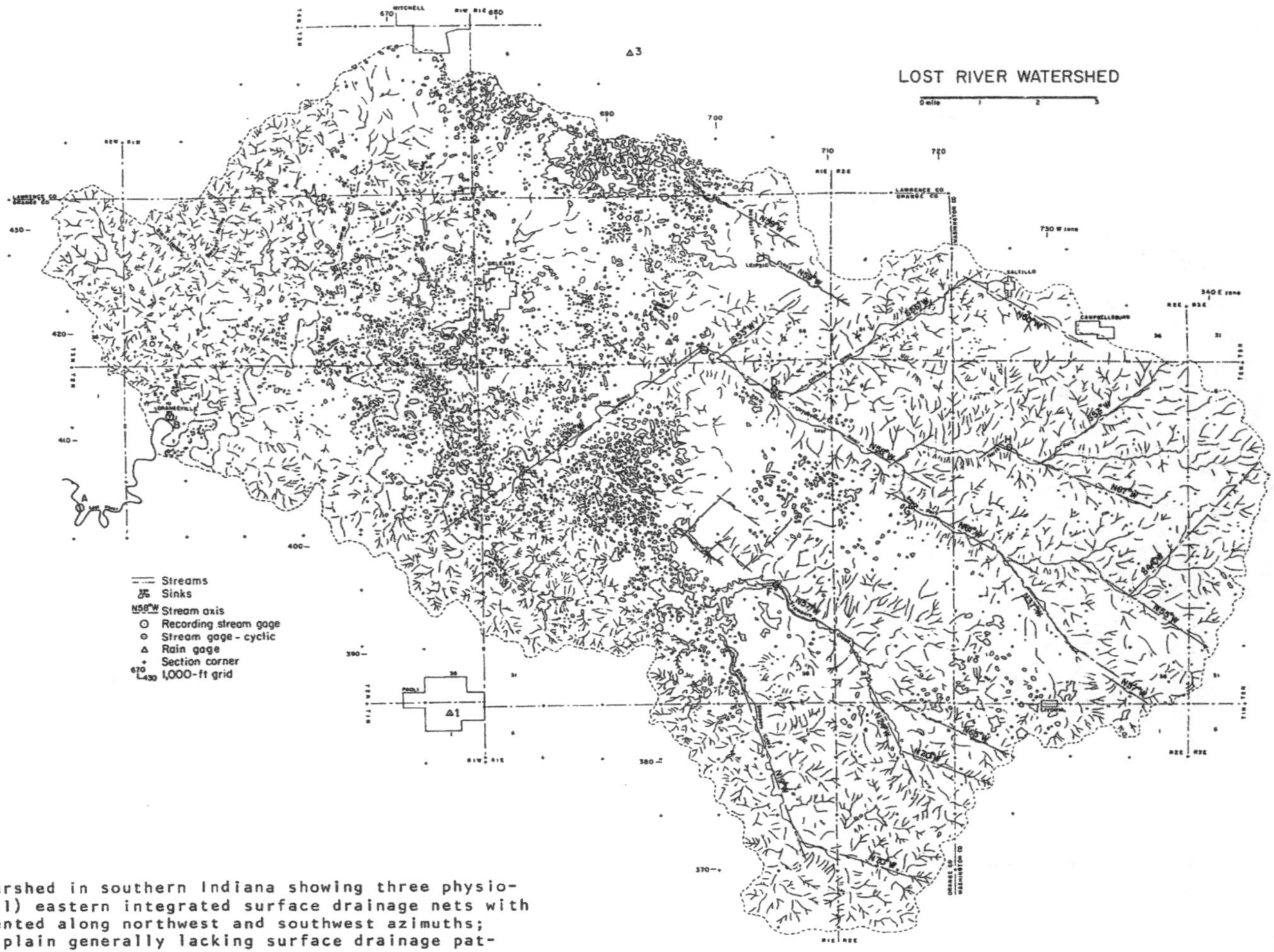

Figure 1
Upper Lost River watershed in southern Indiana showing three physiographic divisions: (1) eastern integrated surface drainage nets with parallel pattern oriented along northwest and southwest azimuths; (2) central sinkhole plain generally lacking surface drainage pattern; and (3) western partially integrated drainage net.

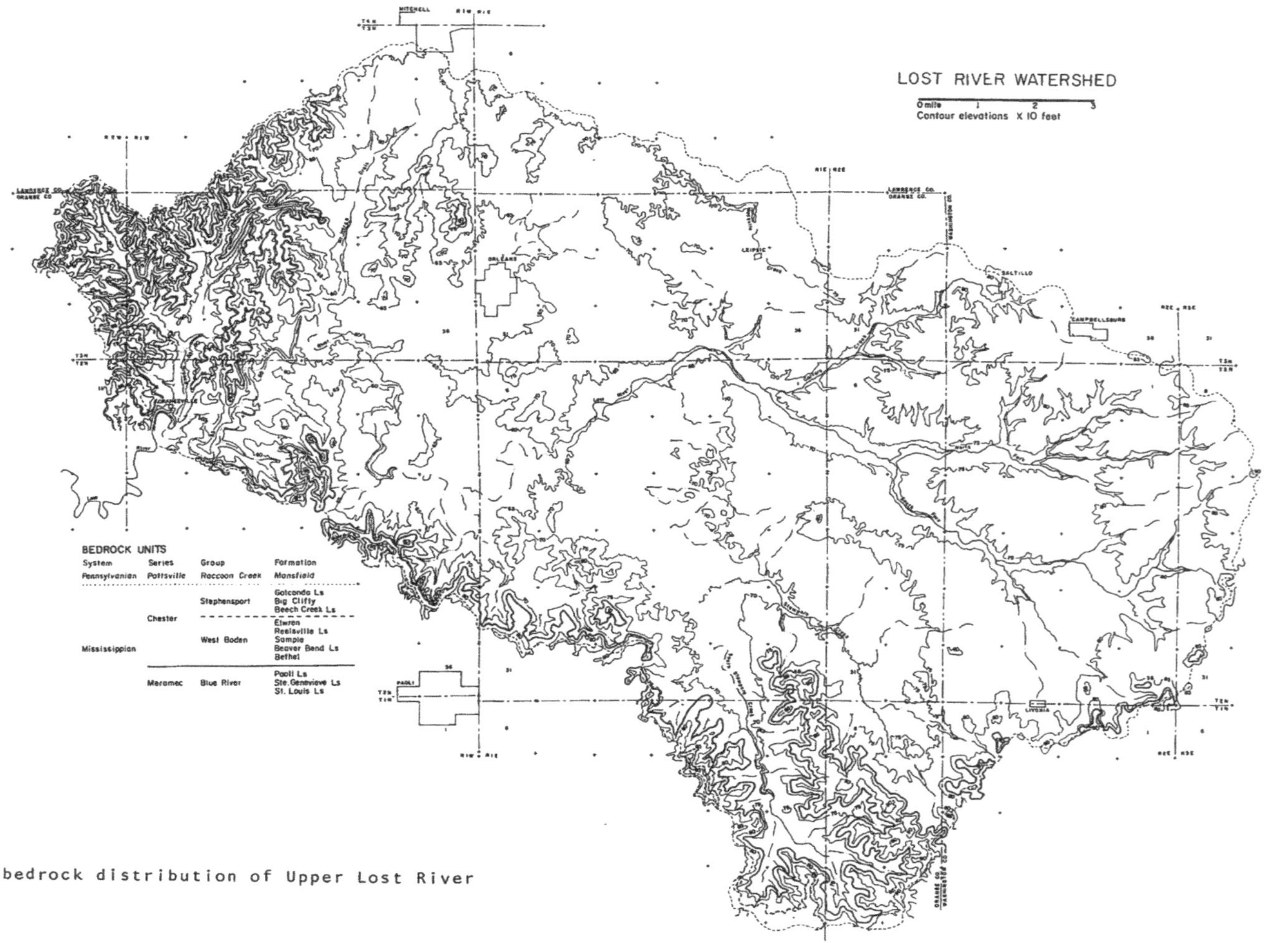

Figure 2
General morphometry and bedrock distribution of Upper Lost River watershed.

(2) The central part of the basin is a broad sinkhole plain where surface drainage nets are essentially absent although Dry Bed of Lost River maintains its angular course across the plain. The drainage basin slopes across the eastern and central areas from about 900 feet in elevation at the eastern divide to about 600 feet at the western margin of the sinkhole plain. These two areas are underlain by rocks of the Mississippian Meramec Series, mainly the St. Louis and Ste. Genevieve Limestones (Perry and Smith, 1958; Sunderman, 1968) which form an essentially uninterrupted limestone sequence some 300 feet thick. The rocks dip westward about 30 feet per mile which is slightly greater than the regional slope. Surface drainage from the eastern part of the watershed terminates abruptly in a series of major swallow holes located along the eastern margin of the sinkhole plain.

(3) The western part of the basin is a rugged hilly area known as the Crawford Upland (Malott, 1922), and underlain by clastic rocks of the Upper Mississippian Chester Series (Fig. 2). Streams in this highland which are incised into the underlying limestone commonly sink at several places along their channels and contribute significant recharge to the subsurface drainage systems. A continuation of the highland to the east forms the southern topographic divide of the Lost River Watershed.

These contrasting physiographic settings present three basic problems in the quantitative geomorphic and hydrologic analysis of the basin and in definition of the basin fluvial morphology: (1) Contribution of discharge of sinking streams in the eastern and western parts of the basin, (2) contribution of discharge of waters from the sinkhole plain, and (3) relation of the surface drainage basin of Lost River to its subsurface drainage basin. Two large karst springs, the Rise of Lost River and the Orangeville Rise are located within the basin and serve as discharge points for drainage from sinking streams and the sinkhole plain.

DEFINITION OF SUBWATERSHEDS

Dye Tracing

Subsurface drainage basins and divides of the Orangeville Rise and Rise of Lost River can and have been defined by fluorescent dye tracing procedures. Dye injected into swallow holes may be detected at spring sites by absorption on charcoal or by quantitative fluorometric measurement of dye concentrations (Dunn, 1957; Wilson, 1968).

In Lost River Watershed previous dye tracing (Murdock and Powell, 1968) has been verified in the present study and additional routing has been defined (Fig. 3). Subbasins may be defined within the Lost River Watershed. Mosquito Creek drains northward underground beyond the topographic divide to Donaldson Cave Spring (Fig. 3, No. 11). South Fork of Stampers Creek drains under the highland, southern topographic divide to Lick Creek (Fig. 3, Nos. 7 - 9). Subsurface drainage divides in karst terrain may or may not correlate with subsurface-surface drainage divides.

Dye tracing shows that the two largest sinking streams in the basin, Stampers Creek and Lost River, drain westward beneath the sinkhole plain 10 and 7 miles respectively, to the Rise of Lost River. The surface drainage from the eastern part of the basin which is diverted into the subsurface follows solution channels completely under the sinkhole plain (Fig. 3, Nos. 5, 6). The sinking streams of the western highland and the northern part of the sinkhole plain (Fig. 3, Nos. 2 - 4, 12 - 14) drain underground about 3 to 6.5 miles to Orangeville Rise (Fig. 3, B). During the course of the present study and in that previously cited (Murdock and Powell, 1968), no sinking stream was found which resurged at both the Rise of Lost River and the Orangeville Rise. A subsurface drainage divide, exists between flow paths 4 and 5 and effectively separates the Lost River Watershed into two components which are hydrologically independent: (1) The eastern integrated surface drainage nets of Lost River above Dry Bed including Carter Creek, North Fork, and South Fork subwatersheds and Stampers Creek subwatershed comprise a surface-subsurface system emerging at the Rise of Lost River. (2) The western highland area and most of the

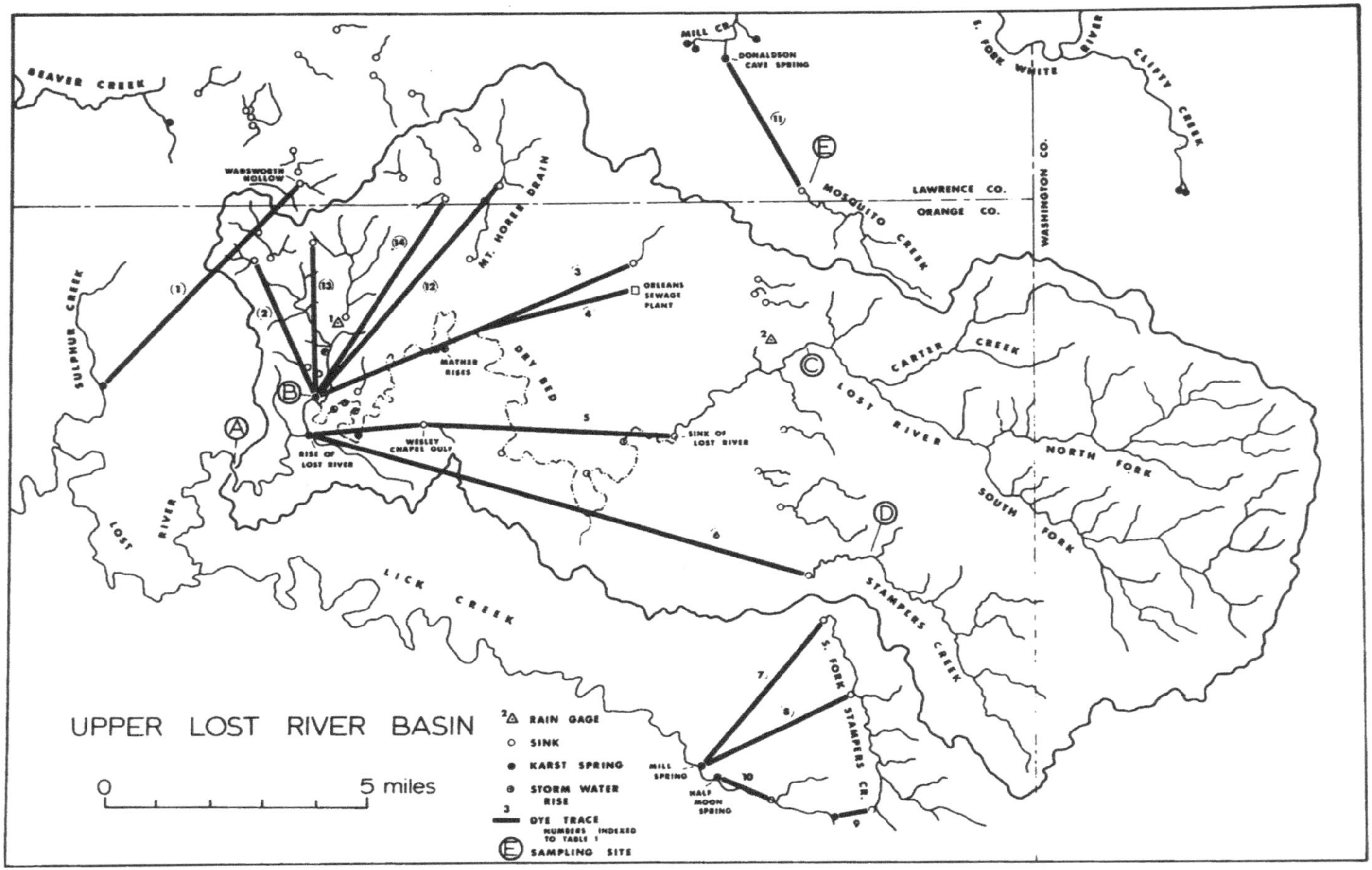

Figure 3 Identification of subsurface flow paths by dye tracing and definition of subwatershed.

Table 1 Underground Flow by Dye Tracing

Path	Swallow Hole Elevation	Rise Elevation	Distance	Travel Time	Hydraulic Gradient
(cf. Fig. 3)	(ft)	(ft)	(mi)	(hrs)	(ft/mi)
1	625	475	5.5	<48	27
2	595	485	3.0	36	37
3	615	485	5.5	36	24
4	625	485	5.9	36	24
5	615	485	7.0	116	24
6	665	485	10.0	120+	26
7	675	610	3.7	<24	18
8	695	610	3.2	--	27
9	715	690	0.8	30	31
10	625	605	1.0	30	20
11	680	600	2.9	<70	28
12	615	485	5.6	<120	23
13	535	485	3.2	<120	16
14	560	485	5.6	<120	14

From Murdock and Powell (1968) and this study.

Table 2 Watershed and Subwatershed Discharges (cfs) During Hydrologic Events

Site	Aug 16 1972	Sep 22 1972	Jan 16 1973	Jan 23 1973	Feb 24 1973	May 24 1973
B Orangeville Rise	14.0	9.8	70	165	67	59
C Sink of Lost River	2.6	.85	25	106	29.8	19.1
D Sink of Stampers Creek	.55	.35	12.6	33.8	10.8	10.0
Total	17.2	11.0	108	305	108	88
A Downstream Gage	17.1	13.7	102	433	109	93

sinkhole plain form a second surface-subsurface system emerging at Orangeville Rise. Drainage areas are about 85 and 48 square miles respectively.

Dye tracing also permits measurement of subsurface flow velocities by timing travel of a dye slug from its point of injection through its rise (Table 1). The procedure is analogous to measuring flow velocity in a surface stream using dye cloud techniques. Of interest here is the discharge from the Orleans Sewage Plant into a sinkhole and its subsurface travel to Orangeville Rise (Fig. 3, No. 4). This tracing has obvious environmental as well as hydrologic significance. During January 1973 seven liters of 20 percent solution of Rhodamine WT dye were injected into the sewage effluent. About 41 hours later the dye slug emerged at the Orangeville Rise. Dye concentrations at sites A and B (Fig. 3) were measured with a filter fluorometer (Fig. 4). The full dye pulse was measured at the recording gage 5.5 miles downstream beginning 46 hours after injection. Discharge at the Orangeville Rise recording gage was 60 cfs during passage of the dye slug.

During a second travel time study when discharge was 148 cfs at Orangeville Rise, dye appeared in 26 hours after injection at Orleans which is 5.9 straightline miles from the Rise. Subsurface flow velocities for the two tests were 13 and 20 feet per minute, and velocity increased with discharge. Based on measured dye concentration, discharge, and amount of dye injected, almost 70 percent of the dye passed through the recording stations as a dye pulse. Such rapid ground water flow velocities indicate that subsurface flow to the Orangeville Rise must be analogous to open pipe flow.

Discharge at Orangeville Rise shows extremely rapid response to rainfall. Comparison of hydrograph response to hourly precipitation data recorded at Bedford, Indiana, 10 miles to the north, indicates a response time of less than 4 hours. This is due to the very rapid flow rates through the major subsurface conduits, and to the direct hydrologic connection of the Rise to several large sinking streams as delineated by dye tracing.

Subwatershed Discharge

Subwatersheds also may be verified in the karst watershed by delineating surface-subsurface discharge systems. A continuous recording gage at Station C (Fig. 3) measures the surface discharge of upper Lost River and its tributaries into the swallow hole above Dry Bed. Non-continuous measurements at Station D give the discharge of Stampers Creek into its swallow hole. Total discharge from the integrated surface drainage net of the eastern area may be considered to be the sum of discharges at these two stations. Their underground conduits jointly emerge at the Rise of Lost River.

The western area of incised drainage discharging to swallow holes and the major part of the sinkhole plain emerge at Orangeville Rise where discharge is continuously recorded (Fig. 3, B).

From the arrangement of the gaging network, the discharge at station A should equal the sum of discharges at stations B, C, and D, plus the water added to the system between the Sinks of Stampers Creek and Lost River and the Rise of Lost River. Note the good agreement between the summed discharge at the downstream gage during various hydrologic events (Table 2). The discrepancy of January 23, 1973 was due to storm discharge bypass of Orangeville Rise through Dry Branch, the main surface drainage from the north (Fig. 3). These hydrologic data verify the subwatershed delineations made by dye tracing and indicate that most of the sinkhole plain is tributary to the Orangeville Rise.

A distinct dissimilarity to the normal fluvial-geomorphic net is the subsurface conduit system beneath the sinkhole plain which has definite upper limits of flow capacity under existing hydraulic gradients. The upper limit of discharge at Orangeville Rise is about 180 to 190 cfs. Gage heights at the Orangeville Rise greater than 5 feet are not associated with significantly increased discharge. When stages are greater than 5 feet at

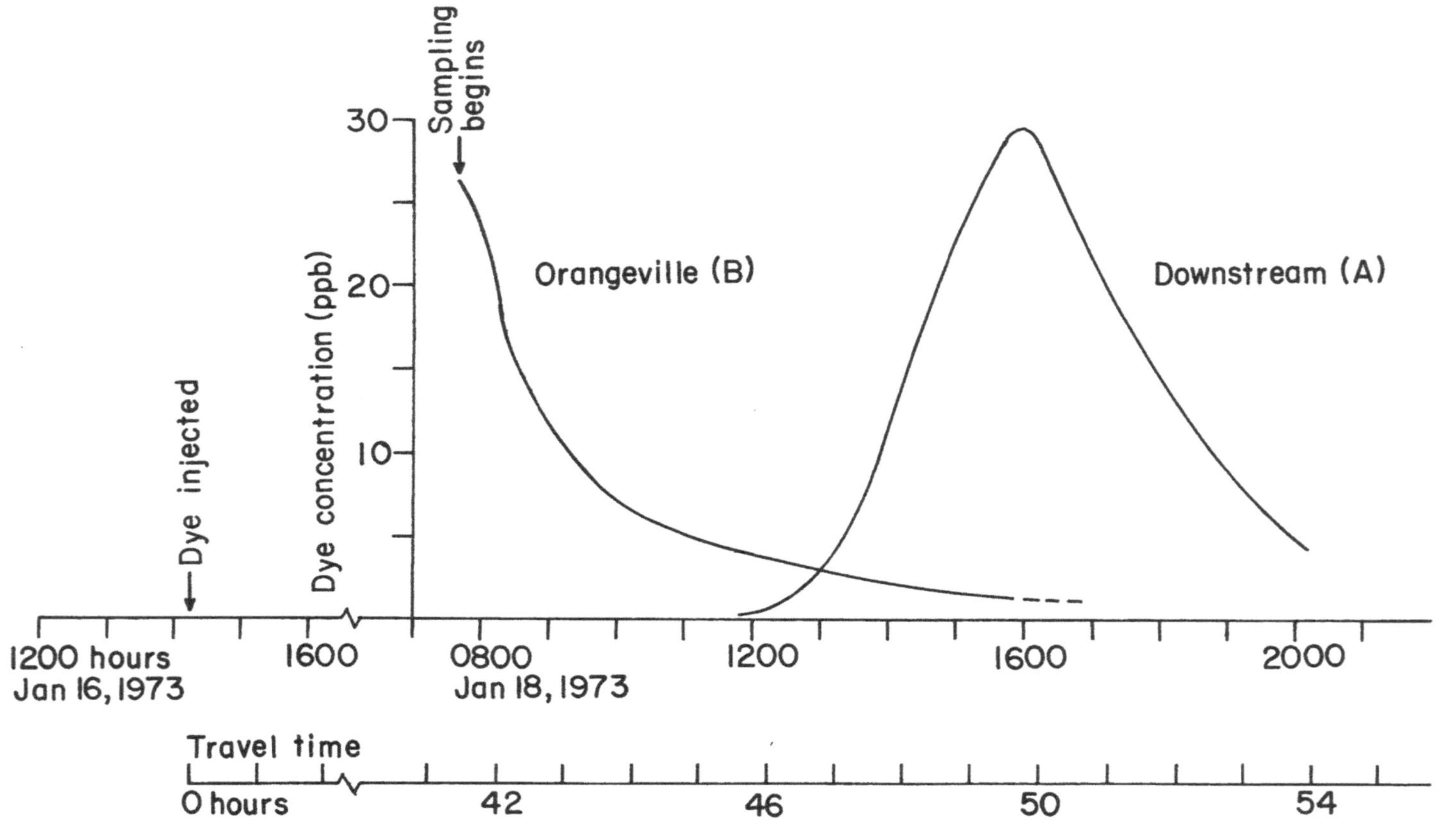

Figure 4 Travel time of dye slug in underground conduit from sinkhole discharge of Orleans Sewage Plant to Orangeville Rise.

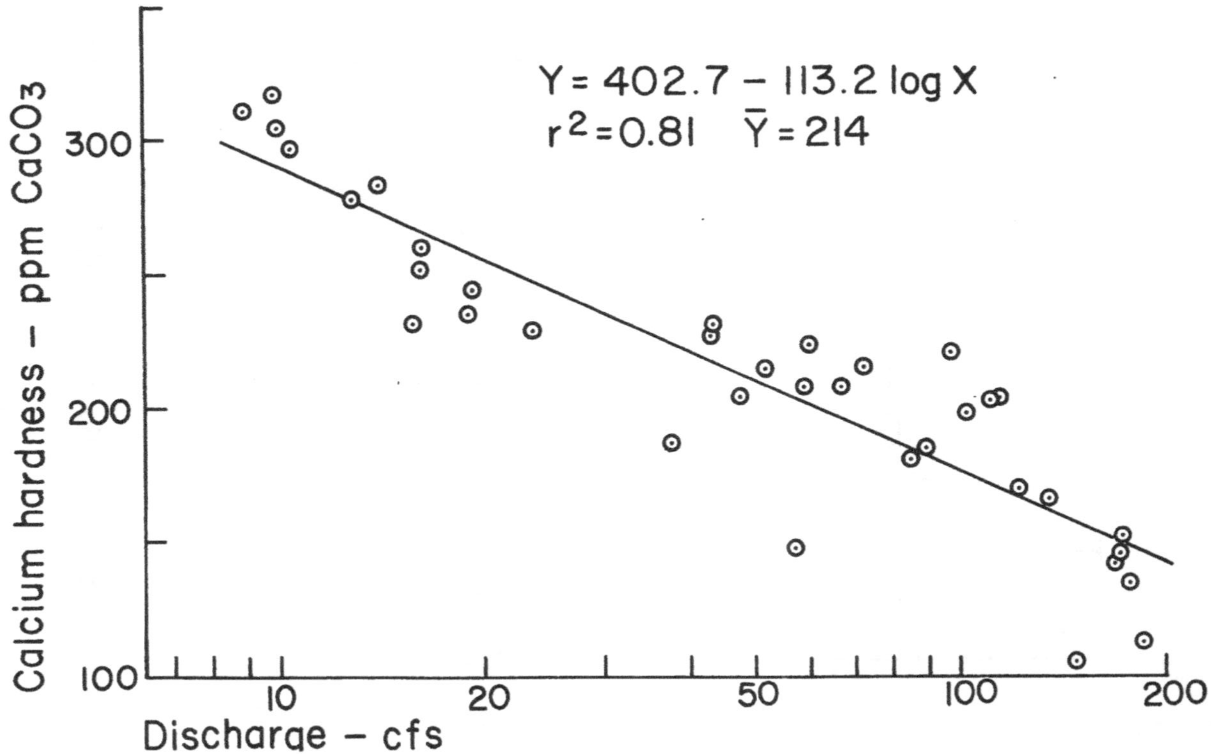

Figure 5 Relation of Ca hardness (ppm $CaCO_3$) to discharge (cfs) of waters at Orangeville Rise.

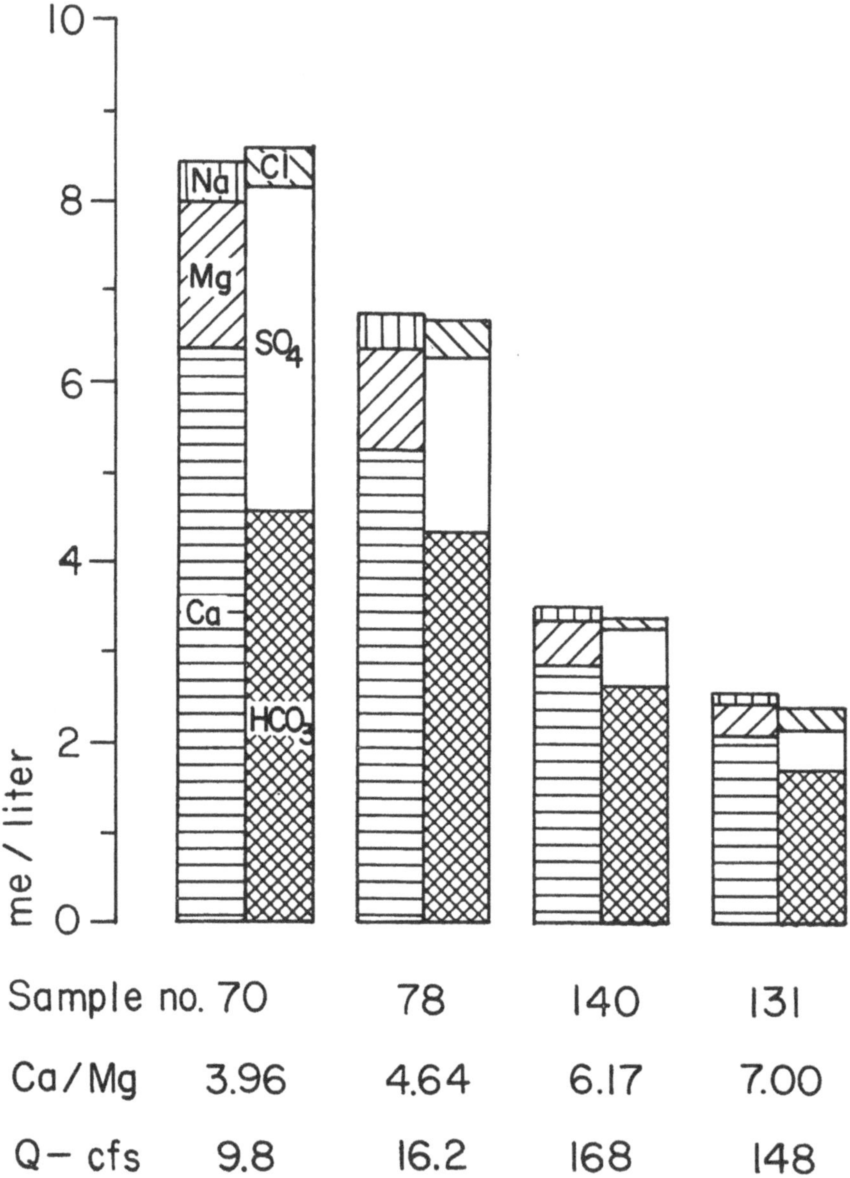

Figure 6
Relation of total dissolved solids (me/liter), Ca/Mg ratios, and discharge (cfs) of waters at Orangeville Rise.

Orangeville Rise, waters emerge at ephemeral storm-water rises such as Mather Rise along Dry Bed of Lost River (Fig. 3). Hydraulic gradient along this underground conduit is about 24 feet per mile. The storm rises act much like a safety valve in a pipe and limit discharge down gradient.

WATER CHEMISTRY AND CORROSION

Within the karst geomorphic framework described previously, estimates of terrain denudation may be made. By sampling and analyzing the chemistry of waters from spring sites with known drainage areas, rates of corrosion within the basin may be estimated. Water samples were collected and routinely analyzed for alkalinity, Ca, Mg, Na, Cl, and SO_4. Total hardness, calcium hardness, and alkalinity are expressed as $CaCO_3$. The geochemistry is a measure of the geomorphic corrosion in a subwatershed, and focus will be on the drainage system emergent at Orangeville Rise (Fig. 3, Nos. 2 - 4, 12 - 14).

Thirty-six water samples were collected during the year beginning June 1972 and represent the full range of discharge at the Rise. Ion concentration relates to discharge. Total hardness is inversely related to discharge as expressed by

$$Y = 516.2 - 154.9 \log X$$

and ranged from 400 ppm $CaCO_3$ at a discharge of 9 cfs to almost 200 ppm at more than 150 cfs. Calcium hardness (Fig. 5) and alkalinity similarly related to discharge as expressed by

$$Y = 402.7 - 113.2 \log X \text{ and}$$

$$Y = 259.9 - 67.2 \log X,$$

respectively. Sulfate varies inversely with discharge as expressed by

$$Y = 542\ X\ 0.636.$$

Na and Cl are present in about molar equivalent concentrations and also vary inversely with discharge.

Ca/Mg ratios are used to determine source of waters in carbonate terrain containing relatively pure limestone and dolomite (Langmuir, 1971). At Orangeville Rise these ratios range from 3.1 to 9.1 and weakly correlate with discharge (Fig. 6). As thin gypsum beds may be included in the limestones underlying the subwatershed, solution of these beds could have pronounced effect on the Ca/Mg ratios of waters at the Rise. If so, waters containing large concentrations of SO_4 should have high Ca/Mg ratios, but the opposite relation exists. Ca/Mg ratios inversely relate to SO_4 as expressed by

$$Y = 10.98 - 3.21 \log X$$

Downstream at French Lick and West Baden where resorters took the baths and purge, SO_4 concentrations in waters from gypsiferous beds are 1000 to 2000 ppm (Blatchley, 1903). The range in Orangeville Rise waters is about 30 to 200 ppm with a mean concentration of 59 ppm. Apparently, solution of gypsum has little effect on the Ca/Mg ratio of waters at the Rise.

Chemical composition of the waters also may be characterized on a milli-equivalent per liter basis (Fig. 6). As discharge increases, total dissolved solids decrease appreciably. As discharge increases above 20 cfs, almost the entire reduction in total dissolved solids is due to decrease in sulfate. As total dissolved solids decrease, Ca/Mg ratios increase greatly. So, with increasing discharge at Orangeville Rise, chemistry of the waters changes from a $Ca\text{-}Mg\text{-}CO_3\text{-}SO_4$ composition to basically a $CaCO_3$ composition.

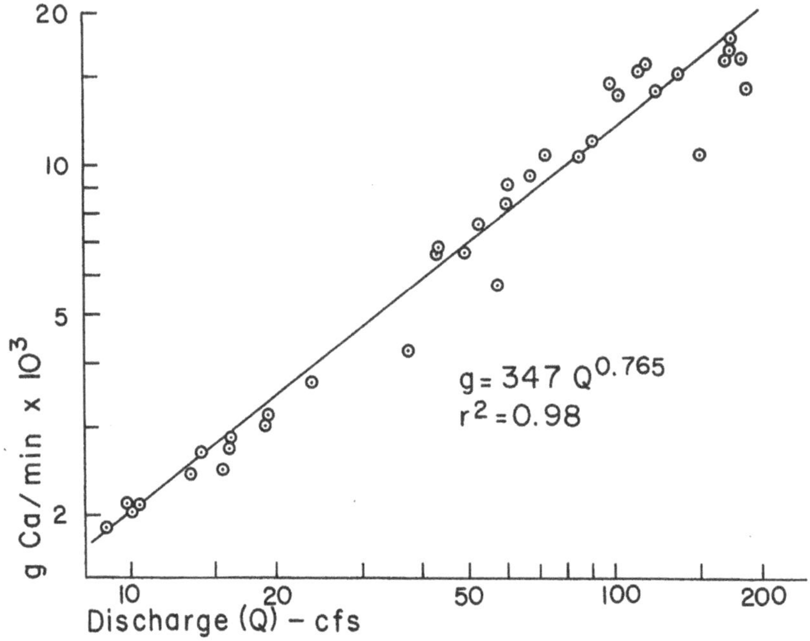

Figure 7 Ca mass flow per unit time in waters of Orangeville Rise.

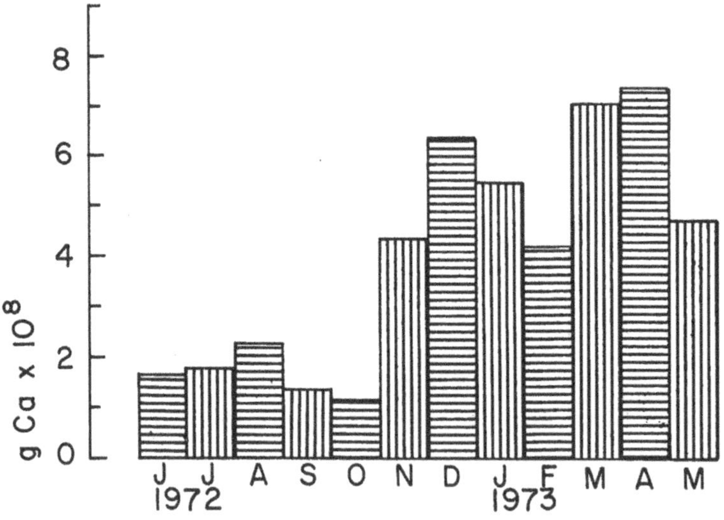

Figure 8 Total Ca discharged per month in waters at Orangeville Rise during year beginning June 1972.

Table 3 Mean Monthly Discharge, Ca Mass Flow Rate, and Monthly Ca Mass Discharge from Orangeville Rise

Month	Mean discharge	Mean Ca flow rate	Total Ca dissolved
	(cfs)	(g Ca/min x 10^3)	(g x 10^8)
June, 1972	22.9	3.81	1.65
July	24.8	4.05	1.81
August	32.4	4.97	2.22
September	18.1	3.18	1.38
October	15.0	2.76	1.23
November	82.1	10.1	4.37
December	130.0	14.4	6.42
January	108.5	12.5	5.59
February	86.8	10.6	4.26
March	147.1	15.9	7.06
April	163.9	17.2	7.41
May, 1973	87.5	10.6	4.74
Total			48.1

Water chemistry--discharge relations are a basis for estimation of corrosion in the subwatershed. From the inverse exponential relation of Ca hardness to discharge (Fig. 5), a tenfold increase in discharge from 10 to 100 cfs results in a 39 percent decrease in calcium content of the water, where the hardness is expressed as $CaCO_3$. More $CaCO_3$ is removed from the subwatershed during periods of higher discharge. From this relation (Fig. 5), instantaneous Ca mass flow may be calculated, and it relates directly to discharge at Orangeville Rise as expressed by the power function

$$\text{g Ca/min} = 347\ Q^{0.765}$$

(Fig. 7). The mass flow rates range from 1.88×10^3 to 1.7×10^4 g/min. A tenfold increase in discharge from 10 to 100 cfs results in a 5.8-fold increase in the amount of $CaCO_3$ removed per unit time.

Using mean monthly discharges, the mean Ca mass flow rate and total amount of Ca discharged at the Rise were calculated for each month during the year June 1972 through May 1973 (Table 3). About 73 percent of $CaCO_3$ was removed from the subwatershed during November through May and coincided with the non-growing season when evapotranspiration was lower (Fig. 8). Assuming a $CaCO_3$ content of 2.45 gm/cm^3 limestone, corrosion in the subwatershed totals 4.9×10^3 cubic meters of rock during the year. This value assumes that all Ca was derived from $CaCO_3$, and that the rock contains, on the average, 10 percent by weight insoluble residue. More refined calculations could increase the total yearly corrosion rate by considering the removal of Ca derived from gypsum.

CONCLUSIONS

Karst terrain presents a specialized and complex case in fluvial geomorphology. Surface drainage channels and nets discharge waters to subsurface conduits and in turn emerge elsewhere to flow again in surface channels. Flow through the subsurface is essentially analogous to closed pipe flow, but discharge may be limited by limited hydraulic gradients. Definition of drainage basins and subbasins requires study of both surficial and subsurface geology and hydrology. After such definition, the impact of geomorphic processes such as corrosion may be evaluated by quantitative analysis of hydrologic and geochemical parameters.

REFERENCES

Blatchley, W. S., 1903, The mineral waters of Indiana: Indiana Geol. Survey 26th Ann. Rept., p. 11-226.

Dunn, J. R., 1957, Stream tracing, Mid-Appalachian region: Nat. Speleo. Soc. Bull., v. 2, p. 7.

Langmuir, D., 1971, The geochemistry of some carbonate ground waters in central Pennsylvania: Geochem. Cosmochem. Act., v. 35, p. 1023-1045.

Leopold, L. B., Wolman, M. G., and Miller, J. P., 1964, Fluvial processes in geomorphology: W. H. Freeman & Co., San Francisco, 522 p.

McConnell, H. and Horn, J. M., 1972, Probabilities of surface karst, in Chorley, R. J. (Ed.) Spatial analysis in geomorphology: Harper and Row, Inc., New York, p. 111-133.

Morisawa, M., 1968, Streams, their dynamics and morphology: McGraw-Hill Book Co., New York, 175 p.

Murdock, S. H. and Powell, R. L., 1968, Subterranean drainage routes of Lost River, Orange County, Indiana: Indiana Acad. Sci. Proc., v. 77, p. 250-255.

Perry, T. G. and Smith, N. M., 1958, The Meramec-Chester and intra-Chester boundaries and associated strata in Indiana: Indiana Geol. Survey Bull. 12, 110 p.

Strahler, A. N., 1968, Quantitative geomorphology, in Fairbridge, R. W. (Ed.) The encyclopedia of geomorphology: Reinhold Book Corp., New York, p. 898-912.

Sunderman, J. A., 1968, Geology and mineral resources of Washington County, Indiana: Indiana Geol. Survey Bull. 39, 90 p.

Thornbury, W. D., 1954, Principles of geomorphology: John Wiley & Sons, Inc., New York, 618 p., also 2d ed., 1969, 594 p.

_____, 1965, Regional geomorphology of the United States: John Wiley & Sons, Inc., New York, 609 p.

Wilson, J. F., 1968, Fluorometric procedures for dye tracing: U.S. Geol. Survey Techniques Water Resourc. Invest., Book 3, Chap. A12.

5

MAGNITUDE-FREQUENCY ANALYSIS AND CHANNEL MORPHOMETRY

G. H. Dury

ABSTRACT

The upper Great Ouse receives as a major tributary the Twin which has the lower floodplain gradient. Testing of the hypothesis that difference in gradient is compensated by differences in other channel characteristics reveals that higher sinuosity on the Ouse, by reducing channel gradient, partly compensates for the difference in flow velocity expectable from floodplain gradient, but also that the effects of sinuosity are offset by differences in hydraulic radius.

Most channel characteristics, and probably some levee characteristics, can be expressed as power functions of drainage area. Magnitude-frequency analysis of the discharge records now available for the Ouse and Nene systems revises an earlier conclusion that discharge at a given recurrence interval is a linear function of drainage area: instead, a power-functional relationship appears. Comparative analysis of channel dimensions on the Nene system supports the findings on the Ouse and the Twin. The results, taken in conjunction with the results of magnitude-frequency analysis, indicate that velocity at a given recurrence interval, in a range that brackets the most probable annual flood, increases significantly in the downstream direction, the increase being far more rapid than the $V \sim ^{0.09}$ required by theory. New evidence from North American rivers confirms that bankfull flow can be equated with $Q_{1.58}$, the most probable annual flood.

Information available for the Ouse includes reaches where water is artificially diverted to the headraces of water mills. The reduced reaches have responded by making slight reductions in mean depth and considerable reductions in width so as to reduce width/depth ratios to values appropriate to the reduced areas of cross-section. Channel slope has undergone little or no change. Within the limits of this study, gradient appears to have little effect on width/depth ratio. The mode of response of the reduced reaches is conceivably relevant to the channel adjustment of streams that become reduced to underfitness.

The flood of March 1947 corresponded, on parts at least of the Nene system, to an event of very great magnitude and very low frequency, corresponding to the thousand-year flood. Its slight to nil geomorphic effectiveness not only serves to confirm that valley meanders and associated features relate to flow regimes very different from those of the present, but also raises queries about the location of catastrophic events on the magnitude-frequency scale.

INTRODUCTION

Despite the brilliant development of empirical and theoretical fluvial morphology during the 1950s and 1960s, much still remains to be done. The present interest in network topology and in the hydrologic effects of urbanization does not mean that further empirical studies are needless--even though Leopold and Langbein (1962) have provided a far-ranging theoretical statement, and Maddock (1969) has signalized disappointing prospects for total hydraulic description. For example, we still know very little about the balance between vertical and lateral accretion of floodplains in the general case, supposing such a case to exist. The relative significance of hydraulic radius, gradient, and roughness in the Manning velocity equation remains to be disentangled, although a start on investigation will have been made by the time that this paper appears. Thresholds of stream behavior, the step functions that these thresholds exemplify, and the relationship of incremental progress up the magnitude-frequency scale to events of catastrophic effectiveness are still

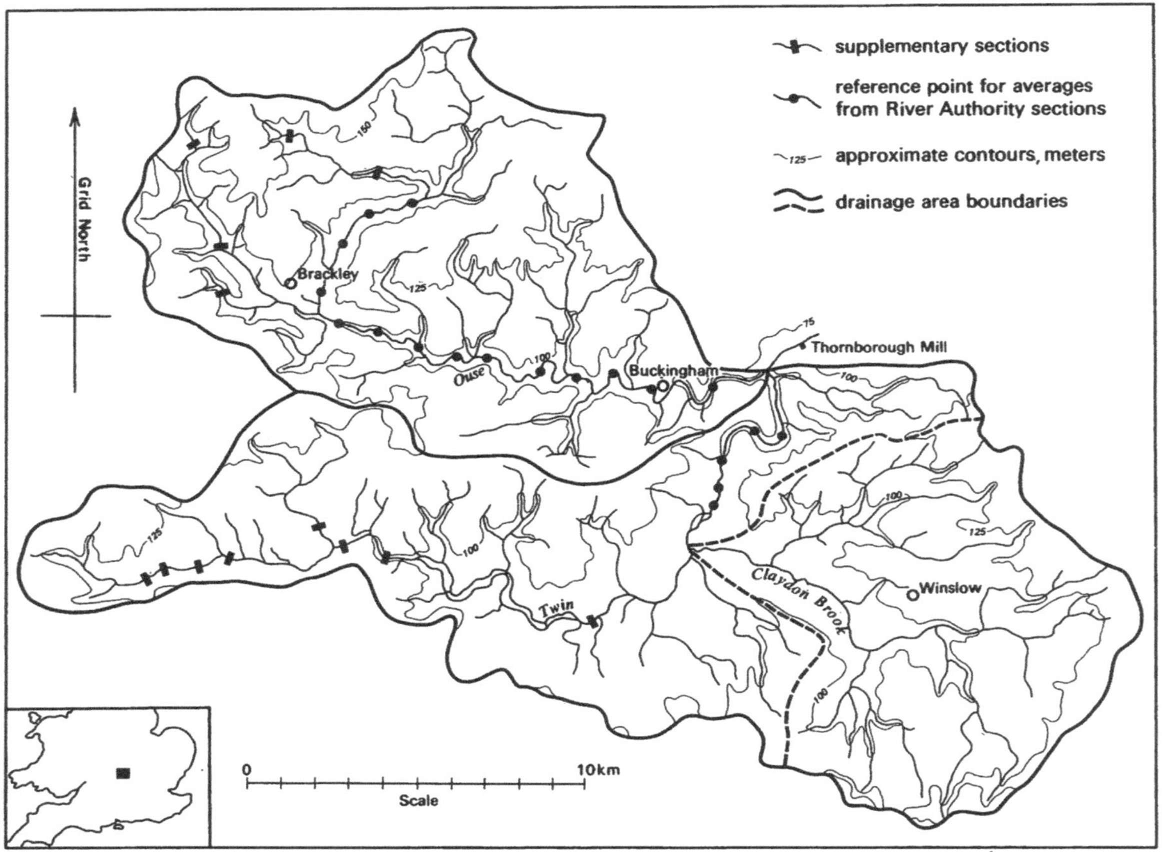

Figure 1   Location map.

less than adequately studied.

Some pertinent questions will be posed, although not necessarily answered in this paper. The work to be described began as a limited enquiry into a minor and apparently self-contained problem of comparative channel gradient. However, it has extended itself far beyond the original limits, into the analysis of channel response to major reduction of discharge, and into the place on the magnitude-frequency scale of bankfull flow and of catastrophic floods. The analysis of bankfull flow is more or less demanded by a quite unexpected finding in the gradient study, that velocity at low recurrence intervals increases markedly downstream on the rivers investigated. The analysis of channel response to reduced discharge uses material incidentally collected for use in the gradient study, while the discussion of catastrophes is made possible by the accidental inclusion in the available records of an event of great magnitude and low frequency.

THE INITIAL HYPOTHESIS

Obliquely traversing the backslope of the main Jurassic cuesta of the English Midlands, the upper Great Ouse receives as a major righthand lateral the Twin. Rising lower down the backslope than does the Ouse, the Twin has the lower floodplain gradient (Figs. 1,2; Dury, 1951, 1952). At the confluence the two streams are of equal order, but the profile downstream of the confluence continues that of the upper Ouse. An hypothesis that differences in floodplain gradient on the two headstreams are compensated by adjustments in other channel characteristics--for example, by increased channel efficiency on the Twin, or by increased sinuosity and decreased channel slope on the Ouse depends in part on the identification of discharge at a given recurrence interval as a function of drainage area (Dury, 1959).

For testing the hypothesis, values were determined for dimensional characteristics; and, since levees complicate the definition of banktop, also for levee characteristics.

Field Data

Channel and floodplain sections, surveyed by the Great Ouse River Authority at intervals of one channel furlong (660 ft) before metrication, and at 0.2km after metrication, extend on the Ouse up to a drainage area of 32km^2 and on the Twin up to 210 km^2. In 1972 I took supplementary sections on the Ouse up to 1.0km^2 and on the Twin up to 9.5km^2. The gap between 76.5km^2 and 210km^2 on the Twin, between my farthest downstream and the Authority's farthest upstream section, is mainly accounted for by the 96km^2 drained by Claydon Brook, a major feeder. The floodplain gap is 5km.

Cross-sections in the River Authority series that relate to meander bends, artificial works, and reaches with major diversions to water mills, were eliminated from analysis, as were recently ditched sections in the supplementary survey. Thirteen of my sections and 37 of the Authority's sections on the Twin and 66 on the Ouse were retained. The close spacing of the Authority's sections makes it convenient to use average values for individual reaches, 8 sets for the Twin downstream of 212km^2, and 14 sets for the Ouse downstream of 33.2km^2 (Table 2). Averages supply 6 sets of levee characteristics for the Twin, that are, however, of limited usefulness since they fall within a 19km^2 range of drainage area, and 7 sets for the Ouse between 37.4 and 130km^2 (Table 3). Regularization of channels notwithstanding, sinuosity can be determined on trains of 2 to 10 bends, or in exceptional cases on discontinuous meanders (Table 2).

On neither stream do profile data consist throughout of floodplain or channel distance and elevation. Channel slope has been read for most of the upper Ouse from the River Authority's diagram (Longitudinal Section, 1965-67 Survey), and for the lowermost Twin from elevations on the Authority's cross-sections. Floodplain slope for the Ouse upstream of 24.27km and for the Twin upstream of 43.05km^2 has been re-determined by reference to measured distances and known elevations. Since sinuosity fails to vary significantly in the downstream direction (Table 4), its mean values, 1.39 on the Ouse and 1.14 on the Twin, have been used to convert between channel and floodplain

Table 1 List of Channel and Levee Characteristics

Channel Characteristic	Symbol
width	w
area of cross-section	A_c
mean depth	$\overline{d}$
width/depth ratio	$w/\overline{d}$
hydraulic radius	R
floodplain slope	S_{fp}
channel slope	S_c
meander wavelength	l
sinuosity	Ω
drainage area	A_d
Levee Characteristic	
crest height above banktop	HL_c
crest height above floodplain	HL_{fp}
banktop height above floodplain	HC_{fp}
width, banktop to crest	WL_c
width, crest to floodplain	WL_{fp}
width, total	WL_{tot}
slope, crest to floodplain	SL_{fp}

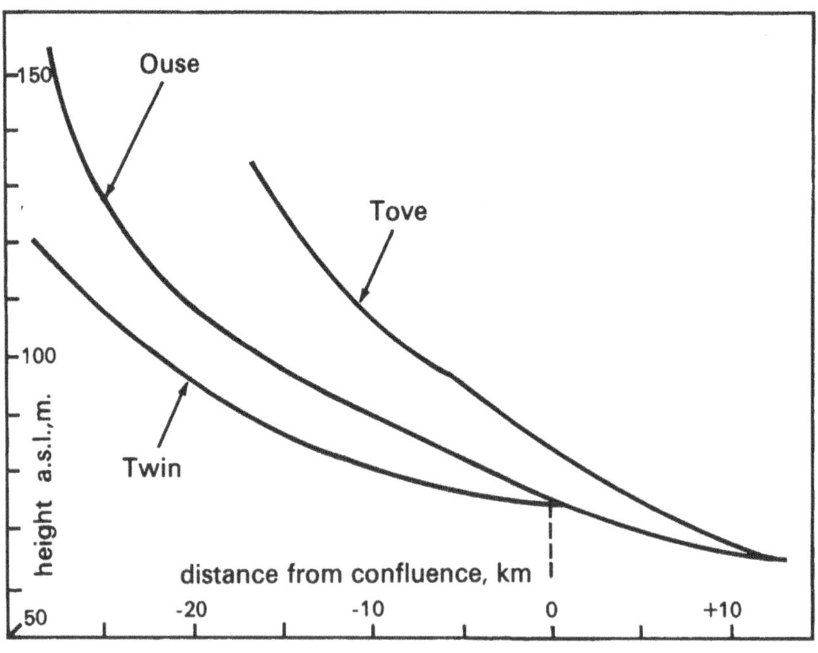

Figure 2 Profile diagram.

Table 2 Values of Channel Characteristics

$A_d$km^2	A_c,m^2	w,m	$\bar{d}$,m	$w/\bar{d}$	R,m	Ω	l,m	10^4xS_{fp}	10^4xS_c
OUSE									
1.00	1.35	2.50	0.54	4.63	0.377	1.32	38.3	-----	-----
2.41	1.54	2.70	0.57	4.75	0.401	1.50	37.3	150.0	107.9
2.67	1.15	3.70	0.31	11.95	0.266	----	----	112.5	80.9
9.55	3.49	4.00	0.87	4.60	0.608	----	----	35.8	25.8
13.75	4.32	5.30	0.82	6.50	0.622	1.59	31.1	54.0	38.8
24.27	----	----	----	----	-----	----	----	31.6	22.7
33.20	8.40	9.90	0.85	11.60	0.724	1.25	68.3	31.4	22.6
37.40	7.10	9.05	0.78	11.50	0.669	----	58.3	26.3	18.9
39.70	8.88	9.30	0.96	9.73	0.791	1.34	75.0	22.2	16.0
43.50	8.47	9.02	0.94	9.60	0.777	1.45	61.1	18.5	13.3
66.00	8.00	9.26	0.87	10.70	0.727	1.39	80.0	18.5	13.3
71.00	7.15	8.10	0.88	9.20	0.725	1.37	77.5	18.5	13.3
73.00	10.33	11.43	0.89	11.85	0.782	1.41	61.7	18.5	13.3
83.80	10.38	10.90	0.95	11.50	0.811	1.17	80.0	18.5	13.3
94.00	12.69	12.20	1.04	11.70	0.889	1.37	120.0	18.5	13.3
100.10	11.80	11.85	1.00	11.85	0.852	1.45	106.0	15.8	11.4
117.00	10.14	10.62	0.96	11.10	0.809	1.54	120.0	14.5	10.4
126.00	15.78	12.86	1.23	10.50	1.030	1.18	80.0	11.8	8.5
128.00	14.35	13.10	1.10	11.95	0.938	----	107.3	----	----
135.25	18.17	12.62	1.44	8.75	1.172	1.54	146.7	11.0	7.9
1446.00	32.84	22.86	1.44	15.88	1.444	----	-----	----	----
3201.00	42.97	26.75	1.61	16.61	1.871	----	-----	----	----
TWIN									
9.50	3.14	5.50	0.57	9.60	0.473	----	-----	34.1	29.9
10.70	2.58	5.90	0.42	14.00	0.383	----	-----	30.3	26.6
12.30	1.44	4.50	0.32	14.00	0.280	----	-----	26.0	22.8
12.50	1.97	5.60	0.35	16.00	0.313	1.08	42.2	35.8	31.4
16.85	3.07	7.70	0.40	19.30	0.361	----	----	22.8	20.0
22.10	----	----	----	-----	-----	1.27	42.5	----	----
36.50	5.37	6.50	0.83	7.85	0.658	1.23	61.3	15.1	13.2
43.50	9.10	9.00	1.01	8.90	0.826	1.33	66.7	14.4	12.6
76.50	12.60	10.60	1.09	9.75	0.986	1.06	67.2	----	----
88.50	-----	-----	----	----	-----	1.18	80.4	----	----
212.00	13.52	13.30	1.02	13.05	0.881	1.23	85.0	7.1	6.2
213.00	10.43	12.40	0.84	14.75	0.741	1.07	80.0	----	----
214.00	15.10	13.25	1.14	11.60	0.972	1.03	85.5	6.4	5.6
216.00	-----	-----	----	-----	-----	----	92.0	----	----
218.00	14.97	13.17	1.14	11.60	0.969	1.02	105.0	7.5	6.6
221.50	13.98	12.72	1.10	11.55	0.937	1.13	125.0	5.2	4.6
222.50	-----	-----	----	-----	-----	----	-----	----	----
232.00	-----	-----	----	-----	-----	----	-----	----	----

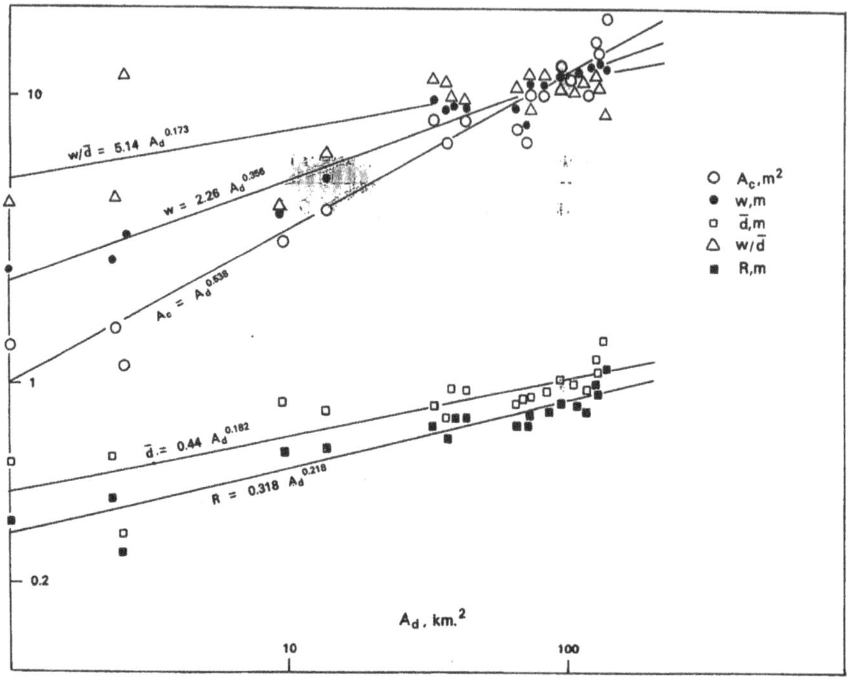

Figure 3 Relation of channel dimensions to drainage area, Ouse.

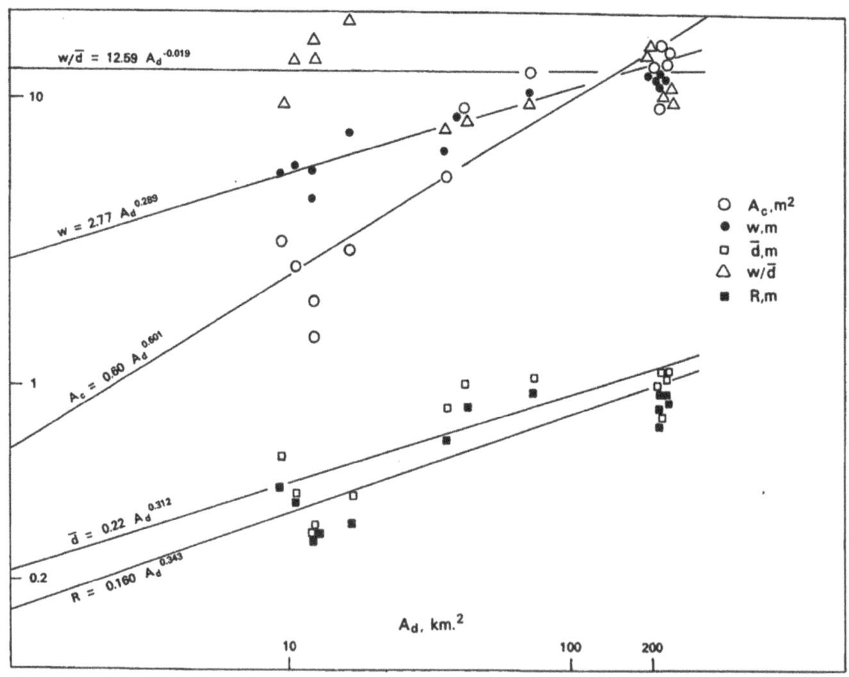

Figure 4 Relation of channel dimensions to drainage area, Twin.

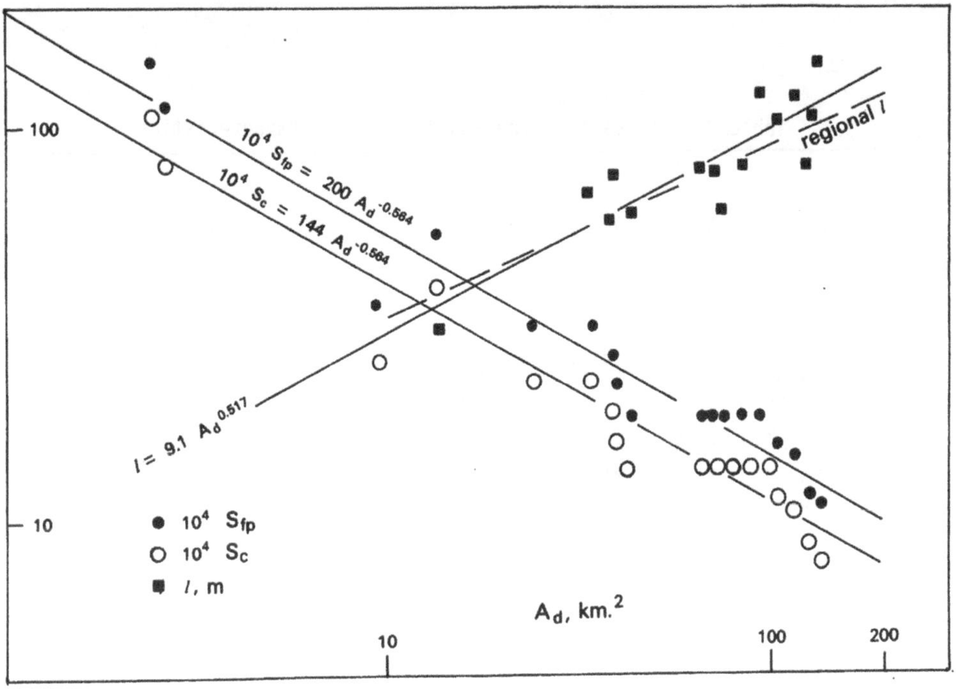

Figure 5 Relation of slope and meander wavelength to drainage area, Ouse.

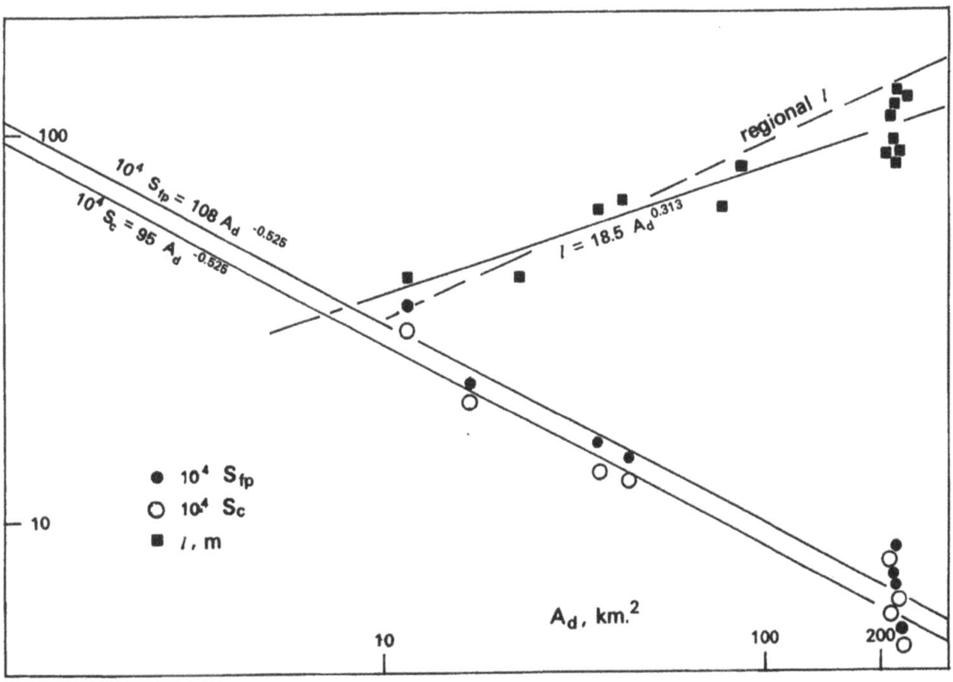

Figure 6 Relation of slope and meander wavelength to drainage area, Twin.

Table 3 Values of Levee and Associated Characteristics

A_d	HL_c, cm	HL_{fp}, cm	HC_{fp}, cm	WL_c, m	WL_{fp}, m	WL_{tot}, m	$10^4 \times SL_{fp}$
OUSE							
37.40	12.08	15.65	3.57	2.83	13.85	16.68	94
43.50	7.02	12.04	5.02	4.21	16.31	20.52	59
71.00	16.70	22.40	5.70	2.31	17.31	19.62	114
95.00	8.33	19.47	11.14	1.41	15.13	16.54	118
117.00	12.93	27.60	14.67	2.54	16.59	19.13	144
126.00	7.47	21.55	14.08	1.57	20.14	21.71	99
130.00	11.10	32.00	20.90	3.10	20.50	23.60	136
TWIN							
213.00	20.00	8.00	--12.00	0.75	-----	-----	---
214.00	11.41	7.89	--3.52	3.99	15.51	19.50	40
218.00	8.64	11.92	3.28	2.42	7.63	10.05	119
221.50	7.38	14.95	7.57	2.64	12.79	15.43	97
222.50	10.40	18.80	8.40	3.54	14.86	18.40	102
232.00	18.80	19.25	0.45	3.36	11.87	15.23	126

Table 4 Coefficients of Correlation Between Log Values of Channel Characteristics and Log Drainage Area

A_c	w	$\overline{d}$	$w/\overline{d}$	R	Ω	l	s_{fp}, s_c
OUSE							
+0.973 0.001>P	+0.985 0.001>P	+0.452 0.02>P >0.01	+0.709 0.001>P	+0.872 0.001>P	--0.104 0.25>P >0.20	+0.880 0.001>P	--0.975 0.001>P
TWIN							
+0.925 0.001>P	+0.916 0.001>P	+0.862 0.001>P	--0.160 0.80>P >0.70	+0.741 0.005>P >0.001	--0.544 0.05=P	+0.918 0.001>P	--0.985 0.001>P
NENE							
+0.953 0.005>P >0.001	+0.979 0.001>P	+0.880 0.02=P	+0.401 0.50>P >0.40	+0.924 0.005>P >0.001		not determined	

slope, as required. The results of conversion are internally consistent.

Levee dimensions are wholly derived from the Authority's sections, which express elevations to 0.01ft (0.25cm). At this degree of detail, it is not everywhere easy to identify the levee crest or the lowest point on the floodplain. From analysis of levee characteristics, overland sections including terraces, artificial structures, and artificial excavations were eliminated.

Results of Analysis: Channel Characteristics

Drainage area provides the basis of reference necessary for comparison. All channel characteristics can be expressed as power functions of drainage area, with varying closeness of correlation (Figs. 3-6, Table 4). A_c, w, and R correlate closely with A_d on both rivers, but the respective relationships between A_c and w ensure for the Ouse a significant downstream increase in $w/\overline{d}$ combined with a slow rate of increase in $\overline{d}$, whereas the more rapid increase in $\overline{d}$ on the Twin is combined with an insignificant downstream change in $w/\overline{d}$. The graphs drawn in Figs. 3 and 4 for $\overline{d}$ and R have been slightly modified from the graphs of regression equations, in order to satisfy the continuity principle. The scale of adjustment involved is exemplified by the alteration of the regression equation $\overline{d} = 0.20A_d^{0.320}$ for the Twin to $\overline{d} = 0.22A_d^{0.312}$.

Meander wavelength correlates positively with drainage area, floodplain and channel slope negatively (Figs. 5,6; Table 4). Although correction for sinuosity reduces slope more on the Ouse than on the Twin, the latter still has the lesser channel gradient. Area for area of drainage, w is similar on the two rivers but A_c runs lower on the Twin; $\overline{d}$ and R are therefore also lower on the Twin. Whether judged by R or by $w/\overline{d}$, the Twin thus appears to fail to compensate for its lower gradient by adjustment in dimensional and form characteristics. The initial hypothesis fails in relation to geometric channel properties.

Results of Analysis: Levee Characteristics

This section will be limited to the Ouse. It is introduced at this juncture, because measurement of cross-sections on reaches with levees involves determination (or selection) of banktop positions. In this study, width has been measured between apparent banktops, the points where bank slope suddenly and markedly decreases upward (Fig. 7). It makes little difference in A_c whether apparent banktops or levee crests are used, but $w, \overline{d}$, $w/\overline{d}$, and R would all be strongly affected by the addition of the 5m or so of width that would result from measurement between levee crests. Sections in the supplementary survey were also taken between apparent banktops. The width values obtained produce rather low values for the wavelength/width ratios l/w, in comparison with the commonly observed and expectable range of 8/1 to 12/1 (Leopold and Wolman, 1960; Bagnold, 1960). Reference of bedwidth to minimum floodplain elevation would mean that, on some sections, bedwidth would be measured within the channel. Reference to mean floodplain elevation (cf. Wolman and Leopold, 1957, p. 89-91) involves the risk of error on reaches where the floodplain is being lowered or raised.

The levee data do not justify a choice of relationship between levee characteristics and drainage area--whether linear, semilogarithmic, or power-functional. In order to permit comparison with the results of analysis of channel characteristics, power-function analysis has been applied, with the results presented in Fig. 8 and Tables 5 and 6. The only correlation significant at better than the 1% level is that of height of channel banktop above floodplain minimum. The regression equations for HL_{fp} and HC_{fp} cannot both be accepted, since their extrapolated graphs would impossibly cross at about 275km^2, predicting a banktop elevation higher than the levee crest. A more likely prospect is that, downstream of about 150km^2, levee crests run about 11cm above banktop level. The tendencies suggested by Fig. 8 include a possible downstream increase in landward levee slope: this tentative result could bear further investigation.

Table 5 Coefficients of Correlation Between Log Values of Levee Characteristics and Log Drainage Area, Ouse

HL_c	HL_{fp}	HC_{fp}	WL_c	WL_{fp}	SL_{fp}
--0.016	+0.846	+0.969	--0.514	+0.726	+0.721
0.80>P >0.70	0.02>P >0.01	0.001>P	0.02>P >0.10	0.10>P >0.05	0.01>P >0.05

Table 6 Ouse Levee Characteristics at 10 and $100 km^2$

A_d, Km^2	HL_{fp}, cm	HC_{fp}, cm	WL_{fp}, m	$10^4 \times SL_{fp}$
10	6.6	0.7	11.2	59
100	23.2	11.9	17.8	130

Slope values are derived from the listed regression values of HL_{fp} and WL_{fp}.

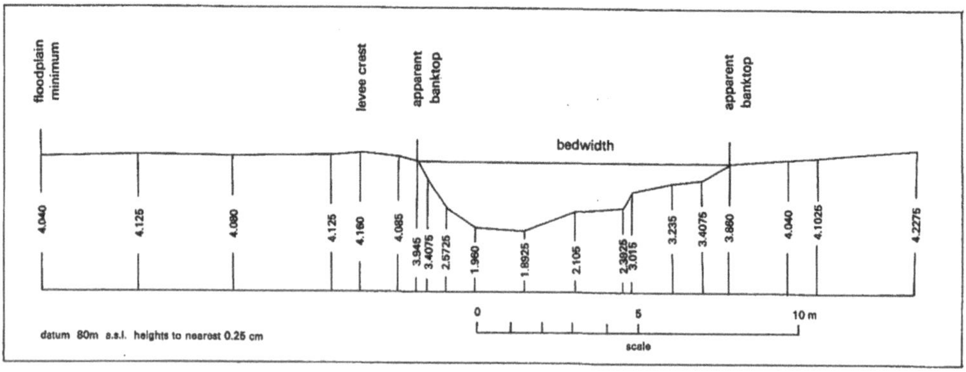

Figure 7 Apparent banktop and bedwidth on a section with a left hand levee: an actual section on the Ouse.

Table 7 Channel characteristics at 10 and 100km^2, from Regression Equations. Ouse(a) and Ouse(b) Values Taken In Downstream Data (Bedford and Brownshill Staunch Points (Case b).

	10km^2	100km^2		10km^2	100km^2
Ouse, A_c, m^2	3.45	11.90	Ouse, R	0.533	0.869
Twin, A_c, m^2	2.45	9.55	Twin, R	0.390	0.776
Ouse(a), A_c, m^2	3.58	10.99	Ouse(a), R	0.538	0.899
Ouse(b), A_c, m^2	4.04	10.68	Ouse(b), R	0.565	0.822
Nene, A_c, m^2	2.73	9.55	Nene, R	0.530	0.950
Ouse, w, m	5.13	11.65	Ouse, $10^4 \times S_{fp}$	55.5	14.53
Twin, w, m	5.39	10.48	Twin, $10^4 \times S_{fp}$	32.5	9.66
Ouse(a), w, m	5.30	11.02	Ouse, $10^4 \times S_c$	39.93	10.45
Ouse(b), w, m	4.04	10.68	Twin, $10^4 \times S_c$	28.3	8.47
Nene, w, m	3.65	7.49	Ouse, l, m (obs.)	29.9	98.3
Ouse, $\bar{d}$, m	0.67	1.02	Twin, l, m (obs.)	38.0	78.2
Twin, $\bar{d}$, m	0.45	0.91	Ouse, l, m (regl.)	32.5	89.5
Ouse(a), $\bar{d}$, m	0.68	0.99	Twin, l, m (regl.)	32.5	89.5
Ouse(b), $\bar{d}$, m	0.70	9.98	Ouse, l/w, (actual)	5.83	8.44
Nene, $\bar{d}$, m	0.75	1.28	Twin, l/w	7.05	7.46
Ouse, w/d	7.66	11.42	Ouse, l/w	6.34	7.68
Twin, w/d	11.98	11.52	Twin, l/w	6.03	8.54
Ouse(a), w/d	7.79	11.13			
Ouse(b), w/d	8.21	11.26			
Nene, w/d	4.87	5.85			

Table 8 Calculated Velocity and Roughness Ratios, Ouse and Twin

	v_O/v_T by:-			
Ad, km^2	$(S_O/S_T)_{fp}^{1/2}$	$(S_O/S_T)_c^{1/2}$	$(R_O/R_T)^{2/3}$	$(R_O/R_T)^{2/3}(S_O/S_T)_c^{1/2}$
10	1.31	1.19	1.23	1.46
100	1.22	1.11	1.08	1.20

	n_O/n_T by:-	
Ad, km^2	A_{c_T}/A_{c_O}	$v_T/v_O(R_O/R_T)^{2/3}(S_O/S_T)_c^{1/2}$
10	0.71	2.056
100	0.80	1.481

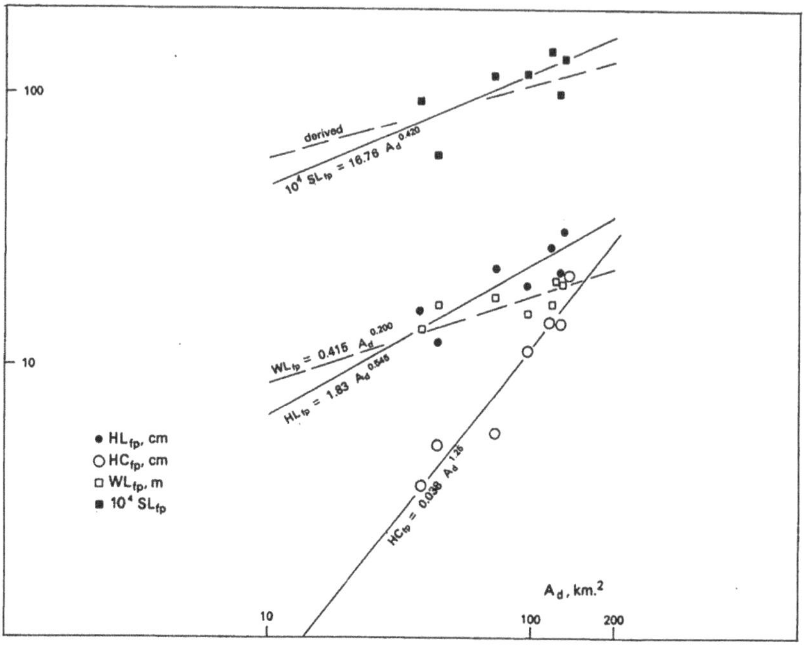

Figure 8 Relation of levee characteristics to drainage area, Ouse.

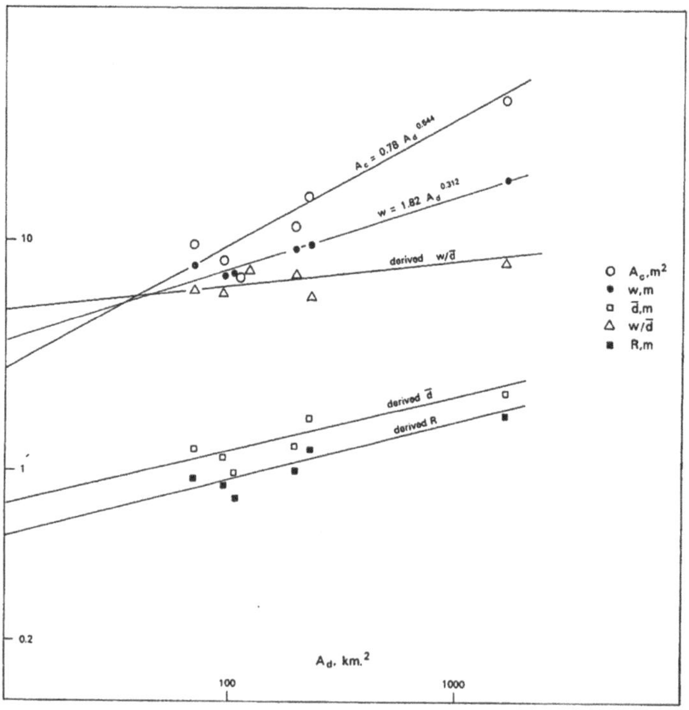

Figure 9 Relation of channel dimensions to drainage area, Nene.

Checks Against Other Data

Despite the elimination of obviously ditched sections, the graph of A_c on the Ouse (Fig. 3) could be strongly affected by the five points of the supplementary survey. A partial check on the regressions of A_c and w for the Ouse can be made with the help of data for natural channels near the Bedford and Brownshill Staunch gauging stations at 1446 and 3021km^2 (Table 2). Extrapolated graphs for the upper Ouse would pass above points plotted for these downstream stations, but inclusion of the downstream data in regressions, or the passing of graphs through $\bar{x},\bar{y}$ for upstream data and through points plotted for Bedford and Brownshill, would have little effect on the values of A_c and w calculated for the Ouse above the confluence with the Twin (Table 7). Artificially enlarged sections in the supplementary survey appear to have been successfully eliminated.

Confirmation comes from an analysis of cross-sectional data for the Nene (Table 7, Fig. 9). This analysis uses surveys by the Welland and Nene River Authority. The regression equations for A_c and w on the Nene run sub-parallel to those on the Ouse, giving closely similar values, area for area.

Channel Characteristics and Velocity Ratios

It is important to establish the basis of cross-sectional measurement as firmly as possible, because the results will now be used in comparative Manning equations. These determine the effects of contrasts in slope, hydraulic radius, and roughness. The outcome of calculations is summarized in Table 8. As shown, floodplain slope difference predicts a velocity on the Ouse 31% greater than that on the Twin at 10km^2, and 22% greater at 100km^2. Channel slope difference reduces the disparity to 19% and 11%. The higher sinuosity on the Ouse seems to compensate for about half the effect of the difference in floodplain gradient. The general conclusion has already been reached, that dimensional and form properties on the Twin fail to compensate for the disparity in channel slope. The Manning equations shows that, in actuality, difference in hydraulic radius offsets the effect of higher sinuosity on the Ouse, almost precisely restoring the velocity difference expectable from floodplain slope at 100km^2, and more than restoring it at 10km^2 (Table 8). This result would not be affected by inclusion of downstream data from Bedford and Brownshill Staunch.

With equivalent discharge/drainage area relationships still assumed, it must follow that velocity through the cross-section, at a given area and for a given recurrence interval, runs lower on the Ouse than on the Twin. The velocity difference must be laid to the roughness factor. Solutions for the relevant equations (Table 8) indicate roughness coefficients about 1 1/2 to 2 times greater on the Ouse than on the Twin.

The initial hypothesis is in this sense sustained, although the tests were expected to reveal compensation in terms of channel gradient, hydraulic radius, and channel form, rather than in terms of the somewhat elusive roughness factor.

DISCHARGE/VELOCITY RELATIONSHIPS

A previous analysis (Dury, 1959) concluded that discharge on the combined Nene and Great Ouse systems, from 69km^2 down to the tidal sluices, is a linear function of drainage area in respect to a fixed recurrence interval in the range 1.2 to 10 years (annual series). Analysis of the longer series of records now available shows discharge per unit area at a given recurrence interval as decreasing downstream (Table 9). Two of the stations on the Nene system are probably suspect, the Willow Brook gauge as being affected by industrial withdrawal, and the Dodford gauge as giving distinctly higher readings after its rebuilding (in WY 1965-6) than before. Records for the Ouse at Thornborough Mill, just below the Ouse-Twin confluence, are expressed as ranges, but can nevertheless be subjected to magnitude-frequency analysis (Fig. 10). The point for $q_{1.58}$ at this station falls almost precisely on the graph drawn for the five retained stations on the Nene, as does the corresponding point for Brownshill Staunch. The point for the Ouse at Bedford plots only slightly high (Fig. 11).

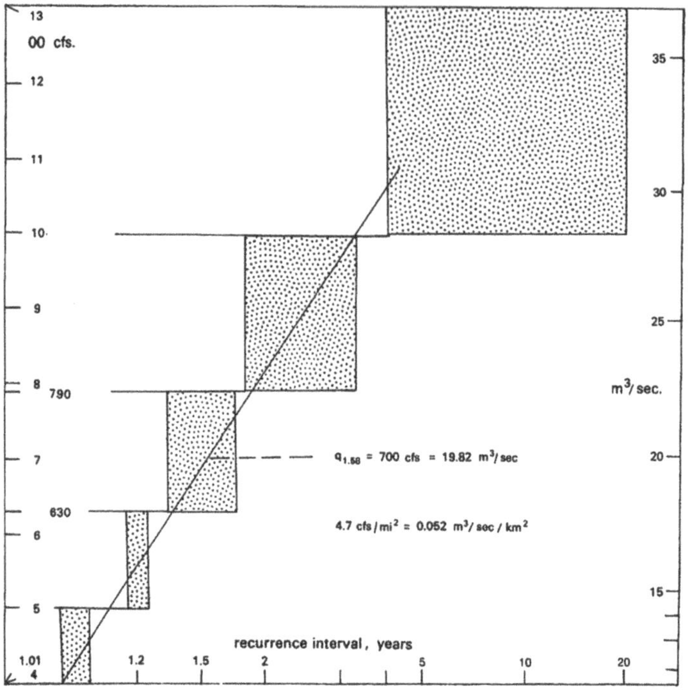

Figure 10 Magnitude-frequency graph for ranges of discharge, Ouse at Thornborough Mill.

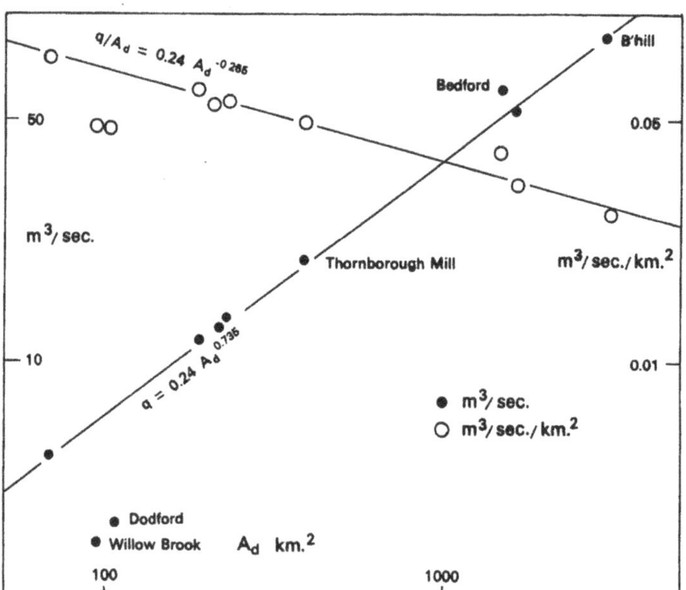

Figure 11 Discharge, and discharge per unit area, at $q_{1.58}$ on the Ouse and Nene systems.

The thirty years of record now available should be capable of defining $q_{1.58}$ within $\pm$ 12%; indeed the very smooth plot of the point values for $q_{1.58}$ against drainage area suggests that the definition is distinctly more accurate than this. The combined regression equations for A_c and $q_{1.58}$ indicate a marked downstream increase in velocity through the cross-section; and the same holds good for other recurrence intervals, e.g., between 1.2 and 5.0 years. In order for the Nene channel at 10km^2 to sustain a velocity equal to that of $q_{1.58}$ at 1000km^2, discharge through the cross-section at 10km^2 would need to equal $q_{5.0}$. It seems necessary to conclude that the inferred downstream increase of velocity at a fixed recurrence interval is a real increase. Table 10 lists comparative values for 10 and 100km^2, plus numerical values of the roughness coefficient for the Ouse and Twin. Although the value obtained for the uppermost Ouse is high, it does not seem impossibly so. Barnes (1967) lists values ranging up to 0.075.

The available information suggests that velocity at the bankfull stage increases in the downstream direction, as rapidly as does velocity at a fixed recurrence interval. The question immediately arising is, of course, that of the place--if any--of bankfull discharge on the magnitude-frequency scale. This question will be treated generally in the following section. Discussion at this juncture will be limited to the study rivers.

It might be supposed that equations for meander wavelength could provide an indirect check on discharge values, and thus on relationships derived from those values. But although wavelengths for the upper Ouse and the Twin plot close to the regional wavelength graph for the Ouse and the Nene (Dury, 1958, Fig. 7; this paper, Figs. 5,6), they prove rather unhelpful. The regional graph leads to $l \propto q_{1.58}^{0.50}$; individual regressions for the Ouse and Twin lead respectively to $l \propto q_{1.58}^{0.70}$ and $l \propto q_{1.58}^{0.43}$. Bedwidth is directly related to discharge, and is probably more easily adjusted than the indirectly-related wave length. The at-a-station relationship $w = 5.43\ q_{1.58}^{0.50}$ (metric units: derived from $l = 30\ q_{1.58}^{0.50}$ in English units and from $l/w = 10$ approx.) predicts widths of 6.2m at 10km^2 and 14.3m at 100km^2, values fairly close to those obtained by regression analysis (Table 7). For what it is worth, the check against bedwidth suggests that the values obtained for $q_{1.58}$ are of the right order for q_{bf}.

A more direct check uses gauging station data. The velocity values in Fig. 12 are derived from values of $q_{1.58}$ and A_c for individual stations, and from estimates of channel capacity on the upper Nene. Stations far downstream, both on the Nene and on the Ouse, are variously unsuitable for the reference to estimated capacity. As shown, velocity at $q_{1.58}$ through actual sections correlates very closely with A_d, increasing significantly in the downstream direction. Velocity at estimated channel capacity runs higher than velocity at $q_{1.58}$ but produces a parallel graph. Its higher values reflect higher discharges that are conceivably affected by artificial dredging and embanking. In any event, the downstream increase in velocity is confirmed, whether at a fixed recurrence interval or at estimated channel capacity; and discharge at estimated channel capacity could certainly be specified by a fixed recurrence interval which would lie above 1.58 but below 2.33 years.

The finding that velocity increases markedly in the downstream direction runs counter to the theoretical proposition (supported as it is by considerable field observation) that, at the bankfull stage, velocity is likely to vary little, or to remain sensibly constant, along the length of the channel (see for instance Leopold, 1953; Leopold and Maddock, 1953, Fig. 32 and text; Leopold, Wolman, and Miller, 1964, Table 7-5; and Leopold and Langbein, 1962). The use of hourly or daily maxima, as opposed to the momentary maxima used here, would serve merely to increase the inferred downstream increase in velocity, on account of the usual contrast of form between upstream and downstream hydrographs; and, in any case, the study rivers typically display flat peaks, with daily and momentary maxima differing little (cf. Dury, 1959).

The theoretical exponents for downstream relationships, as listed by Leopold and Langbein (1962), are compared in Table 11 with the exponents produced by the present study. Only the slope exponents, averaging --0.75 for the Ouse-Twin system, come close to the theoretical value, here --0.74. The rate of downstream increase in width and depth on the Ouse-Twin, and thus

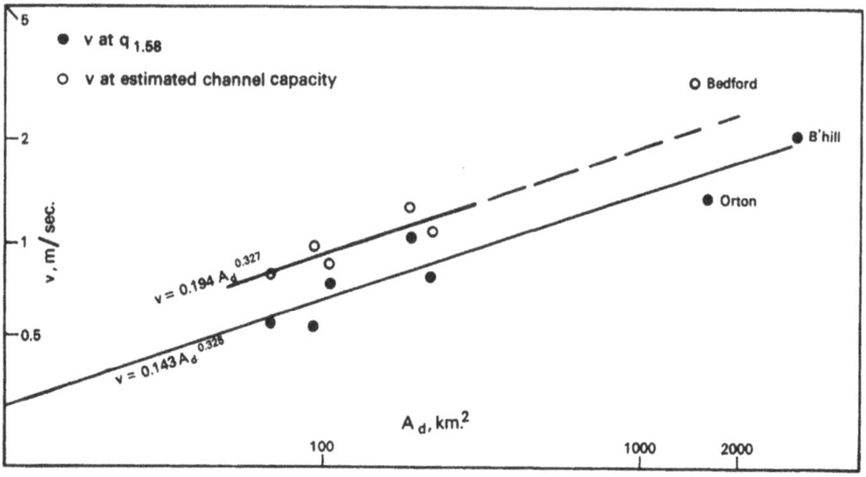

Figure 12 Velocity at $q_{1.58}$ and at estimated channel capacity.

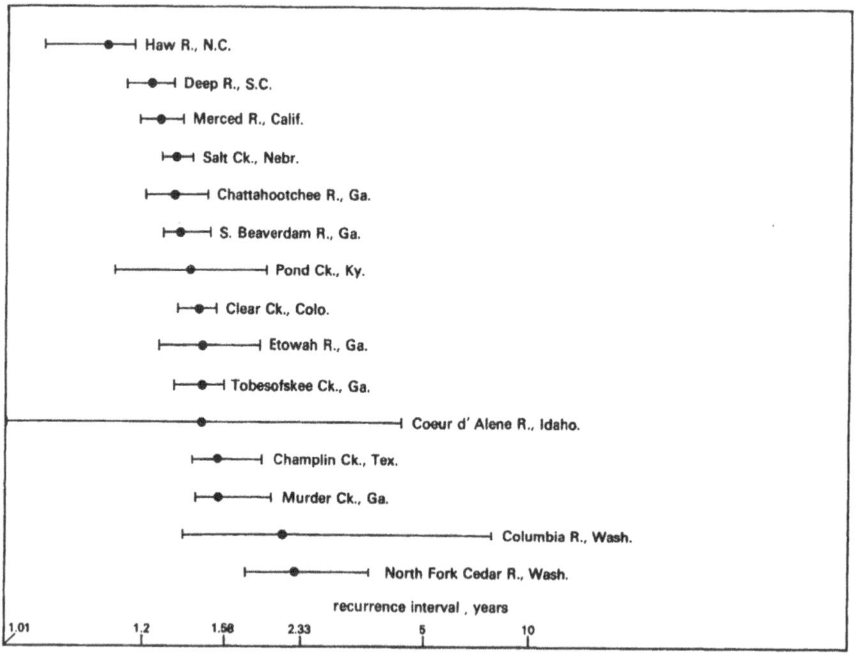

Figure 13 Results of magnitude-frequency analysis for usable stations in WSP 1849, below-bank to bankfull discharges.

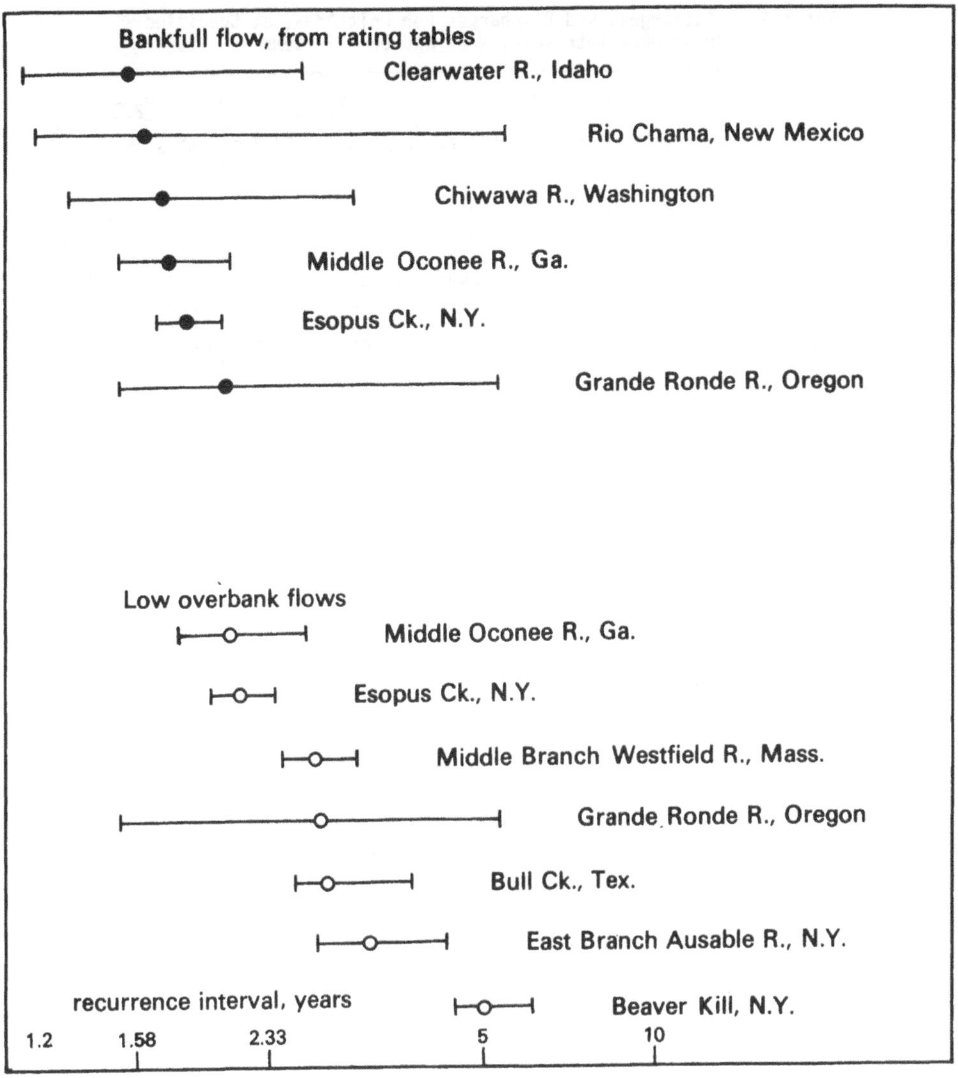

Figure 14 Results of magnitude-frequency analysis for usable stations in WSP 1849, bankfull discharges from rating curves and low overbank discharges.

Table 9 Discharges, and Discharges Per Unit Area, at Specified Recurrence Intervals, Nene and Ouse Systems

	A_d, km^2	$q_{1.2}$ m^3/s	$q_{1.2}$ m^3/s/km^2	$q_{1.58}$ m^3/s	$q_{1.58}$ m^3/s/km^2	$q_{2.33}$ m^3/s	$q_{2.33}$ m^3/s/km^2
NENE							
Harper's Brook	69	2.97	0.043	5.24	0.075	7.36	0.107
Willow Brook	93	3.11	0.033	4.39	0.047	5.66	0.061
Nene/Dodford	106	3.40	0.032	5.10	0.048	6.65	0.063
Ise	192	9.49	0.049	11.61	0.060	13.87	0.072
Nene/Upton	220.5	10.90	0.049	12.32	0.056	13.59	0.062
Nene/St Andrews	230	9.91	0.043	13.03	0.057	16.28	0.071
Nene/Orton	1615	41.77	0.026	53.80	0.033	66.54	0.041
OUSE							
Thornbgh Mill	385	16.14	0.042	19.82	0.051	23.78	0.062
Bedford	1446	42.47	0.029	62.29	0.043	82.11	0.058
Brownshill Stch	3201	65.13	0.020	87.78	0.027	109.01	0.034

	$q_{5.0}$ m^3/s	$q_{5.0}$ m^3/s/km^2	$q_{10.0}$ m^3/s	$q_{10.0}$ m^3/s/km^2
NENE				
Harper's Brook	11.04	0.160	14.02	0.203
Willow Brook	7.93	0.085	9.63	0.104
Nene/Dodford	9.20	0.087	11.33	0.107
Ise	?17.56	?0.091	?	?
Nene/Upton	15.71	0.071	17.56	0.080
Nene/St Andrews	21.52	0.094	25.91	0.113
Nene/Orton	83.36	0.052	102.64	0.064
OUSE				
Thornbgh Mill	30.30	0.079	35.39	0.092
Bedford	113.26	0.078	138.74	0.096
Brownshill Stch	145.12	0.045	175.55	0.055

of cross-sectional area, is distinctly less than theory predicts, so that the rate of downstream increase in velocity is distinctly greater, as has been seen. For the present, no explanation of the discrepancies can be offered.

THE FREQUENCY RELATIONSHIPS OF BANKFULL FLOW: SOME NEW EVIDENCE

The comparative use of discharge at a given frequency has obvious advantages, since actual streams may be affected by tendencies to cut or fill. For example, some of the bankfull stages listed for U.S. rivers are inundation stages on considerably incised streams. Artificial enlargment or embankment of channels can greatly affect stage-discharge relationships, including the frequency of bankfull discharge (cf. Dury, 1961). If natural bankfull discharge on poised streams can be tied to a fixed recurrence interval, comparison among existing streams or between present and former streams would be greatly facilitated.

The theoretical justification for using a fixed recurrence interval comes from the work of Gumbel (1958) and Langbein (1949). The most probable annual flood, $q_{1.58}$ on the annual series, is identical in magnitude with $q_{1.0}$ on the partial-duration series, but the time-specifications differ between the two; $q_{1.0}$ on the partial-duration series is expectable at an average interval of one year. Empirical justification comes from various studies (Wolman and Leopold, 1957; summaries of additional work in Leopold, Wolman, and Miller, 1964, p. 319-321; Fig. 7-10 and Text) that approximately identify bankfull discharge with $q_{1.50}$ (annual series).

Kennedy (1972), in a recent criticism of the identification of bankfull (= dominant, channel-forming) discharge with discharge at a fixed recurrence interval, has overlooked one complete block of evidence (cf. Dury, 1973). She objects to the concentration of one group of study sites in a single midwestern State, and quite erroneously credits Nixon (1959) with obtaining for British rivers a recurrence interval of 0.5 years. Nixon actually obtained a per cent flow duration of 0.6, equivalent to a little over 2 days/year (his p. 162-3). Such a duration does not mean two separate events a year; and in any event, per cent duration is not a suitable comparative measure (cf. Dury, 1961). Kennedy is however correct in emphasizing the implications of using a fixed recurrence interval in the interpretation of underfit streams.

To anticipate the conclusion of this section: $q_{1.58}$ (annual series) is a suitable measure of natural bankfull discharge. This conclusion confirms the preceding inference about a downstream increase in velocity on the Nene and Ouse, and supports other work that uses $q_{1.58}$ (or the close approximation $q_{1.50}$) as a measure of flow at the bankfull stage.

Basic data for analyzing the recurrence interval of bankfull flow come from Barnes's roughness study (Barnes, 1967: U.S.G.S. Water-Supply Paper 1849) and the tabulations of flood peaks in WSP 1671-7, 1679-80, 1682, and 1686-9. These tabulations include the hydrologic areas that include Barnes's sites, and span the times of the roughness surveys. The peak discharges in WSP 1849 could be expected to approximate bankfull, for, among the criteria of site and event selection, Barnes lists a major flood discharge just prior to survey, plus the confinement of the peak within channel banks and the absence of extensive flow in floodplains. The tabulations of annual floods in the WSP 1600 series were used to construct magnitude-frequency graphs, and $q_{1.58}$ was defined on these graphs before the magnitudes of the floods in the roughness survey were checked.

WSP 1849 contains information on 50 sites, but two of these lie close together on a single stream and were studied in relation to a single event. Six cases for which no records appear in the tabulations of flood peaks were omitted from analysis. Also omitted were 10 stations affected during the relevant period by damming, regulation, or major diversion, and one stream where the peak in the roughness study was far below the base of the partial-duration series. Despite the specification of lack of extensive flooding, the peak used in the roughness study was on some streams the 1st to 3rd highest flood of record: that on the Columbia River at The Dalles, Oregon, went 17.5m above listed bankfull stage and involved 3 to 6m depth of over-

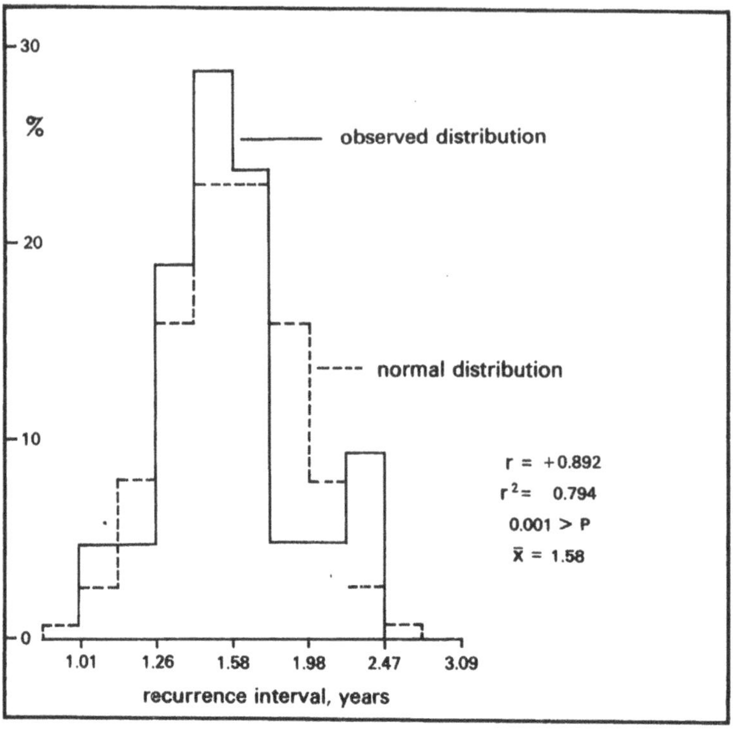

Figure 15 Log-normal distribution of frequencies for near-bankfull discharges at usable stations in WSP 1849.

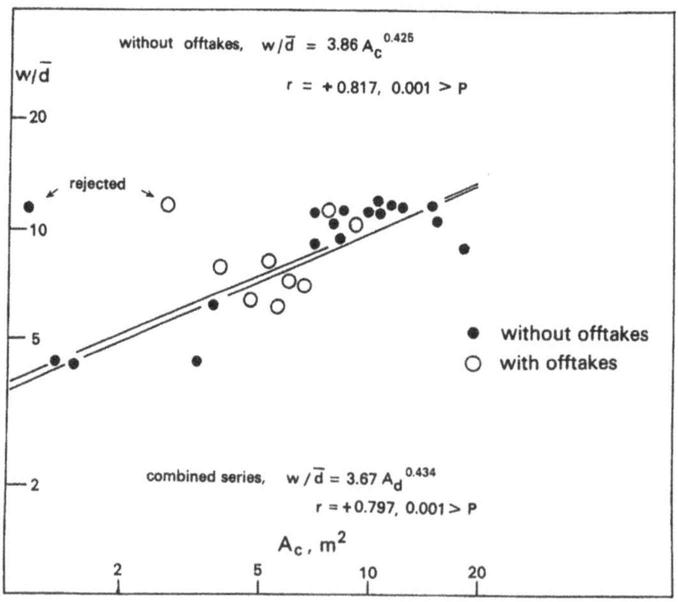

Figure 16 Comparison of width/depth ratio with cross-sectional area.

Table 10 Discharge and Velocity at the Most Probable Annual Flood, and Roughness Coefficients, at 10 and 100km^2

	A_d, km^2	A_c, m^2	$q_{1.58}$, m^3/sec	v m/sec	n
Ouse		3.45		0.377	0.110
Twin	10	2.45	1.30	0.531	0.054
Nene		2.73		0.476	-----
Ouse		11.90		0.592	0.050
Twin	100	9.55	7.05	0.738	0.033
Nene		9.55		0.738	-----

Table 11 Exponents in the Downstream Relationships $v \propto q^m$, $d \propto q^f$, $w \propto q^b$, $s \propto q^z$, and $n \propto q^y$.

	m	f_x	b	z	y
Theoretical (Leopold and Langbein, 1962)	0.09	0.36	0.55	--0.74	--0.22
This study, Ouse	0.27	0.25	0.48	--0.79	--0.39
This study, Twin	0.19	0.42	0.43	--0.79	--0.39
This study, Nene	0.26	0.32	0.42	------	------

Table 12 Results of Magnitude-Frequency Analysis for Streams Described in WSP 1849

	Years of record	Definition of $q_{1.58}\pm$	$q_{1.58}$, central	$q_{1.58}$, min.	$q_{1.58}$, max.	study flood	study flood r.i., central	study flood r.i., minimum	study flood r.i., maximum
Haw River near Benaja, N.C.	31	12%	39.64	34.88	44.40	28.32	1.12	1.03	1.19
Deep River at Ramseur, S.C.	37	10%	283.15	269.00	311.47	253.02	1.25	1.17	1.34
Merced River near Yosemite, Calif.	50	8%	67.25	61.87	72.63	55.21	1.28	1.22	1.38
The Salt Creek at Roca, Nebr.	12	22%	84.95	66.26	103.63	52.67	1.35	1.28	1.43
Chattahootchee River near Leaf, Ga.	22	14%	169.89	146.11	193.68	144.41	1.35	1.23	1.52
South Beaverdam R. at Dewy Rose, Ga.	19	15%	26.96	22.91	31.01	23.22	1.37	1.30	1.55
Pond Creek near Louisville, Ky.	17	16%	43.89	36.81	50.97	41.91	1.42	1.14	2.00
Clear Creek near Golden, Colo.	64	5%	40.35	38.31	42.39	39.08	1.48	1.37	1.58
Etowah R. near Dawsonville, Ga.	23	14%	67.25	57.82	76.68	63.99	1.50	1.27	1.95
Tobesofkee Creek near Macon, Ga.	24	14%	76.45	65.75	87.15	71.92	1.50	1.35	1.65
Coeur d'Alene R. near Pritchard, Idaho	9	30%	328.46	229.92	426.99	319.96	1.50	1.01	4.40
Champlin Creek near Colorado City, Tex.	12	22%	63.71	49.69	77.73	67.67	1.60	1.45	1.98
Murder Ck. near Monticello, Ga.	11	22%	23.22	18.09	28.26	23.78	1.60	1.46	2.07
Columbia River at Vernita, Wash.	21	15%	10,477	8,905	12,034	11,496	2.20	1.40	8.00
N. Fork Cedar R. near Lester, Wash.	12	22%	19.82	15.46	24.18	28.20	2.33	1.82	3.60
q_{bf} defined by means of rating curve									
Clearwater River at Kamiah, Idaho	47	5%	1,345	1,278	1,412	1,330[1]	1.55	1.23	1.60
Rio Chama at Chamita, New Mex.	15	19%	133.08	118.44	158.37	135.91[1]	1.60	1.27	5.50
Chiwawa River near Plain, Wash.	20	15%	70.79	60.17	81.41	73.62[1]	1.67	1.35	3.10
Middle Oconee River near Athens, Ga.	24	14%	138.74	116.94	158.57	150.07[1]	1.70	1.50	2.05
Esopus Creek at Coldbrook, N.Y.	30	12%	261.92	235.72	293.35	288.82[1]	1.80	1.65	2.00
Grande Ronde R. at La Grande, Oreg.	10	25%	83.53	62.65	104.41	104.77[1]	2.00	1.50	5.30
Overbank discharges									
Middle Oconee River near Athens, Ga.	24	14%	138.74	116.94	158.57	173.00	2.05	1.75	2.60
Esopus Creek at Coldbrook, N.Y.	30	12%	261.92	235.72	293.35	393.58	2.10	1.92	2.35
Middle Br. Westfield R. at Goss Hts, Mass	50	5%	67.96	67.39	71.35	96.27	2.70	2.40	3.10
Grande Ronde R. at La Grande, Oreg.	10	25%	83.53	62.65	104.41	130.81	2.75	2.00	5.20
Bull Creek near Ira, Tex.	7	30%	25.48	17.84	33.13	91.18	2.80	2.50	3.80
E.Br. Ausable R. at Ausable Forks, N.Y.	38	8%	152.90	140.73	165.08	218.03	3.25	2.70	4.30
Beaver Kill at Cooks Falls, N.Y.	48	5%	229.35	218.03	240.68	438.89	5.00	4.40	6.00

[1]Discharge at bankfull stage, from rating curves (author)

flow. It seems clear that, at some of the sites listed in WSP 1849, extensive flooding at times of high discharge is inhibited only by the narrowness of valley bottoms and the steepness of valley walls. Where no bankfull stage is listed, streams for which the study flood was of great magnitude and low frequency were eliminated.

Twenty-five sites of the original list in WSP 1849 survived elimination. Only 16 of these escaped overbank flow during the study peaks. One of the 16 non-flooding rivers crested below bankfull, but bankfull discharge can be read from a rating curve. Among the 9 flooding rivers, rating curves give bankfull flow in 5 cases. Three of the flooding rivers experienced low overbank flow of modest recurrence interval, data on which have been retained for comparison; and data on 4 additional streams, that also experienced low overbank flow but for which no bankfull stage is listed, have been similarly retained (Table 12).

As Benson (1960) has shown, the length of available record controls the probability of defining a flood of given recurrence interval within a given percentage and with a given probability (95 out of 100 trials) of being correct. In the present study, a working nomogram gave the range of definition and set appropriate envelope curves on the magnitude-frequency graphs. The central values for $q_{1.58}$ in Table 12 result from the initial analysis; maximum and minimum values have been read on the envelope curves. Where bankfull discharge depends on a rating curve drawn from the stage/discharge data in the WSP 1600 series, and where a shifting rating exists, the range of recurrence interval for q_{bf} has been appropriately widened. The results of the magnitude-frequency analysis are summarized in Table 12 and illustrated in Figs. 13 and 14.

The central values of recurrence interval for below-bank to bankfull flow range from 1.12 to 2.33 years (annual series) on the streams in the roughness survey, for events where overbank flow is not in question. For streams where bankfull flow has been obtained from rating curves, the range is 1.55 to 2.00 years. Central values for modest overflow range from 2.05 to 5.00 years.

Among the rivers in the roughness survey that provide usable records, the Columbia at Vernita, Wash., and the Clearwater at Kamiah, Idaho, are far larger than the others. When these are left out, as tending to produce spuriously high correlation coefficients, a regression of the non-flooding crest discharges in 15 cases and of bankfull discharge defined by rating tables in 4 cases on the central values of $q_{1.58}$ produces $r = +0.979$, $r^2 = 0.958$, $0.001 > P$, and study (or bankfull) flood = $0.972 q_{1.58}$. More importantly, when the range of recurrence interval from 1.01 to 1.58 years is used as a logarithmic base, and when the within-bank discharges in Table 12 are analyzed accordingly, these discharges are found to be log-normally distributed about the recurrence interval of 1.58 years, with $r = +0.891$, $r^2 = 0.794$, and $0.001 > P$ (Fig. 15). That is to say, the relevant data in the roughness survey of WSP 1849, combined with the magnitude-frequency analyses based on the WSP 1600 series, strongly support the proposition that natural bankfull discharge can be specified by $q_{1.58}$ on the annual series or $q_{1.0}$ on the partial-duration series.

The limited conclusions that bankfull flow on the Nene and Ouse can be specified in terms of recurrence interval, and that velocity at bankfull on those streams increases downstream, are sustained. More generally, the reconfirmation that natural bankfull discharge is equivalent to $q_{1.58}$ (annual series) sustains the interpretations of underfitness that depend on discharge at a given recurrence interval.

CHANNEL RESPONSE TO REDUCTION OF DISCHARGE

The general theory of underfit streams (Dury, 1964a, 1964b, 1965) deals mainly with responses to reduced discharge in terms of channel plan. The present study provides information on responses in terms of cross-section.

Diversion of water from a stream channel into millraces might, on general grounds, be expected to influence channel characteristics downstream

Table 13 Comparative Data on Offtakes, River Ouse

Reaches without offtakes						Reaches with offtakes					
Ad, km^2	Ac, m^2	w,m	$\bar{d}$,m	$w/\bar{d}$	R,m	Ad, km^2	Ac, m^2	w,m	$\bar{d}$,m	$w/\bar{d}$	R,m
33.2	8.40	9.90	0.85	11.60	0.724	35.3	3.94	5.69	0.71	8.02	0.5
37.4	7.10	9.05	0.78	11.50	0.669	38.5	6.14	6.71	0.91	7.40	0.720
39.7	8.88	9.30	0.96	9.73	0.791	41.5	5.23	6.59	0.80	8.25	0.639
71.0	7.15	8.10	0.88	9.20	0.725	72.0	6.72	6.94	0.97	7.15	0.757
83.8	10.38	10.90	0.95	11.50	0.811	89.0	9.20	9.89	0.93	10.60	0.783
94.0	12.69	12.20	1.04	11.70	0.889	97.0*	7.85*	9.58*	0.82	11.68	0.700
100.4	11.80	11.85	1.00	11.85	0.852	109.0	2.75	5.80	0.48	12.20	0.407
117.0	10.14	10.62	0.96	11.10	0.809	121.5	4.87	5.72	0.85	6.74	0.656
127.0*	15.07*	12.98*	1.16	11.19	0.984	130.0	5.67	6.00	0.95	6.32	0.718

*means of two values, for 96 and 98 km^2 and for 126 and 128 km^2.

Table 14 Comparative Effect of Offtakes, River Ouse: Ratios between Reaches with Offtakes and Reaches without Offtakes

Ad, km^2	Ac ratio	w ratio	$\bar{d}$ ratio	$w/\bar{d}$ ratio	R ratio
34.25	0.469	0.575	0.816	0.705	0.765
37.95	0.864	0.741	1.166	0.636	1.076
40.60	0.589	0.709	0.830	0.854	0.808
71.50	0.940	0.857	1.097	0.781	1.044
86.40	0.886	0.907	0.977	0.928	0.965
95.50	0.619	0.785	0.789	0.995	0.787
104.70	0.233	0.489	0.476	1.027	0.478
119.25	0.480	0.539	0.891	0.605	0.811
128.5	0.376	0.462	0.814	0.568	0.730
means	0.606	0.673	0.872	0.789	0.829
derived means			0.900	0.748	0.856

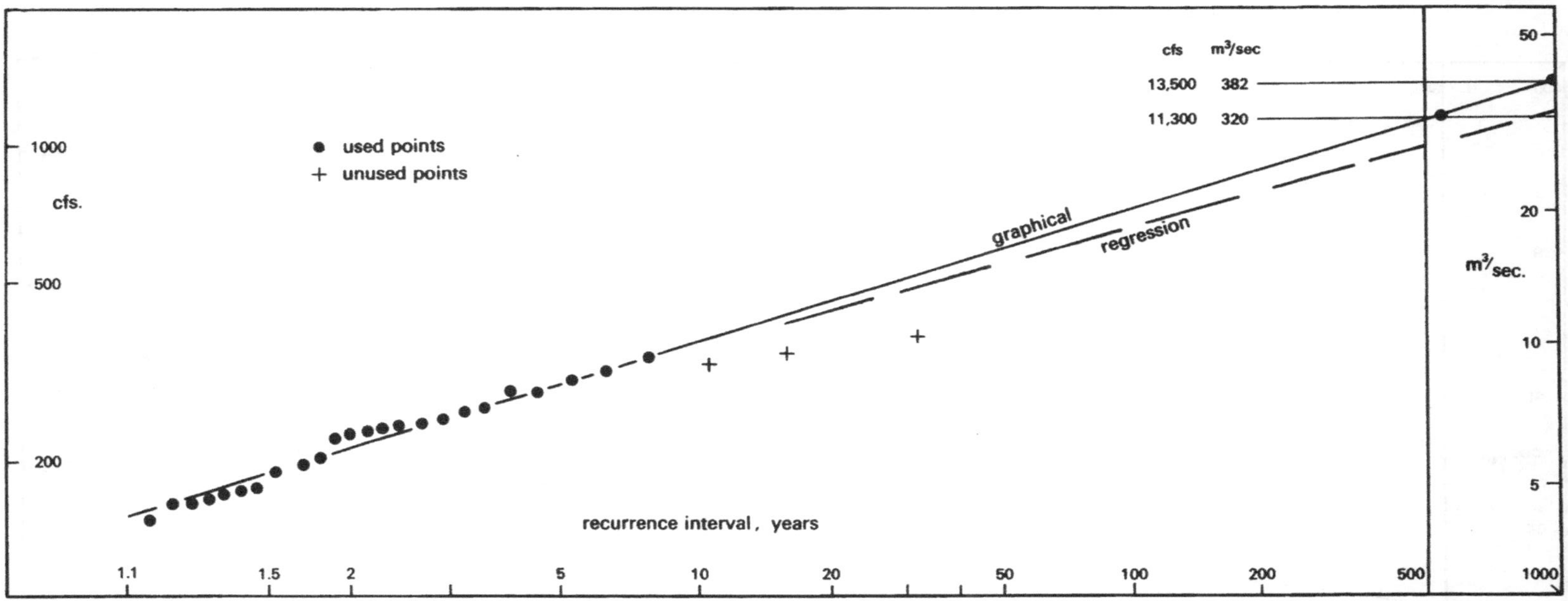

Figure 17 Log-Gumbel graph for the Nene at Orton, 1615km^2: possible location of the March 1947 event.

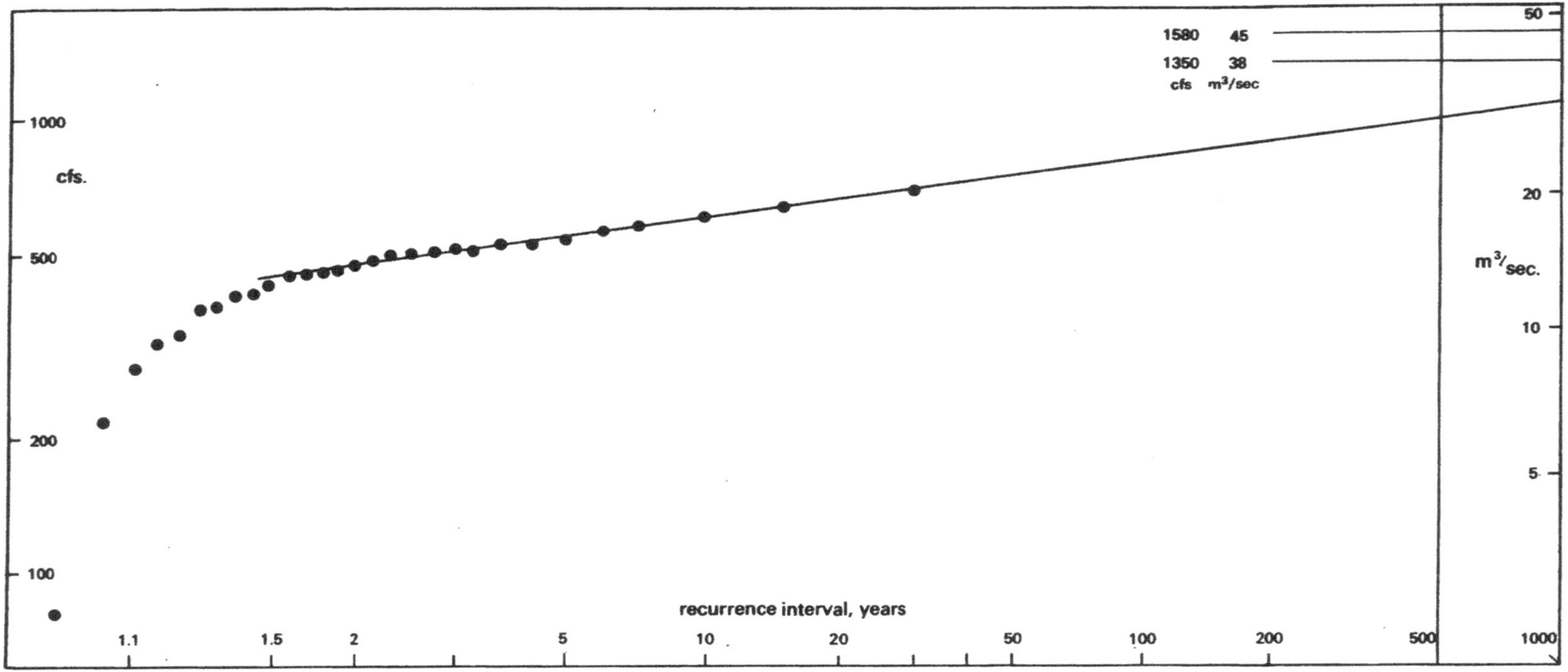

Figure 18 Log-Gumbel graph for the Nene at Upton, 220km^2: possible location of the March 1947 event.

of the offtakes. Information available for the upper Ouse, upstream of the confluence with the Twin, shows that values of A_c, w, $\overline{d}$, and R on reaches downstream of offtakes run generally below the values predicted by regression equations for unaffected reaches. No systematic distribution of the effects of reduced discharge can be identified in relation to drainage area; but a general inference is nevertheless possible.

Table 13 lists comparative data for reaches upstream of offtakes (= without offtakes) and downstream of offtakes (= with offtakes), the values for reaches without offtakes being determined as closely as possible upstream of the paired reaches without. Table 14 lists ratios calculated from the comparative data in Table 13. In no case does the ratio value vary significantly along the length of the stream. Accordingly, the means of 0.606 and 0.673 for the A_c and w ratios can be taken as fairly representing the reduction in cross-sectional area and bedwidth that results from offtaking. Derived means in Table 14 are developed from these values. Reduction in mean depth on reaches with offtakes is seen to be distinctly less, proportionately speaking, than reduction in cross-sectional area. In consequence, width/depth ratio is greatly reduced. That is to say, the reaches reduced in volume by diversion have responded by reducing their $w/\overline{d}$ values. Since there is little or no scope for increase of channel slope, response to reduced discharge must be expressed largely, or entirely, in terms of channel size and shape; and the reduced streams have assumed new form ratios that indicate increased channel efficiency.

When A_c is used as a basis of reference, values of $w/\overline{d}$ for channels with and without offtakes can be comparatively plotted (Fig. 16). The $w/\overline{d}$ values for reaches with offtakes fall into the same series as values for reaches without offtakes. In other words, the reduced channels have assumed form ratios appropriate to their reduced cross-sections. Variance in cross-sectional area explains about two-thirds of the variance in form ratio. The slope factor has little effect: multiple regression shows that it explains only 8.4% of the variance in form ratio that is left unexplained by variance in cross-sectional area.

These results confirm a downstream decrease of channel efficiency on the Ouse, insofar as that efficiency is measured by the form ratio. More importantly, they show that the channels reduced by offtaking have responded by reducing their width rather than their depth. Such information as is available on the former channels of streams that are now underfit suggests that a similar response has been made by streams reduced in volume by natural means.

THE THOUSAND-YEAR FLOOD AND ITS LACK OF EFFECT

From time to time, discussions of underfit streams and valley meanders produce the proposition that valley bends are produced, and eroded, by events of great magnitude and low frequency: that is to say, by unusually high floods. The records now available permit a check on, and a denial of, this proposition. Magnitude-frequency analysis produces, for several stations on the Nene system, Gumbel graphs with strong rectilinear tendencies between about 1.5 and 30 years of recurrence interval. Although extrapolation of magnitude-frequency graphs is never defensible--even when it must be undertaken in circumstances of exigency--the graphs actually obtained suggest that, at some stations, the very high discharges of March 1947 related to very great recurrence-intervals indeed. Thus, in 33 years of record, 1939-40 through 1971-72, the Upton gauge on the Nene recorded as the second highest annual flood a discharge of 19.59m^3/sec; the highest annual flood, a daily mean of 38.23m^3/sec in March 1947, plots somewhere off the usual graph. It can be taken to correspond to a momentary peak of 44.74m^3/sec (see Dury, 1959, Fig. 7). During the 32 years 1940-41 through 1971-72, the second highest annual flood at the Orton gauge was 109.01m^3/sec; the highest, in March 1947, was 382.26m^3/sec (estimate revised from the original 319.96m^3/sec). This value also plots somewhere off the graph. These two events appear to approximate the 1000-year flood.

The antecedent circumstances were certainly unusual. The winter of 1946/47 in much of the British Isles brought considerable snowfall to begin with, then a persistent blocking high, which spread westward from the mainland, bringing prolonged severe frost. When the blocking pattern broke down, snowmelt and runoff were rapid. Extensive flooding occurred on the Ouse and the Nene, among other eastern rivers.

Figures 17 and 18 are log-Gumbel graphs for the Nene at Orton and Upton, the form being selected in accordance with the recommendation of Benson (1962). The extrapolated graph for Upton is consistent with a recurrence interval of more than 1000 years for the March 1947 event; and it is very difficult to read a recurrence interval even as low as 500 years. The extrapolated graph for Orton is conservative, in taking no account of the three events ranking next in the series: allowance for these events would lower the slope of the graph, and predict for the March 1947 event a recurrence interval greater than 1000 years, instead of the 1000-year interval that almost precisely matches the rest of the scatter. Once again, whatever the recurrence interval, or range of interval, that might be adopted, it is difficult to see the March 1947 event at Orton as corresponding to a recurrence interval of less than 500 years.

This being so, the very slight geomorphic effects of the March 1947 floods, throughout the Nene and Ouse systems, are instructive. No widespread erosion or deposition took place. In this area at least, even the 1000-year event does not seem great enough to be a catastrophe. One wonders how far up the magnitude-frequency scale a castrophe would lie.

Neo-Catastrophism

The likely rise of a neo-catastrophic school has already been adumbrated (Dury, 1972). It deserves to be taken seriously. The foregoing observations about the thousand-year flood on the Nene are seriously meant. Paleomorphology, and its sister (or mother) study, paleoclimatology, teach us that geomorphic effects occur on one or other side of physical thresholds. That is to say, there are limits beyond which a steady-state system passes from equilibrium into disequilibrium, before settling into a new steady-state condition of different equilibrium. But the threshold effect, the jump, the step function, are completely unprovided for in the usual analysis of magnitude-frequency. There seems to be plenty of scope for investigation of the bounds of operation inside a given system. A possible criterion with respect to stream action would be that of geomorphically effective discharge at a given recurrence interval. If this criterion were adopted, then the constraints of progress up the magnitude-frequency scale would be loosened. But, in that case, we should be back to the definition of channel-forming discharge in terms of recurrence interval. Alternatively, one might speculate that large-scale erosion of the channel banks and/or floodplains on the Nene and Ouse systems is unlikely to occur until floodwater depth approximates the mean depth of the channel at bankfull, at, say, a drainage area of 1000km^2. In that case we should be thinking about a depth of floodwater of the order of 2m or more. It should not be impracticable to obtain data on depth of inundation, to plot these on a scale of magnitude-frequency, and at the same time to construct velocity profiles for the floodwater currents. In this way it might become possible to indicate, very roughly, the likely recurrence interval <u>in present conditions</u> for catastrophic floodplain or bank erosion.

CONCLUSIONS

1. Channel characteristics on the Ouse, its tributary the Twin, and the neighbouring Nene can be expressed as power functions of drainage area. Since discharge at a given recurrence interval can also be expressed as a power function of drainage area, channel characteristics can be expressed as power functions of discharge.

2. Levee characteristics on the Ouse can probably also be expressed, in the main, as power functions of drainage area or of discharge.

3. Increased sinuosity on the upper Ouse largely compensates for the steeper floodplain gradient on that river than on the Twin, but its compensatory effect is offset by difference in hydraulic radius. In consequence, greater channel roughness on the Ouse than on the Twin must be postulated.

4. For a range of recurrence intervals (annual series) from 1.2 to 5 years, discharge varies with the 0.735 power of drainage area. This conclusion differs from an earlier conclusion that discharge is a linear function of drainage area.

5. Velocity through the cross-section at a given recurrence interval, and specifically at $q_{1.58}$, increases in the downstream direction. The exponents obtained from this study range from 0.19 to 0.26, distinctly greater than the 0.09 indicated by theory. Confirmation of the downstream increase comes from a specific analysis of gauging station data, while estimates of natural channel capacity demonstrate a parallel downstream increase in velocity at actual bankfull stage.

6. Analysis of estimated channel capacity on the Nene supports the proposition that bankfull discharge can be specified in terms of recurrence interval. Analysis of data contained in U.S.G.S. Water-Supply Papers strongly supports the equating of bankfull discharge with $q_{1.58}$ (annual series).

7. Diversion of water to millraces on the upper Ouse has caused the channel downstream of offtakes to reduce its mean depth slightly and its width greatly. The reduced channels have assumed $w/\bar{d}$ ratios that match the ratios of unreduced channels of comparable cross-sectional area. The observed response is generally similar to that of channels on streams reduced to underfitness by natural means.

8. For some stations on the study rivers, log-Gumbel graphing indicates that the event of March 1947 approximated the thousand-year flood. The very slight geomorphic effects of this event raise questions about the place of catastrophic events on the magnitude-frequency scale and of the probability-forecasting of fluvial catastrophes.

ACKNOWLEDGMENTS

I owe thanks to a number of individuals and organizations for support and assistance. Miss K. P. Kittson of the Sydney Teachers' College, N.S.W., Australia, and my wife, M. H. Dury, acted as unpaid field assistants during the summer of 1972. My own fieldwork was aided by grants from the Whitbeck Fund, Department of Geography, UW-Madison, and from the Graduate School, UW-Madison (Project No. 130171). Officers of the Great Ouse River Authority (in particular, Mr. G. Pearson) and of the Welland and Nene River Authority (in particular, Mr. B. Gooding) helped me to secure cross-sectional data, and responded to elaborate follow-up enquiries through the mail. Illustrations have been prepared in the University of Wisconsin Cartographic Laboratory, with funding from the UW Graduate School (Project No. 140128). Mr. K. S. Deakin, of Magdalen College School, Brackley, Northants., England, kindly took the site photographs used in the oral presentation of this paper.

REFERENCES

Bagnold, R.A., 1960, Some aspects of the shape of river meanders: U.S. Geol. Survey Prof. Paper 282-E, 10p.

Barnes, H.H., 1967, Roughness characteristics of natural channels: U.S. Geol. Survey WSP 1849, 213p.

Barnes, H.R., and Golden, H.G., 1966, Magnitude and frequency of floods in the United States, Pt. 2-B: U.S. Geol. Survey WSP 1674, 409p.

Benson, M.A., 1960, Characteristics of frequency curves based on a theoretical 1000-year record: U.S. Geol. Survey WSP 1543-A, p. 51-74.

Benson, M.A., 1962, Evolution of methods for evaluating the occurrence of floods: U.S. Geol. Survey WSP 1580-A, 30p.

Bodhaine, G.L., and Thomas, D.M., 1964, Magnitude and frequency of floods in the United States, Pt. 12: U.S. Geol. Survey WSP 1687, 337p.

Dury, G.H., 1951, Some aspects of the geomorphology of part of the Midland Jurassic belt: U. of London, unpubl. Ph.D. thesis.

Dury, G.H., 1952, Some long-profiles of the Great Ouse system: Journ. Northants. Nat. Hist. Soc. F.C. v. 32, p. 135-140.

Dury, G.H., 1958, Tests of a general theory of misfit streams: Inst. Brit. Geog., Trans. and Papers no. 25, p. 105-118.

Dury, G.H., 1959, Analysis of regional flood frequency on the Nene and the Great Ouse: Geog. Journ. v. 125, p. 223-229.

Dury, G.H., 1961, Bankfull discharge, an example of its statistical relationships: Internat. Assoc. Sci. Hydrol., Bulletin vi^e année, no. 3, p. 48-55.

Dury, G.H., 1964a, Principles of underfit streams: U.S. Geol. Survey Prof. Paper 452-A, p. A1-A67.

Dury, G.H., 1964b, Subsurface exploration and chronology of underfit streams: U.S. Geol. Survey Prof. Paper 452-B, p. B1-B56.

Dury, G.H., 1965, Theoretical implications of underfit streams. U.S. Geol. Survey Prof. Paper 452-C, p. C1-C43.

Dury, G.H., 1973, Channel habit, underfit streams, and the conduct of technical debate: Area V. 5, p. 152-154.

Green, A.R., 1964, Magnitude and frequency of floods in the United States, Pt. 1-A: U.S. Geol. Survey WSP 1671, 260p.

Gumbel, E.J., 1958, Statistical theory of floods and droughts. Inst. Water Engrs., Journ. v. 12, p. 157-184.

Husling, H., and Kallio, N.A., 1964, Magnitude and frequency of floods in the United States, Pt. 14: U.S. Geol. Survey WSP 1689, 320p.

Kennedy, B.A., 1972, 'Bankfull' discharge and the meander form: Area v. 4, p. 209-212.

Langbein, W.B., 1949, Annual floods and the partial-duration series: Amer. Geophys. Union, Trans. v. 30, p. 879-881.

Leopold, L.B., 1953, Downstream change of velocity in rivers: Amer. J. Sci. v. 251, p. 606-624.

Leopold, L.B., and Langbein, W.B., 1962, The concept of entropy in landscape evolution: U.S. Geol. Survey Prof. Paper 500-A, 20p.

Leopold, L.B., and Maddock, T., 1953, The hydraulic geometry of stream channels and some physiographic implications: U.S. Geol. Survey Prof. Paper 252, 57p.

Leopold, L.B., and Wolman, M.G., 1960, River meanders: Geol. Soc. Amer., Bulletin v. 71, p. 769-794.

Leopold, L.B., Wolman, M.G., and Miller, J.P., 1964, Fluvial Processes in Geomorphology. W.H. Freeman, 522p.

Maddock, T., 1969, The behavior of straight open channels with moveable beds: U.S. Geol. Survey Prof. Paper 622-A, 70p.

Matthai, H.F., 1968, Magnitude and frequency of floods in the United States, Pt. 6-B: U.S. Geol. Survey WSP 1680, 491p.

Nixon, M., 1959, A study of the bank-full discharges of rivers in England and Wales: Inst. Civil Engrs., Proc. v. 12, p. 157-173.

Patterson, J.L., 1965, Magnitude and frequency of floods in the United States, Pt. 8: U.S. Geol. Survey WSP 1682, 506p.

Patterson, J.L., 1966, Magnitude and frequency of floods in the United States, Pt. 6-A: U.S. Geol. Survey WSP 1679, 471p.

Speer, P.R., and Gamble, C.R., 1964a, Magnitude and frequency of floods in the United States, Pt. 2-A: U.S. Geol. Survey WSP 1673, 329p.

Speer, P.R., and Gamble, C.R., 1964b, Magnitude and frequency of floods in the United States, Pt. 3-B: U.S. Geol. Survey WSP 1676, 340p.

Speer, P.R., and Gamble, C.R., 1965, Magnitude and frequency of floods in the United States, Pt. 3-A: U.S. Geol. Survey WSP 1675, 630p.

Thomas, C.A., Broom, H.C., and Cummans, J.E., 1963, Magnitude and frequency of floods in the United States, Pt. 13: U.S. Geol. Survey WSP 1688, 250p.

Tice, R.H., 1968, Magnitude and frequency of floods in the United States, Pt. 1-B: U.S. Geol. Survey WSP 1672, 585p.

Wiitala, S.W., 1965, Magnitude and frequency of floods in the United States, Pt. 4: U.S. Geol. Survey WSP 1677, 357p.

Wolman, M.G., and Leopold, L.B., 1957, River flood plains, some observations on their formation: U.S. Geol. Survey Prof. Paper 282-C 23p.

Young, L.E., and Cruff, R.W., 1967, Magnitude and frequency of floods in the United States, Pt. 11: U.S. Geol. Survey WSP 1686, 308p.

6

EROSIONAL FORMS AND PROCESSES FOR THE CATASTROPHIC PLEISTOCENE MISSOULA FLOODS IN EASTERN WASHINGTON

Victor R. Baker

ABSTRACT

Periodic catastrophic flooding from Lake Missoula produced an amazing assemblage of erosional forms in the Columbia Plateau region of eastern Washington. Lithologic and structural irregularities in the Yakima Basalt were significant in localizing the plucking-type erosion by high velocity flood water. The anticlinal structures of the western region resulted in numerous constricted channels. The phenomenal flow velocities achieved in these constrictions produced the most spectacular scabland features, including rock basins, potholes, and abandoned cataracts. In contrast, the eastern scabland region was characterized by more uniform flow conditions associated with more subdued erosional topography.

Secondary flow phenomena, including various forms of vortices and flow separations, are considered to be the principal erosive agents. The intense pressure and velocity gradients of vortices along the irregular channel boundaries produced the plucking-type erosion. The great depth of the flood flows probably considerably reduced the effectiveness of cavitation as an erosional agent.

A sequence of erosional bed forms can be seen in many of the scabland divide crossings. The sequence seems to proceed in time as follows: (1) longitudinal grooves on the tops of basalt flows, (2) linear rock basins, grooves, and potholes, and (3) eventual formation of a deeply incised inner channel of the "Dalles" type. This sequence is analogous to the results of recent flume experiments on simulated bedrock by S. A. Schumm and R. G. Shepherd.

Flood erosion of sediments on the Columbia Plateau has produced such diverse features as scour marks, armored channels, scour holes, and streamlined loess hills. The hydrodynamic similarity between the loess hills and airfoils is demonstrated by calculating equivalent lemniscate loops, as suggested by R. J. Chorley's study of glacial drumlins.

INTRODUCTION

Recent studies of large-scale modern and ancient flooding (e.g. Birkeland, 1968; Bradley and others, 1972; Malde, 1968; Scott and Gravlee, 1968) have reintroduced an awareness of the importance of catastrophic events in fluvial processes. When J.Harlen Bretz began his studies of the Channeled Scabland in eastern Washington over 50 years ago, the intellectual climate was not so receptive. Olson (1969, p. 503) states, "during its not always calm history, the story of the development of the Channeled Scabland was thought by some to have brushed beyond the dividing line in flaunting catastrophe too vividly in the face of the uniformity that had lent scientific dignity to the interpretation of the history of the earth."

The Channeled Scabland region (Fig. 1) coincides with the portion of the Columbia Plateau that was subjected to periodic catastrophic flooding during the late Pleistocene (Bretz, 1969; Baker, 1973). Bretz (1923) introduced the term "scabland" to refer to the chaotically eroded tracts of bare basalt that occupy the floors of channels cut through the thick loessial cover of the plateau. Bretz (1923, 1928, 1932) described the amazing assemblage of

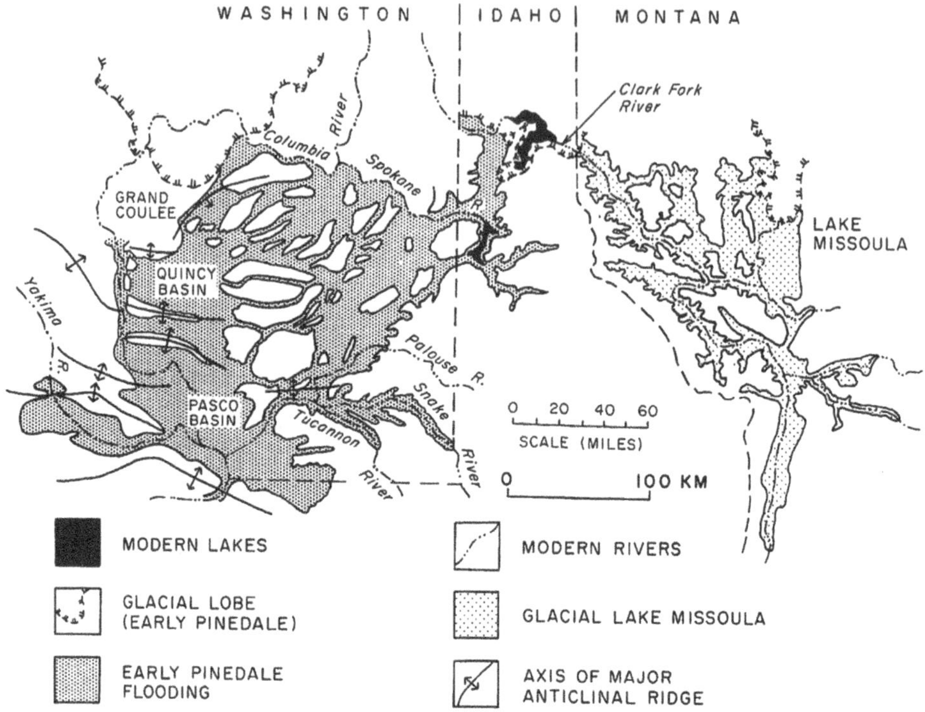

Figure 1 Modern and ancient drainage on the Columbia Plateau and in western Montana (Baker, 1973).

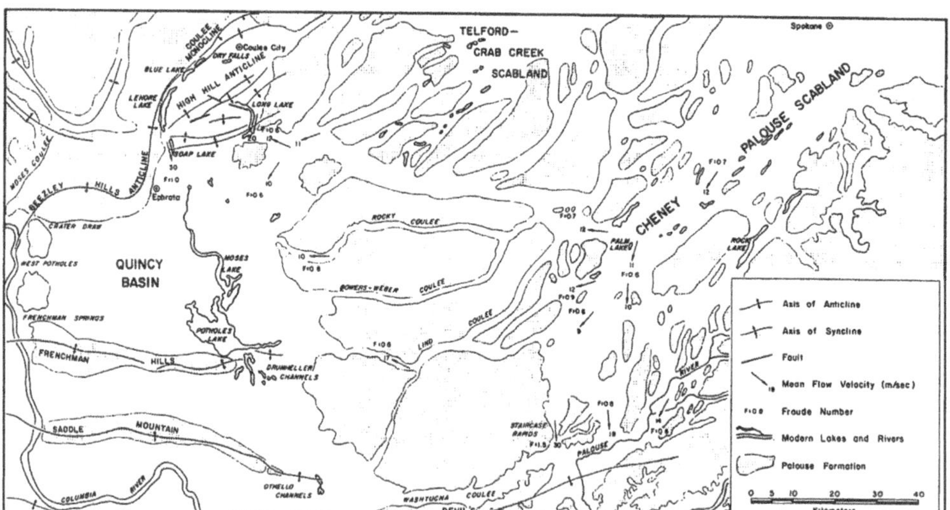

Figure 2 Structural and paleohydraulic features of the Channeled Scabland. Structural data from Grolier and Bingham (1971) and Newcomb (1970). Mean flow velocities were determined from high water mark evidence and channel geometry according to procedures described by Baker (1973).

erosional and depositional forms in the scablands, including rock basins, anastomosing channels, cataracts, gravel bars, and coulees. Numerous features, especially the huge gravel bars, multiple high-level divide crossings, and rock basins as much as 40 m deep led him to postulate that a catastrophic flood of enormous proportions had swept across the Columbia Plateau. The source of the flood was eventually shown to be glacial Lake Missoula, which was dammed by a glacier in the vicinity of modern Lake Pend Oreille (Bretz, 1930).

Objections to the flood hypothesis (Allison, 1933; Flint, 1938a) were answered in meticulous detail by Bretz and others (1956). The presence of remarkable giant current ripples on the scabland bars and evidence for the rapid emptying of Lake Missoula (Pardee, 1942) strongly supported the flood hypothesis. Recent studies of Quaternary geology in the regions of Missoula flooding have accepted this reasoning (Baker, 1973; Fryxell and Cook, 1964; Richmond and others, 1965; Trimble, 1963). Nevertheless some reluctance to accept even Bretz's (1969) modified hypotheses for the origin of the Channeled Scabland persists to the present day. Newcomb and others (1972) have held to Allison's (1933) hypothesis for the explanation of deep water ponding in the western scablands, and have interpreted the late Pleistocene history of the Hanford area without reference to any of Bretz's work.

Following the considerable criticism of his ideas in the late 1920's (see Bretz, 1927a, 1927b; McKnight, 1927; Bretz and others, 1956), Bretz (1932, p. 82) agonized over his flood hypothesis. He stated, "somewhere must lurk an unrecognized weakness. Where is it? If it exists, it probably lies in the hydraulics of the concept." Bretz asserted that the turbulence of the glacial flood and the jointing of the basalt may be important, an idea gained from his earlier studies of the Columbia River (Bretz, 1924). He added (Bretz, 1932, p. 83), "we do not know enough about great flood mechanics to make any conclusions valid.....Hydraulic competency must be allowed the glacial streams, however much it may differ from that of stream floods under observation."

The present paper will attempt to show that the erosional features first described by Bretz are an expected hydrodynamic consequence of the exceedingly high flow velocities in very deep flood water, as calculated by Baker (1973). Moreover, it will be shown that scabland erosional forms seem to follow an orderly progression in time that is a characteristic of deep-swift flows which generate large-scale turbulence along irregular boundaries. The details of this process are markedly influenced by the lithologic and structural characteristics of the rock.

EROSIONAL CHARACTERISTICS OF THE YAKIMA BASALT

The Miocene plateau lavas known collectively as the Yakima basalt of the Columbia River Group (Waters, 1961) underlie most of the Channeled Scabland region. The monotonous appearance of the basalt flows belies a detailed stratigraphy that can be traced for scores of kilometers. Bingham and Grolier (1966), Mackin (1961), Schmincke (1967), Waters (1961), Wright and others (1973) have described the flow by flow stratigraphy in terms of such criteria as size and shape of columns, vesicle types, pillow zones, mineralogy, spiracles, and chemical composition. Local interbeds of tuffaceous sediments and buried soil zones are distinctive.

Structural Deformation

The basalt surface has a general regional slope of 6-8 m/km southwestward from elevations of 850 m near Spokane to 120 m near Pasco. Bretz (1923, 1928) described the development of a consequent stream pattern along this dipslope prior to the late Pleistocene flooding. Deformation of the dipping plateau surface

Figure 3 Structural control of hydraulic plucking erosion:

a erosion at the crest of High Hill anticline south of Coulee City, Washington

b joint-controlled rock basins and channels on the Palouse-Snake River divide.

increases to the west (Fig. 2). The great topographic ridges, such as Frenchman Hills, Saddle Mountain, and Horse Heaven Hills, are interpreted to be anticlines which grew during basalt deposition (Waters, 1955).

Bretz recognized that the large-scale structural features of the Columbia Plateau were important in localizing flood erosion. He suggested that the structural step of the Grand Coulee monocline allowed the initiation of a 250 m high cataract, whose northward recession from near Coulee City created Grand Coulee (Bretz, 1932). Erosion in Lenore Canyon was aided by fractured rock on the steep eastern limb of Beezley Hills anticline. Scabland divide crossings at High Hill anticline (Bretz, 1932) and Frenchman Hills (Bretz, 1928) were shown to be eroded most intensely at the tension-jointed anticlinal crest (Fig. 3a). Joint control of rock basins and scabland channels (Fig. 3b) is well displayed on the Palouse-Snake divide crossing (Trimble, 1950, 1954).

Lithology and Jointing

Planes of weakness within the basalt bedrock were an important influence on fluvial erosional forms produced by Missoula flooding. The individual basalt flows average 25-60 m thick and are characterized by a variety of depositional and cooling features which allowed variable resistance to flood erosion (Fig. 4). Most flows have a scoriaceous upper portion grading into the rubbly top typical of aa lava. Some flows exhibit the classic Tomkieff (1940) sequence of 2-tier columnar jointing. Columns in the lower colonnade may be 1-2 m in diameter. An entablature of long slender columns and hackly fragments is present in all flows. In some flows an upper colonnade of much smaller columns occurs above the entablature. The cooling history of the flows controls the nature of columnar jointing (Spry, 1962). Irregular cooling surfaces developed because of erosion during the intervals between basalt emplacement. This often resulted in flaring columns in the next higher Yakima basalt flow. Rapid cooling produced hackly fragmented lava and the "brickbat" zones described by Mackin (1961, p. 10).

Between the outpourings of basalt, enough time elapsed to permit weathering, growth of forest cover, and the formation of lakes. Local sedimentary intercalations in the basalts include conglomerate beds, clay layers, and fresh-water diatomite. The lavas which overrode these lake beds formed pillow-palagonite complexes and spiracles (gas chimneys) as described by Waters (1960).

Observations of basalt boulders transported by Pleistocene flooding (Baker, 1973) revealed that the largest boulders were always portions of hackly jointed entablature. Fragments of individual columns were common in scabland bars, but boulders greater than 1-2 m in diameter were not produced from lower colonnades because of the well-developed cooling-contraction joints in these zones.

Erosional Forms

Bretz (1928) pointed out that the Channeled Scabland is a complex anastomosis involving perhaps 100 glacial streamways, 50 of which bifurcate in a downstream direction. Flood flows in many of these channels spilled over into adjacent valleys, crossing pre-flood divides. The pre-flood valleys simply did not have the capacity to convey the enormous flood discharges without overflowing.

Perhaps the most prevalent topographic form in the scabland tracts is butte-and-basin topography (Fig. 5). The usual development is small anastomosing channels and rock basins surrounding

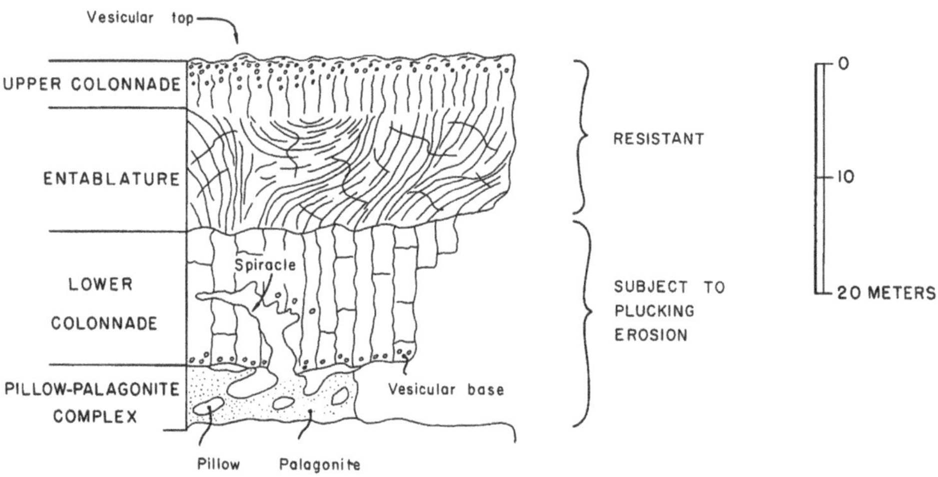

Figure 4

Cross section of an idealized Yakima basalt flow showing structural features important to flood erosion processes. The upper colonnade and the pillow-palagonite complex are only present in some basalt flows. Diagram is modified from Swanson (1967) and Schmincke (1967), and uses the terminology of Tomkeieff (1940), Waters (1960), and Spry (1962).

a

b

Figure 5 Butte-and-basin topography near Long Lake (a) and Blue Lake (b). Potholes show undercutting of their entablature rims particularly on the downstream end.

a

b

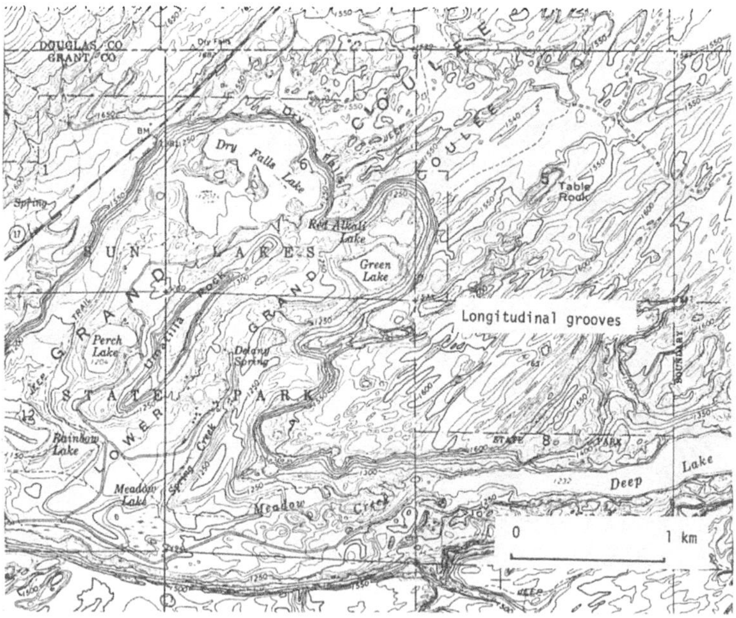

Figure 6

Dry Falls cataract and mouth of upper Grand Coulee (a). Longitudinal grooves just upstream from Dry Falls (b) probably resulted from the erosive action of longitudinal vortices aligned parallel to the flow direction.

buttes and mesas, with a total relief of 30-100 m. The rock basins range in size from shallow saucers to the scale of Rock Lake, 11 km long and 30 m deep (Fig. 2). Bretz (1932, p. 26-28) described this combination of features as follows: "The channels run uphill and downhill, they unite and they divide, they deepen and they shallow, they cross the summit, they head on the back-slopes and cut through the summit; they could not be more erratically and impossibly designed.....Channeled scabland is river bottom topography magnified to the proportion of river-valley topography."

The most impressive erosional forms created by the Missoula floods are probably the abandoned cataracts, largest of which is Dry Falls (Fig. 6), 5-1/2 km wide and 120 m high. Bretz and others (1956, p. 1029) recognized that many of the cataracts were formed subfluvially rather than by the plunge-pool undercutting classically illustrated by Niagara Falls. This contention is supported by the high-water mark evidence left by the early Pinedale flood (Baker, 1973, Fig. 7). Bretz (1932, 1933) described in detail the initiation of a 250 m cataract near Coulee City, Washington, and its 32 km upstream recession to create the upper Grand Coulee. The unique capacity of vertically jointed basalt to maintain the lip of a recessional cataract was held to be the primary consideration in this type of erosion.

In contrast to the upper Grand Coulee, Lenore Canyon and Moses Coulee (Fig. 7) show the existence of pre-flood stream valleys by the hanging valleys of tributary streams draining uneroded Palouse-soil topography. The floors of such valleys were deepened by flooding, often with the production of a distinct inner channel (Fig. 7).

Significant scabland features never described by Bretz are the longitudinal grooves (Fig. 6) which locally cover flood-exposed basalt entablature. The grooves generally parallel flood flow directions and are oblique to structural features. In the lower Grand Coulee they mark the otherwise flat summits of mid-channel buttes which have been isolated by rock basins and anastomosing channels. Near Dry Falls the grooves average 5 m deep and 50 m in width. They are relatively evenly spaced with about 12 grooves over a kilometer width. Four grooves, 300 m wide, mark the approach to Potholes cataract.

HYDRAULICS OF A CATASTROPHIC FLOOD

By studying the extensive high-water mark evidence left by the last major scabland flood, which may have occurred about 22,000 years B. P., it is possible to reconstruct high-water surfaces and flow depths. Combining this evidence with the channel geometry, Baker (1973) utilized hydraulic engineering procedures to calculate discharges and mean flow velocities. Additional paleohydraulic information is provided by the numerous giant current ripples that occur throughout the region. From this kind of evidence it has been shown that discharges as great as 21.3×10^6 m^3/sec were conveyed through the Channeled Scabland (Baker, 1971).

The distribution of mean flow velocities in the Channeled Scabland (Fig. 2) is important in understanding flood erosion processes. Constricted channels in the western scablands, such as Lenore Canyon, Othello Channels, and Drumheller, achieved flood flow velocities as high as 30 m/sec. Such high velocities were reached only because of the unique combination of great flow depth (60-120 m) and very steep water surface gradients (2-12 m/km) that characterized the Missoula Flood. The constrictions contain the best developed butte and basin topography, cataracts, and inner channels. Adjacent basins such as the Hartline, Quincy, and Pasco Basins, produced ponded water and the accumulation of sediment eroded from the constrictions (see Baker, 1973, p. 15).

a

b

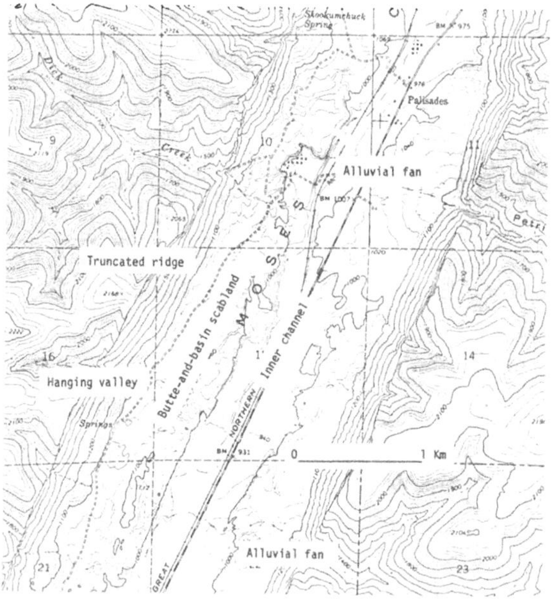

Figure 7 Inner channel development in Moses Coulee. The channel margins show grooved and potholed scabland that characterized the channel bottom prior to subfluvial incision. Giant current ripples occur below the inner channel (a). The extent of the preflood valley is shown by the truncation of tributary ravines and their interfluves (b). Postflood alluvial fans (b) locally overlie butte-and-basin scabland.

In contrast, the Cheney-Palouse and Telford-Crab Creek scabland tracts on the eastern plateau (Fig. 2) conveyed flows that averaged 12-15 m/sec. Butte and basin topography is much less intense in these tracts than in the constrictions of the western scablands. Only a few cataracts are found and these are much smaller than in the west. Giant current ripples, absent in constricted reaches, are common adjacent to eroded scabland in the Cheney-Palouse and Telford-Crab Creek tracts. As shown in Fig. 2, these areas all had Froude numbers less than 1.0. Froude numbers were determined from

$$F = \overline{V}/\sqrt{g\overline{D}} \qquad (1)$$

where $\overline{V}$ is the mean flow velocity calculated by slope-area or critical depth hydraulics (Chow, 1959), $\overline{D}$ is the mean channel depth, and g is the acceleration of gravity. Engineers concerned with canal design make a practice of avoiding supercritical flow ($F > 1.0$) because of its great erosive power. As pointed out by Koloseus (1971, p. 3-49), the higher stagnation pressures of supercritical flow give rise to uplift forces of such a magnitude as to remove the lining of a canal.

A fundamental hydraulic characteristic of very deep, high gradient flood flows is the development of secondary circulation, flow separation, and the birth or decay of vorticity around obstacles and along irregular boundaries. Such three-dimensional flow phenomena in rivers are poorly understood even by hydraulic engineers. Indeed, Simons and Gessler (1971) have suggested that fluvial morphologic studies can make a valuable contribution to engineering river mechanics by describing examples of such phenomena in nature. Matthes (1947) has termed these phenomena collectively as "macroturbulence." The most important erosive form of macroturbulence is the "kolk," a Dutch term which Matthes used to designate intense energy dissipation by upward vortex action. The intense pressure and velocity gradients of the vortex produce a phenomenal hydraulic lift force along the filament of the vortex (Fig. 8). The precise magnitude of this suction effect is unknown but it certainly is greatly in excess of normal hydraulic lift forces.

The conditions necessary for the generation of kolks include the following:

(1) A steep energy gradient,
(2) A low ratio of actual sediment transported to potential sediment transport,
(3) An irregular, rough boundary capable of generating flow separation.

These conditions are especially well displayed by the constricted reaches of the western scablands. Bretz and others (1956) and Bretz (1969), who likened the kolk to a subfluvial tornado, considered it to be the major erosive agent in the scablands.

BEDROCK EROSIONAL PROCESSES

Cavitation

The enormous flow velocities achieved in many scabland channels caused considerable reduction of absolute pressure, perhaps to the fluid vapor pressure. Shock waves produced by collapse of the air bubbles which form in such situations are recognized as intense erosional agents by civil engineers (Symposium, 1947). Embleton and King (1968, p. 271) suggested that this process of cavitation may have been responsible for the excavation of scabland rock basins. The bizarre "p-forms," potholes, and other erosional features which occur on rock surfaces in Scandinavia are interpreted as the result of cavitation erosion by subglacial melt-water streams flowing in concentrated routes or tunnels (Dahl, 1965). However, Schumm and Shepherd (1973) have suggested that such topography need not be subglacial.

Engineering experiments have shown that bubble collapse near materials such as steel or concrete produce hammer-like blows

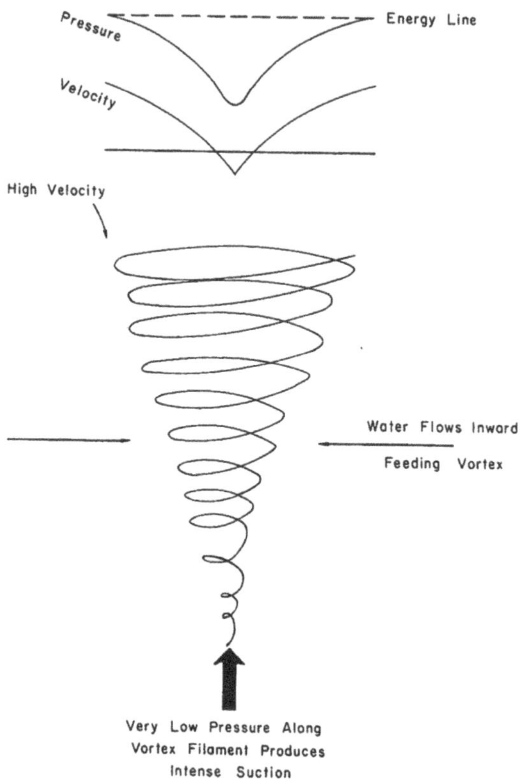

Figure 8

Characteristics of a hydraulic vortex ('kolk"). Such macroturbulent effects are set up by flow obstructions. The intense pressure and velocity gradients of the vortex produce a hydraulic lift.

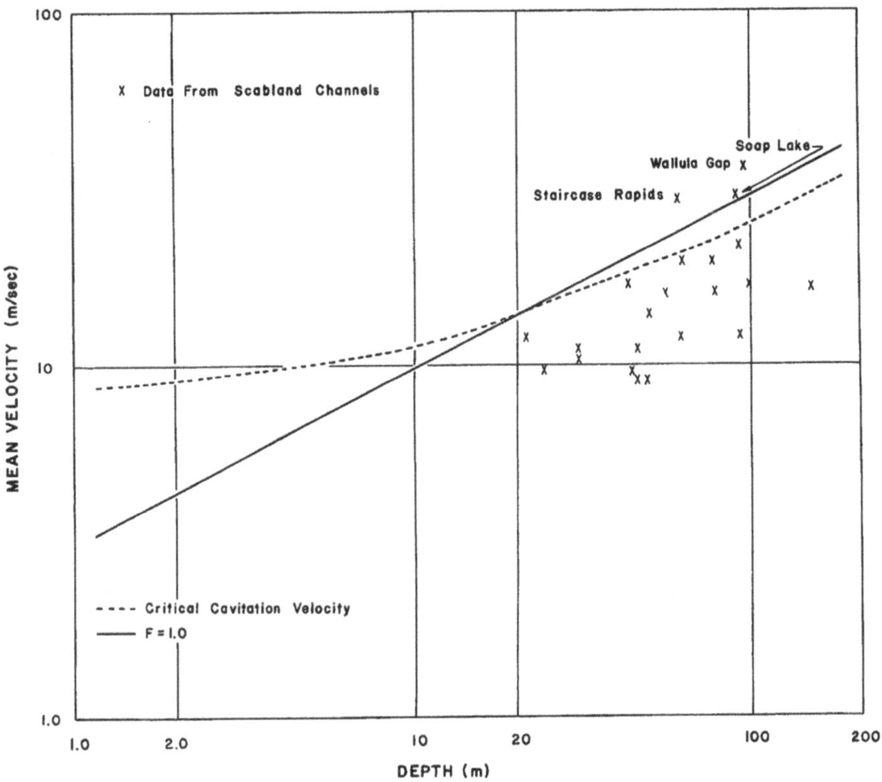

Figure 9 Influence of flow depth on the critical mean flow velocity necessary for cavitation to occur according to Barnes (1956, p. 499). Hydraulic data from scabland channels are plotted for comparison. Most of the flood flows were subcritical and below the critical depth-velocity combinations necessary to produce cavitation erosion.

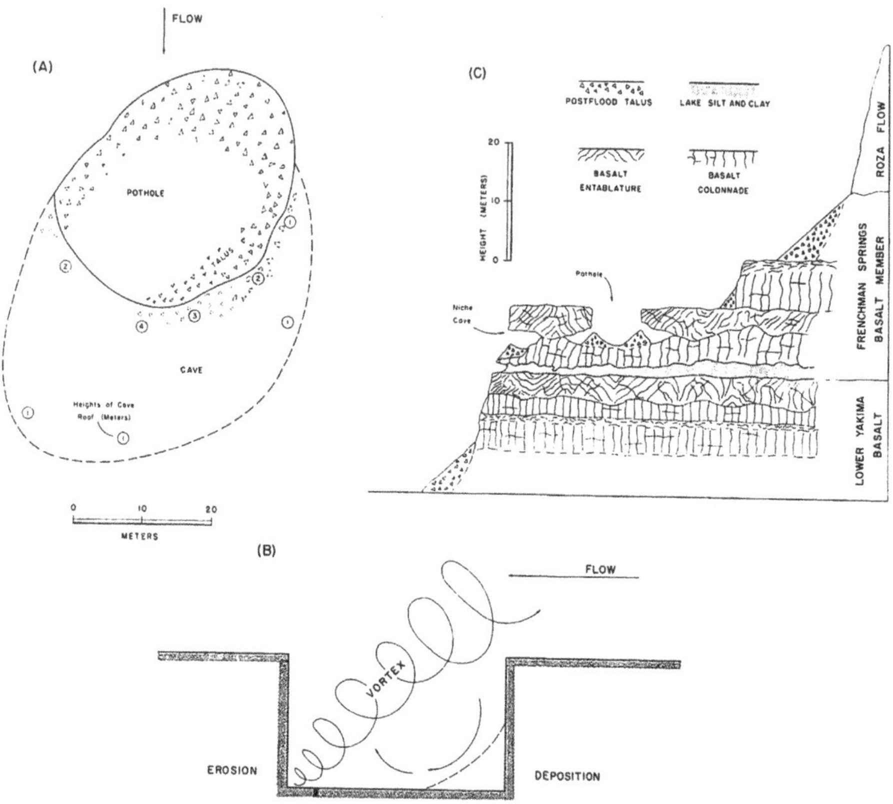

Figure 10 Pothole erosion at Lenore Caves, 15 km north of Soap Lake in Lenore Canyon. The sketch map (a) shows undercutting of the pothole rim on the downstream side. This is consistent with experimental observations of vortex scour (b), as described by Moore (1972). The relationship of pothole to the local basalt section is shown in (c).

with local pressures as high as 30,000 atmospheres (Barnes, 1956, p. 494). These very intense local pressures are capable of shattering the surface of nearly any solid material. The fractured surface layers are quickly carried away by turbulent water and fracturing continues. Barnes (1956, p. 503) proposed that such processes in high velocity glacial streams may initiate the formation of potholes. He added, however, that cavitation forms are very short-lived. Once the potholes form, they would then trap abrasive particles of bedload and continue to enlarge by the rotary action of the water (Alexander, 1932).

Cavitation can only occur for certain critical conditions. Hjulstrom (1935, p. 311) noted that the minimum velocity necessary for cavitation to occur in a river is about 12 m/sec. This figure is somewhat misleading, however, because it applies only to relatively shallow, swift streams. Barnes (1956, p. 499) showed that depth is also an important factor, according to the equation

$$\overline{V}_c = 4.6\ (33.1 + D)^{0.5} \qquad (2)$$

where $\overline{V}_c$ is the critical mean flow velocity necessary for cavitation to occur at any particular depth D under normal fluvial flow conditions. This equation indicates that the critical conditions necessary for cavitation were only achieved in a few scabland channels (Fig. 9).

Hydraulic Plucking and Pothole Erosion

Bretz (1924) first suggested that scabland-type erosion was a result of plucking rather than abrasion. He observed that eddies generated by the irregular channel walls at The Dalles area of the Columbia River are responsible for the plucking of basalt columns. He stated that inner channels of "The Dalles" type, potholes, and other scabland features could only be produced by a river of high discharge and steep gradient eroding close and vertically jointed rock. It was later suggested by Bretz and others (1956, p. 1028) that the hydraulic lift necessary for plucking action was provided by the kolk action of Matthes (1947). Bretz (1969) added, "plucking action thus was greatly augmented wherever submerged basalt ledges with the proper form appeared during flooded regimens."

The morphology of pothole erosion at Lenore Caves (Fig. 10) provides some insight into the details of the hydraulic plucking process. Like many scabland potholes, the one at Lenore Caves is asymmetrical in cross section parallel to the flood flow direction. Columns were preferentially plucked on the downstream side from beneath the resistant pothole rim of entablature. The result was a cave extending 20 m under the basalt (Fig. 10a). The symmetry of erosion is consistent with Moore's (1972) experimental flume studies of vortex scour. These studies indicate that at a flood peak, vertical vortices will scour at the downstream end of a hole on a stream bed and that the vortex axis will slant upstream (Fig. 10b).

Sequence of Erosional Forms

Recent experimental studies of fluvial erosion utilizing simulated bedrock (Shepherd, 1972; Shepherd and Schumm, in press) indicate that a sequence of erosional bed forms may develop in bedrock as a function of time. First to appear in these experiments were the faint streaks of longitudinal lineations associated with potholes and transverse erosional ripples. The lineations then became enlarged into prominent longitudinal grooves. Eventually the grooves decreased in number, and finally one narrow, deep inner channel formed. In the experiments the inner channels were incised below base level, and nickpoints migrated headward upstream (see Schumm and Shepherd, 1973, p. 7, for a longitudinal profile). Deposition of bedload occurred downstream from the headcut, culminating in a low-gradient inner channel with a sand bed and simulated bedrock banks. The lower part of this channel was actually an elongate basin, incised below base level.

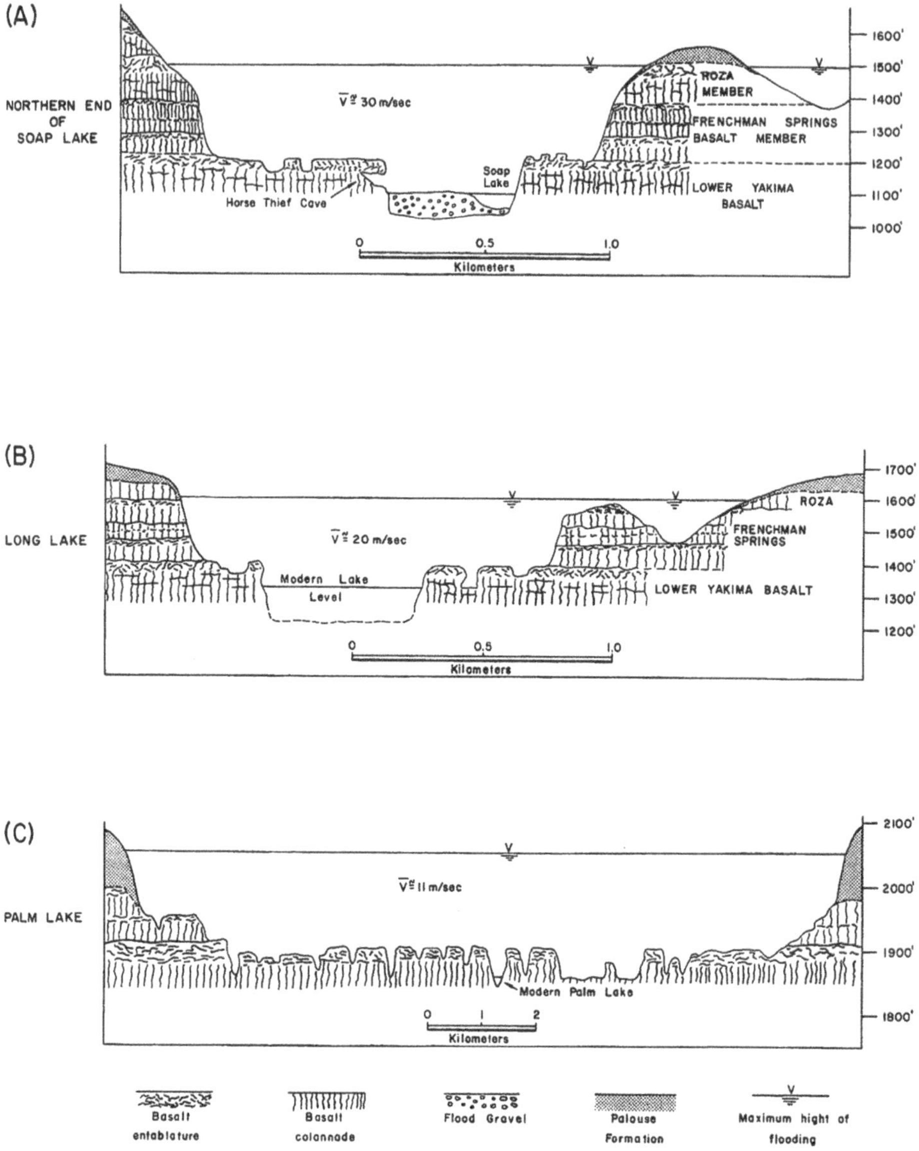

Figure 11

Representative scabland channel cross sections showing probable extent of flooding and maximum mean flow velocities. Soap Lake (a) and Long Lake (b) show inner channel development with marginal butte-and-basin topography. Palm Lake (c) shows grooved scabland.

Shepherd (1972) observed that some of the scabland erosional topography described by Baker (1973) was analogous to features produced in the flume experiments. The present study supports that contention. Scabland channel cross sections (Fig. 11) show numerous inner channels (Fig. 7), grooves (Fig. 6), and potholes (Figs. 5, 10). Moreover, the occurrence of eroded scabland divide crossings, analogous to all the stages observed in the flume experiments, suggests that the scabland channels may have followed a similar sequence of development. However, the peculiar erosional characteristics of the jointed basalt induced many differences in the details of the erosional forms.

Longitudinal grooves are less common in the scablands than might be indicated by the flume experiments. Where formed (Fig. 6), they are developed on resistant basalt entablature. They probably developed prior to the destruction of that surface by the headward migration of a cataract and associated development of an inner channel. Shepherd (1972) suggested that powerful roller vortices with filaments parallel to the flow direction probably formed the experimental grooves. A similar phenomenon was observed in laboratory experiments by Allen (1971), who noted the very regular spacings of the vortices in uniform bed materials.

Once formed in the well-jointed basalt, scabland inner channels seem to have propagated headward by cataract recession. Bretz and others (1956) envisioned powerful kolks at the plunge pool locations, undermining the cataract lip. Sediment, usually basalt columns and large blocks of entablature, were transported away from the cataract lip in a state of quasi-suspension, buoyed by the intense macroturbulence (Baker, 1973, p. 26-29). The competency of the kolks is difficult to estimate theoretically. Boulders up to 30 m in diameter (estimated from aerial photography) were transported from the lip of West Potholes cataract. Certainly the vertical vortices (kolks) were exceedingly powerful agents of lift and transport.

Scabland channels tend to be relatively straight rather than sinuous. A striking example of the channel straightening tendency of the great flood flows is provided by the Crab Creek area (see Bretz and others, 1956, Pl. 8; Baker, 1973, Fig. 22). Prior to flooding this area possessed well-formed valley meanders with a wavelength of 2000 meters, indicating a probable normal bankfull stage of 850 m^3/sec as estimated from the work of Dury (1965, Fig. 5). The Missoula flood flows in this area were approximately 2,800,000 m^3/sec (Baker, 1971, Pl. 1). Bretz (1928) noted that the flooding could not tolerate the leisurely pre-flood curves of the incised stream. Slip-off slopes were vigorously attacked producing what he called "trenched spur buttes." As shown by Ippen and others (1962) this effect is a predictable consequence of the high shear stress gradient that occurs on the inside of curves in an experimental trapezoidal channel.

A hypothetical sequence of progressive erosion of a typical scabland divide crossing (Fig. 12) is envisioned as follows. The first flood water to overtop the divide encountered soft Palouse loess and Ringold formation (Phase I). The high velocity water quickly exposed the underlying basalt, leaving an occasional streamlined loess hill as a remnant of the former cover (Phase II). The entablature of the uppermost basalt flow was then encountered. This probably yielded to groove development, possibly associated with longitudinal roller vortices. The first exposure of well-developed columnar jointing, perhaps at the top of a flaring colonnade along the irregular cooling surface, introduced a very different style of erosion (Phase III). Large sections of columns could now be removed at this site with the simultaneous development of vertical vortices (kolks). With the enlargement and coalescence of the resultant potholes, the surface assumed the bizarre butte and basin topography that characterizes much of the Channeled Scabland (Phase IV). The eventual topographic

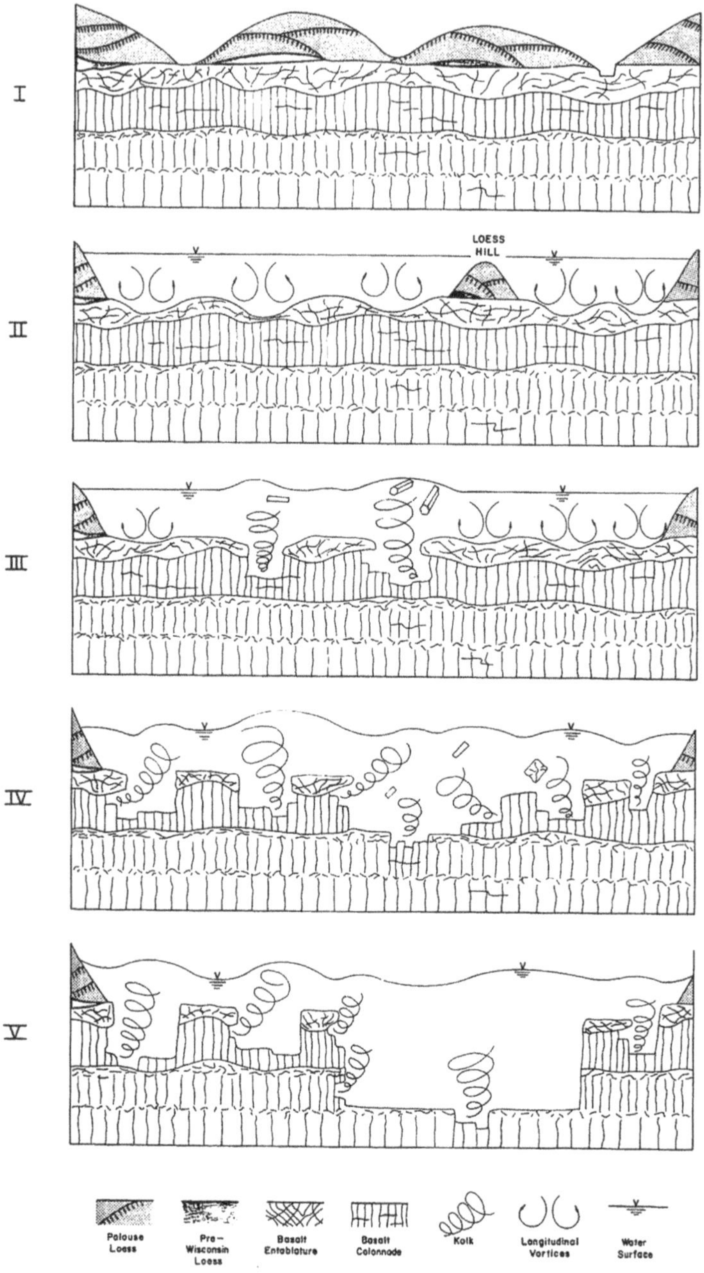

Figure 12

Hypothetical sequence of flood erosion for a typical scabland divide crossing.

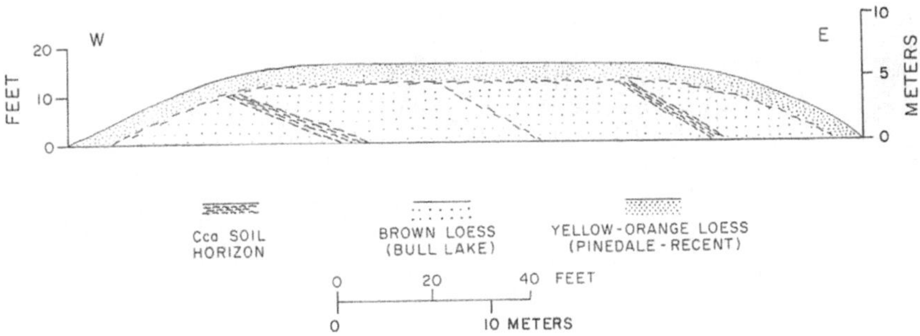

Figure 13a Typical loess stratigraphy as exposed by a road cut 3 km east of La Crosse, Washington.

Figure 13b Streamlined loess hills on the Palouse-Snake divide crossing. Note the prominent prows formed by convergent hillslopes pointing upstream for the flood flows. The scarps average 40 m in height. Flow depths averaged 60 m (Baker, 1973, fig. 8).

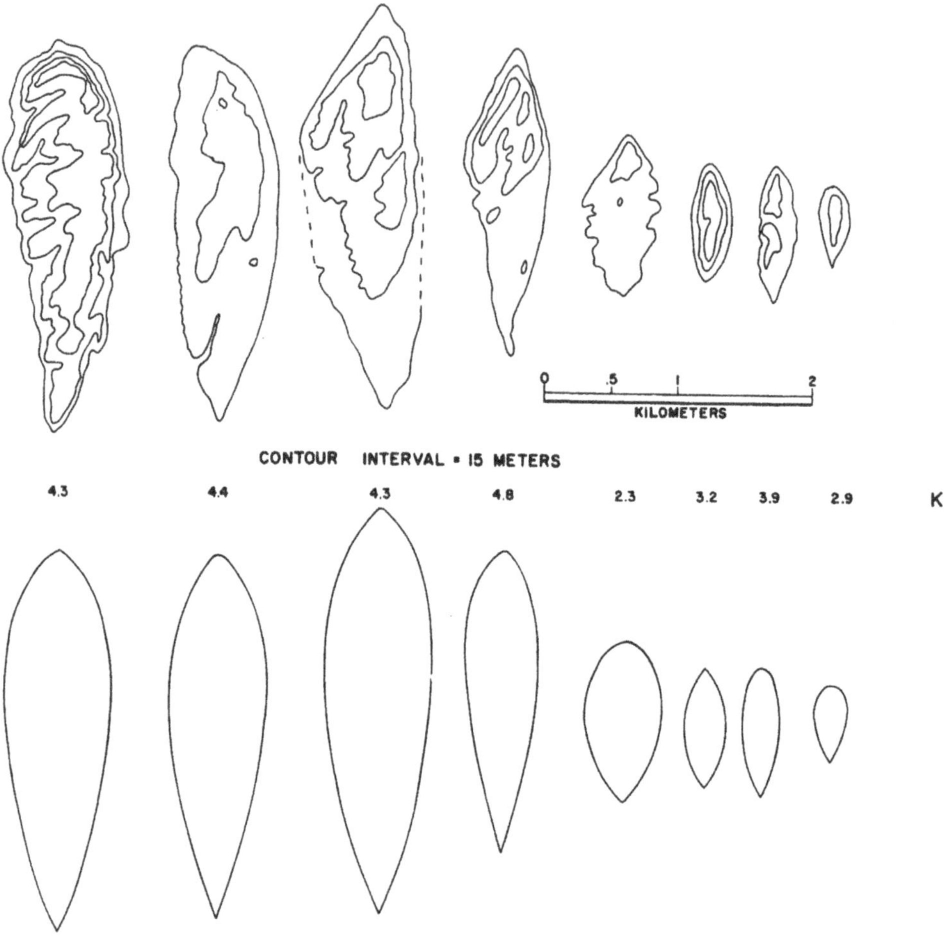

Figure 14
Comparison of typical streamlined scabland loess hills to equivalent lemniscate loops as calculated according to Chorley's (1959) analysis of drumlin shapes. Water flowing over and around the easily scoured loess caused the streamlined shapes, permitting less resistance to flow. Subsequent gulleying of the soft loess has somewhat modified the streamlined shapes.

form was the development of a prominent inner channel (Phase V). Such inner channels may have been initiated at downstream structural steps in the basalt, and then migrated headward by cataract recession. The lateral enlargement of inner channels probably proceeded by the undercutting of resistant entablature as columns were plucked out by kolks. Horsethief cave, north of Soap Lake, is an excellent example of this type of erosion.

SCOUR OF SEDIMENTS

Several sedimentary units overlie the Yakima basalt in the Channeled Scabland. During the early Pleistocene, the great structural basins of the western scablands accumulated a sequence of partly consolidated silt, gravel, and clay known as the Ringold Formation (Flint, 1938b; Newcomb, 1958). Overlying these sediments is the Palouse Formation, a complex blanket of loess sheets that locally reaches thicknesses of 75 m (Ringe, 1970). Richmond and others (1965) consider the Palouse loess to be Bull Lake in age. The Palouse Formation contains several buried soils and is often mantled by younger loess (Fig. 13a). Bretz (1928) recognized that these soft sediments overlying the basalt produced a rather distinct erosional assemblage of their own. He identified 3 major types: (1) loess scarps as high as 60 m with slopes as steep as 35°, (2) loess islands, and (3) loess-bounded channels. Erosional forms also occur in the flood gravels, resulting from scour by either subsequent floods or the waning stages of the flood which deposited the gravel.

Streamlined Loess Hills

Bretz (1923, p. 624-626) first recognized that the hundreds of isolated loessial hills of the eastern scablands possessed remarkably steep, ungullied marginal hillslopes. These slopes converge to form a definite prows that point up the local scabland gradient (Fig. 13b). Although Bretz and others (1956) interpreted these hills as fluvially-eroded loessial "islands," high-water mark reconstruction (Baker, 1973) has shown that many of the hills were eroded subfluvially. The streamlined shapes of these hills bear a close resemblance to airfoils, to which they are probably hydrodynamically similar. Chorley's (1958) study of drumlins proposes that airfoils and similar features be described by a dimensionless parameter K,

$$K = \frac{l^2 \pi}{4A}, \qquad (3)$$

where l is the length of the long axis of the feature, and A is the area. The equation for a lemniscate loop, equivalent to the natural form, is given by:

$$p = l \cos K \theta, \qquad (4)$$

where p and θ are polar coordinates. The close relationship of loess hill shapes to lemniscate loops (Fig. 14) strongly supports Bretz's contention that these hills were streamlined by a rapidly flowing fluid. Water velocities averaged 12-15 m/sec for depths of 30-60 m in these areas.

Scour Marks

Large boulders on scabland bars and bedrock projections along scabland channels produced a deformation of flood flow streamlines that resulted in distinctive erosional scour marks. Engineers have studied the hydrodynamics of such scour in order to protect bridge piers during floods (Laursen, 1960). The scour is generated by two basic systems of vortices (Shen, 1971): the horseshoe-vortex system and the wake-vortex (Fig. 15a).

The prominent crescentic scour hole on the northwest (upstream) side of the boulder in Figure 15b was probably caused by the hydrodynamic stretching and accumulation of vortex filaments in the front of the boulder. Richardson (1968) described this process as a characteristic effect of a blunt-nosed obstacle on an approaching

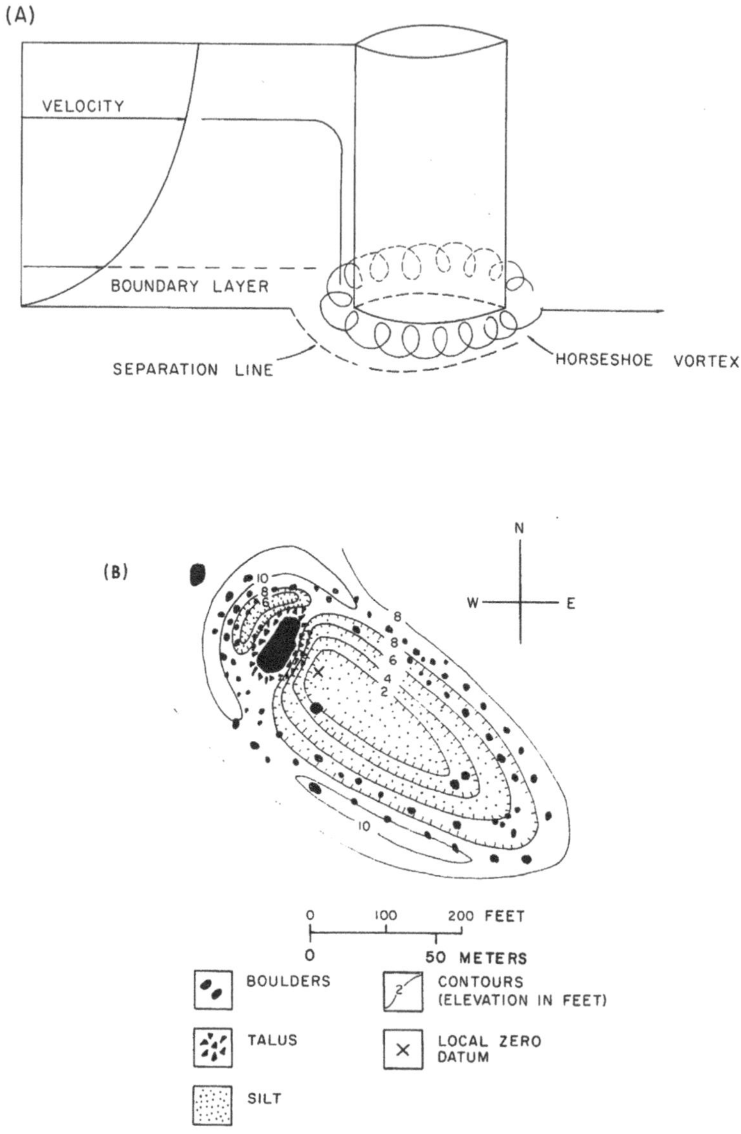

Figure 15a Formation of a horseshoe-vortex system at the front of a vertical cylinder mounted on an experimental flume bed (after Moore and Masch, 1963).

Figure 15b Scour hole development near a 18 x 11 x 8 m boulder 2-1/2 km west of Rocky Ford Creek Fish Hatchery, Ephrata, Washington (Baker, 1973).

two-dimensional velocity field. The strong pressure field produced by the blunt obstacle causes a separation of the boundary layer which then rolls up ahead of the obstacle to form a horseshoe vortex. Karcz (1968) suggested that this mechanism is responsible for the current crescents that commonly occur upstream from obstacles.

The large elliptical scour hole that formed on the downstream side of the boulder in Figure 15b formed as a result of a wake-vortex system generated by flow separation in the rear of the boulder. Engineering experiments have shown that the wake vortex system is a function of Reynolds number (velocity of approach times obstacle diameter times fluid density, divided by dynamic viscosity of the fluid). With other variables in the Reynolds number held constant, increasing velocity results first in a pair of vertical vortices. When the flow changes from laminar to turbulent, vortices form and migrate downstream. Karcz (1968) notes that the average velocity in the wake region is quite low at this stage and deposition is likely on the lee of the obstacle. Indeed the scablands contain many examples of pendant bars that were deposited as elliptical forms in the lee of obstacles (Baker, 1973). Malde (1968) described similar features that occurred during the Bonneville flood in the Snake River Plain of Idaho.

At very high flow velocities, vortex intensity grows and the sucking action of kolks dominates. Shen (1971, p. 23-25) observed that the wake-vortex system then acts like a vacuum cleaner in removing bed material at this stage. Thus, the scour hole in the lee of the boulder may indicate high flow velocities. Baker (1973, p. 41-42) suggested that the high flow velocities were generated during the draining of the Quincy Basin in the waning stages of the flood. This explanation is also consistent with the prominent armoring that characterizes the Quincy Basin fill.

DISCUSSION

Most bedrock river valleys are created over relatively long periods of time. Channel walls are weakened by slope processes, producing colluvium which is later removed by an occasional flood event. The details of this interaction vary in different climatic-geologic settings, but in no case is fluvial erosion alone thought to create the entire valley. Perhaps the contradiction of this line of reasoning was the most startling of Bretz's revelations concerning the Channeled Scabland. Not only did he document as much as 300 m of incision occurring solely by fluvial processes, but he also showed the likelihood that nearly all the erosion had occurred in the course of a few discrete floods.

Of course, the reason for the immense scale of scabland erosion is the immense scale of those floods. Rather than valley erosion, the scablands experienced subfluvial channel scour by high gradient flowing water that reached depths as great as 200 m. The water flowed down the steep regional dip slope of the plateau, covering interfluves and reducing small valleys to mere bottom roughness elements. There is no acceptable modern analog for the energy expenditure of so much water, moving at so high a velocity. The engineering experiments and procedures referred to in this paper were developed for features orders of magnitude smaller in scale than the Channeled Scabland. The extrapolation of known hydraulics to the scabland erosion problem has left many problems unsolved, including the following:

1. Three-dimensional flow phenomena and vortices characterized the flood flows. What is the hydraulics of these phenomena? What is the competence of flow characterized by these features?

2. The accuracy of the velocity determinations for scabland flooding is not known (see discussion by Baker, 1973, p. 22). It may be that energy dissipation as kolks in channels with highly irregular boundaries exerts a major influence on the magnitude of downchannel velocities. How can existing hydraulic theory be

modified to account for such extreme conditions?

3. The lithologic and structural peculiarities of the Yakima basalt greatly aided the process of hydraulic plucking. How would other rock types respond to the type of flow phenomena produced in catastrophic floods?

4. The discharges, flow velocities, and erosional forms produced by the Missoula floods are probably the largest and most spectacular that can be documented in the geologic record. What is the implication of this record for interpreting the geomorphic influence of infrequent flood events of great magnitude on "normal" river systems?

ACKNOWLEDGMENTS

Field work in eastern Washington was supported by National Science Foundation Grant GA-21478. Professor William C. Bradley, University of Colorado, was principal investigator to the project. The author benefited from his wise counsel as well as from discussions with H. E. Malde, G. E. Neff, and R. G. Shepherd. Report preparation was aided by a University of Texas Research Institute grant.

REFERENCES

Alexander, H. S., 1932, Pothole erosion: Jour. Geology, v. 40, p. 305-337.

Allen, J. R. L., 1971, Transverse erosional marks of mud and rock: their physical basis and geological significance: Sedimentary Geology, v. 5, p. 167-385.

Allison, I. S., 1933, New version of the Spokane flood: Geol. Soc. America Bull., v. 44, p. 675-722.

Baker, V. R., 1971, Paleohydrology of catastrophic Pleistocene flooding in eastern Washington: Geol. Soc. America Abstracts with Programs, v. 3, p. 497.

__________, 1973, Paleohydrology and sedimentology of Lake Missoula flooding in eastern Washington: Geol. Soc. America Special Paper 144, 79 p.

Barnes, H. L., 1956, Cavitation as a geological agent: Amer. Jour. Sci., v. 254, p. 493-505.

Bingham, J. W., and Grolier, M. J., 1966, The Yakima Basalt and Ellensburg Formation of south-central Washington: U.S. Geol. Survey Bull. 1224-G, p. G1-G15.

Birkeland, P. W., 1968, Mean velocities and boulder transport during Tahoe-age floods of the Truckee River, California-Nevada: Geol. Soc. America Bull., v. 79, p. 137-141.

Bradley, W. C., Fahnestock, R. K., and Rowekamp, E. T., 1972, Coarse sediment transport by flood flows on Knick River, Alaska: Geol. Soc. America Bull., v. 83, p. 1261-1284.

Bretz, J H., 1923, The channeled scablands of the Columbia Plateau: Jour. Geology, v. 31, p. 617-649.

__________, 1924, The Dalles type of river channel: Jour. Geol., v. 32, p. 139-149.

__________, 1927a, Channeled scabland and the Spokane flood: Wash. Acad. Sci. Jour., v. 17, p. 200-211.

__________, 1927b, The Spokane flood; a reply: Jour. Geology, v. 35, p. 461-468.

__________, 1928, The channeled scabland of eastern Washington: Geog. Rev., v. 18, p. 446-477.

__________, 1930, Lake Missoula and the Spokane flood (Abstract): Geol. Soc. America Bull., v. 41, p. 92.

__________, 1932, The Grand Coulee: American Geog. Soc. Spec. Publ. 15, 89 p.

__________, 1933, The channeled scabland: International Geological Congress, XVI session, Guidebook 22, 16 p.

__________, 1969, The Lake Missoula floods and the channeled scabland: Jour. Geology, v. 77, p. 505-543.

Bretz, J H., Smith, H. T. U., and Neff, G. E., 1956, Channeled scablands of Washington; new data and interpretations: Geol. Soc. America Bull., v. 67, p. 957-1049.

Chorley, R. J., 1959, The shape of drumlins: Jour. Glaciology, v. 3, p. 339-344.

Chow, V. T., 1959, Open-channel hydraulics: McGraw-Hill, New York, 680 p.

Dahl, R., 1965, Plastically sculptured detail forms on rock surfaces in northern Nordland, Norway: Geogr. Annlr., v. 47, p. 83-140.

Dury, G. H., 1965, Theoretical implications of underfit streams: U.S. Geol. Survey Prof. Paper 452-C, 43 p.

Embleton, C., and King, C. A. M., 1968, Glacial and periglacial geomorphology: New York, St. Martin's Press, 608 p.

Flint, R. F., 1938a, Origin of the Cheney-Palouse scabland tract: Geol. Soc. America Bull., v. 49, p. 461-524.

__________, 1938b, Summary of later-Cenozoic geology of southeastern Washington: Amer. Jour. Sci., v. 35, p. 223-230.

Fryxell, R., and Cook, E. F., 1964, A field guide to the loess deposits and channeled scablands of the Palouse areas, eastern Washington: Washington State Univ., Lab. Anthropology, Rept. Inv. 27, 32 p.

Grolier, M. J., and Bingham, J. W., 1971, Geologic map and sections of parts of Grant, Adams, and Franklin Counties, Washington: U. S. Geol. Survey Misc. Geol. Invest. Map I-598.

Hjulström, F., 1935, Studies in the morphological activity of rivers as illustrated by the River Fyris: Geol. Inst. Univ. Upsala, Bull., v. 25, p. 221-528.

Ippen, A. T., Drinker, P. A., Jobin, W. R., and Shemdin, O. H., 1962, Stream dynamics and boundary shear distribution for curved trapezoidal channels: Mass. Inst. Tech. Report No. 47, 81 p.

Karcz, I., 1968, Fluviatile obstacle marks from the wadis of the Negev (southern Israel): Jour. Sed. Petrology, v. 38, p. 1000-1012.

Koloseus, H. J., 1971, Rigid boundary hydraulics for steady flow, in H. W. Shen, ed., River mechanics: H. W. Shen, Ft. Collins, Colo., p. 3-1-3-51.

Laursen, E. M., 1960, Scour at bridge crossings: Jour. Hydraulics Div., Proc. Amer. Soc. Civil Eng., v. 86, no. HY2, p. 39-54.

Mackin, J. H., 1961, A stratigraphic section in the Yakima Basalt and the Ellensburg Formation in south-central Washington: Washington Dept. Conserv., Div. Mines and Geology Rep. Inv. 19, 45 p.

Malde, H. E., 1968, The catastrophic late Pleistocene Bonneville Flood in the Snake River Plain, Idaho: U. S. Geol. Survey Prof. Paper 596, 52 p.

Matthes, G. H., 1947, Macroturbulence in natural stream flow: Am. Geophys. Union Trans., v. 28, p. 255-262.

McKnight, E. T., 1927, The Spokane flood: a discussion: Jour. Geology, v. 35, p. 453-460.

Moore, B. R., 1972, Theoretical considerations in the application of vortex scour action in suction dredge holes to the control of sedimentation on river bars, Ohio River, southern Illinois: Geol. Soc. America Abstracts with Programs, v. 4, no. 7, p. 600.

Moore, W. L., and Masch, F. D., 1963, Influence of secondary flow on local scour at obstructions in a channel: Federal Inter. Agency Sedimentation Conference Proc., 1963, U. S. Dept. of Agriculture, Misc. Publ. 970, p. 314-320.

Newcomb, R. C., 1958, Ringold formation of Pleistocene age in type locality: Am. Jour. Sci., v. 256, p. 328-340.

__________, 1970, Tectonic structure of the main part of the basalt of the Columbia River Group, Washington, Oregon, and Idaho: U. S. Geol. Survey Misc. Geol. Inv. Map I-587.

Newcomb, R. C., Strand, J. R., and Frank, F. J., 1972, Geology and ground-water characteristics of the Hanford Reservation of the U.S. Atomic Energy Commission, Washington: U. S. Geol. Survey Prof. Paper 717, 78 p.

Olson, E. C., 1969, Introduction to J Harlen Bretz's paper on "The Lake Missoula Floods and the Channeled Scabland": Jour. Geol., v. 77, p. 503-504.
Pardee, J. T., 1942, Unusual currents in glacial Lake Missoula, Montana: Geol. Soc. America Bull., v. 53, p. 1569-1600.
Richardson, P. D., 1968, The generation of scour marks near obstacles: Jour. Sed. Petrology, v. 38, p. 965-970.
Richmond, G. M., Fryxell, R., Neff, G. E., and Weiss, P. L., 1965, The cordilleran ice sheet of the northern Rocky Mountains, and related Quaternary history of the Columbia Plateau, in H. E. Wright and D. G. Frey, ed., The Quaternary of the United States: Princeton Univ. Press, Princeton, N. J., p. 231-242.
Ringe, D., 1970, Sub-loess basalt topography in the Palouse Hills, southeastern Washington: Geol. Soc. America Bull., v. 81, p. 3049-3060.
Schmincke, H. U., 1967, Stratigraphy and petrography of four upper Yakima Basalt flows in south-central Washington: Geol. Soc. America Bull., v. 78, p. 1385-1422.
Schumm, S. A., and Shepherd, R. G., 1973, Valley floor morphology; evidence of subglacial erosion: Area, v. 5, p. 5-9.
Scott, K. M., and Gravlee, G. C., Jr., 1968, Flood surge on the Rubicon River, California--hydrology, hydraulics, and boulder transport: U. S. Geol. Survey Prof. Paper 422-M, 38 p.
Shen, H. W., 1971, Scour near piers, in H. W. Shen, ed., River mechanics: Ft. Collins, Colorado, H. W. Shen, p. 23-1 - 23-25.
Shepherd, R. G., 1972, A model study of river incision: Unpub. M.S. thesis, Colorado State Univ., Fort Collins, Colo., 135 pp.
Shepherd, R. G., and Schumm, S.A., in press, An experimental study of river incision: Geol. Soc. America Bull.
Simons, D. B., and Gessler, J., 1971, Research needs in fluvial processes, in H. W. Shen, ed., River mechanics: Ft. Collins, Colo., H. W. Shen, p. 32-1 - 32-13.
Spry, Alan, 1962, The origin of columnar jointing, particularly in basalt flows: Geol. Soc. Australia Jour., v. 8, p. 191-216.
Swanson, D. A., 1967, Yakima Basalt of the Tieton River area, south-central Washington: Geol. Soc. America Bull., v. 78, p. 1077-1110.
Symposium, 1947, Cavitation in hydraulic structures; a symposium: Am. Soc. Civil Eng. Trans., v. 112, p. 1-124.
Tomkeieff, S. I., 1940, The basalt lavas of the Giant's Causeway district of northern Ireland: Bull. Volcanologique, v. 6, p. 90-143.
Trimble, D E., 1950, Joint controlled channeling in the Columbia River basalt: Northwest Sci., v. 24, p. 84-88.
____________, 1954, Geology of the Haas quadrangle: U. S. Geol. Survey Geologic Quadrangle Map GQ-103.
____________, 1963, Geology of Portland, Oregon and adjacent areas: U. S. Geol. Survey Bull. 1119, 119 p.
Waters, A. C., 1955, Geomorphology of south-central Washington, illustrated by the Yakima East quadrangle: Geol. Soc. America Bull., v. 66, p. 663-684.
____________, 1960, Determining direction of flow in basalts: Amer. Jour. Sci., v. 258-A, p. 350-366.
____________, 1961, Stratigraphic and lithologic variations in the Columbia River Basalt: Amer. Jour. Sci., v. 259, p. 583-611.
Wright, T. L., Grolier, M. J., and Swanson, D. A., 1973, Chemical variation related to the stratigraphy of the Columbia River basalt: Geol. Soc. Amer. Bull., v. 84, p. 371-386.

7

REFLECTIONS ON THE ORIGIN OF SMALL-SCALE LONGITUDINAL STREAMBED SCOURS

Iaakov Karcz

ABSTRACT

Flow-aligned elements of streambed-scour may be regarded as a product of secondary, helicoidal motions induced by flow instability. In spite of the implied uniformity of origin, they vary greatly in size, shape, and distribution-density within the channel. Field evidence and experiments suggest that this variability reflects possible differences in the parent kinematic flow patterns. Three main groups of structures appear, irrespective of channel geometry and apparently unrelated to meandering tendencies:

a. Structures induced by "defects", such as obstacles, cavities etc., with features of scour and deposition emanating from the individual defects as a result of flow separation phenomena in front and rear of the obstacle.

b. Patterns of continuous longitudinal ridges and troughs, attributed to flow-configurations characterized by streamwise-persisting, rhythmic spanwise variation in velocity, suggestive of a system of longitudinal vortex rolls with alternating sense of rotation. The finest of such structures is current lineation, which appears to represent a manifestation of the fundamental structure of the flow at the boundary. Such lineations are present at velocities both lower and higher than those required to produce ripples. Larger sand ribbons are best visible at, or close to, transition to upper plane-bed phase of sediment transport.

c. Patterns of elliptical, arcuate and spindle-shaped scours, which appear in association with large bed undulations or along otherwise planar streambeds. These structures are interpreted in terms of flow separation behind skewed negative steps. It is possible however, that in some cases they are triggered by downstream-migrating, hairpin-shaped vorticity systems, with secondary flows pointing towards and away from the bed. Some larger scale scours, which appear scattered along segments of the channel, are probably the result of secondary currents induced by spatial or temporal changes in flow parameters or in hydraulic geometry.

All these structures show a considerable variation in scale. This variation may be simply an indication as to whether the structure is a manifestation of the kinematic flow pattern within the sublayer, within the boundary layer or within the main flow.

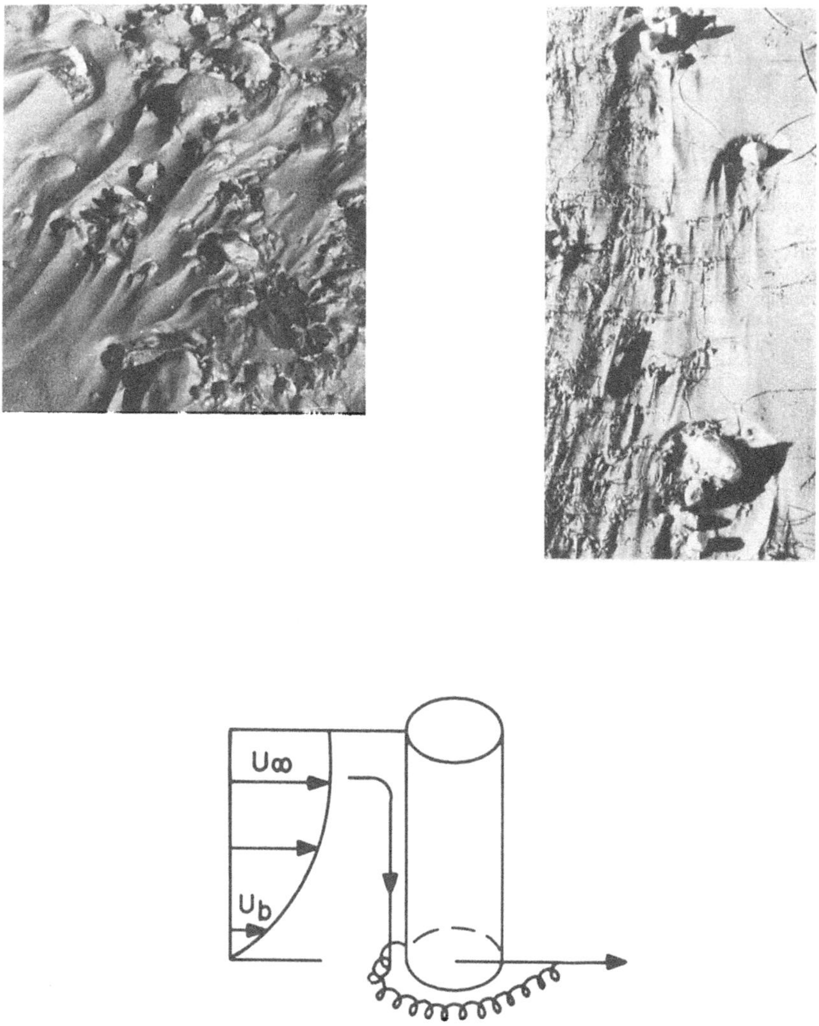

Fig.1 a,b, current crescents and sand shadows
c, formation of a horseshoe vortex

INTRODUCTION

Flow-generated streambed configurations manifest the instability of the original interface between the bed and the flow (for references see: Allen, 1968, 1969a,b; Dzulynski, 1966; Engelund and Hansen, 1966; Karcz, 1970). Usually these features are aligned either parallel or normal to the direction of flow, and are geometrically similar to instability-induced flow patterns, reported from other media by astronomers, meteorologists, oceanographers, and chemical engineers, who studied their origin employing sophisticated mathematical tools, and concepts ranging from hydrodynamics to non equilibrium thermodynamics.

In spite of the intimate sequential relationship between longitudinal and transverse flow patterns predicted by modern stability theory (for references: Chandrasekhar, 1962; Stuart, 1965) and in spite of the common occurrence of longitudinal elements of streambed sculpture, repeatedly emphasized by field geologists and geomorphologists, experimental hydraulic research has focused almost entirely on transverse bed forms. This was due mainly to the early recognition of the hydraulic importance of ripples, dunes and antidunes, and in part also to the wishful assumption that these structures are more amenable to elegant linear mathematical treatment. A strange discrepancy has therefore developed in treatment of natural bed forms. The more common in alluvial channels, ripples, dunes, plane beds and antidunes, were interpreted with marked self assurance, usually in terms of numerical relationships derived under largely idealized experimental conditions, whereas other features, such as current lineations, sand ribbons, flutes, and elliptical scours, were discussed only in rather vague terms. The recent work of Allen (1964, 1969a, 1971), Dzulynski and Walton (1963, 1965) and others, refocused attention on longitudinal structures, indicating possible mechanisms of origin.

This paper summarizes some results and concepts, which emerged from an ongoing study of: obstacle marks, sand ribbon and troughs, and elliptical scours. The study is in progress, and some of the ideas presented here have yet to withstand the trial of reality.

OBSTACLE MARKS

Obstacle-induced bed forms along streambeds and beaches range in size from 10^{-3} to 10^{1} m and include structures such as current crescents and obstacle shadows (Fig. 1), produced by separation of flow upstream and downstream of obstacles. Upstream, adverse pressure gradient causes the fluid to curl into a horseshoe-shaped vortex, with horns extending downstream along the sides of the obstacle. The vortex filaments are stretched, the fluid particles rotate and advance helically, diving components of motion inducing scour. The structure and strength of the rear system depends mainly on the obstacle Re number and on the turbulence level. Close to the obstacle, in the near-wake [Fig. 2], separation leads to two free-shear layers which envelope a slow-recirculating low-pressure region, believed to be site of deposition of the sand shadow. Near its closure, the free shear layers become unstable and roll into a pair of larger vortices. At Re numbers below 50 the vortices are virtually stationary, at higher values they grow and are shed alternately, in a regular fashion known as the von Karman street. With increase in Re, shedding and diffusion of vortices becomes more rapid and less regular, and at Re > 3000 this vortical far-wake region has the appearance of a turbulent wedge. However, some evidence suggests the presence of von Karman street-like configuration also in high Re number flows (Roshko, 1961; Wu, 1972). Of lesser importance in generation of obstacle marks is the possible appearance

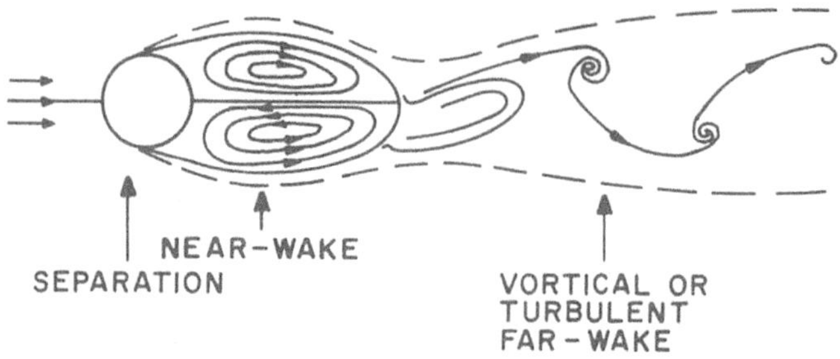

Fig.2 Schematic structure of a wake system

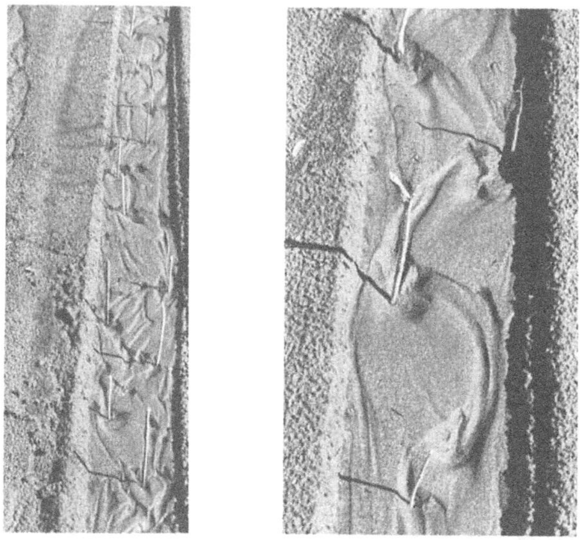

Fig.3 Development of thalweg sinuosity

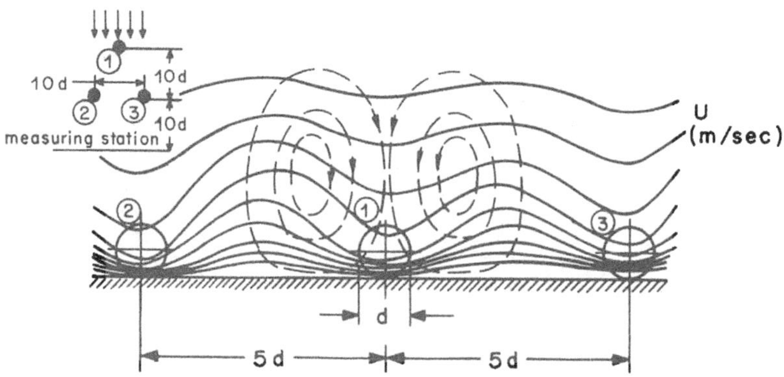

Fig.4 Secondary currents behind a row of spheres (Schlichting, 1968)

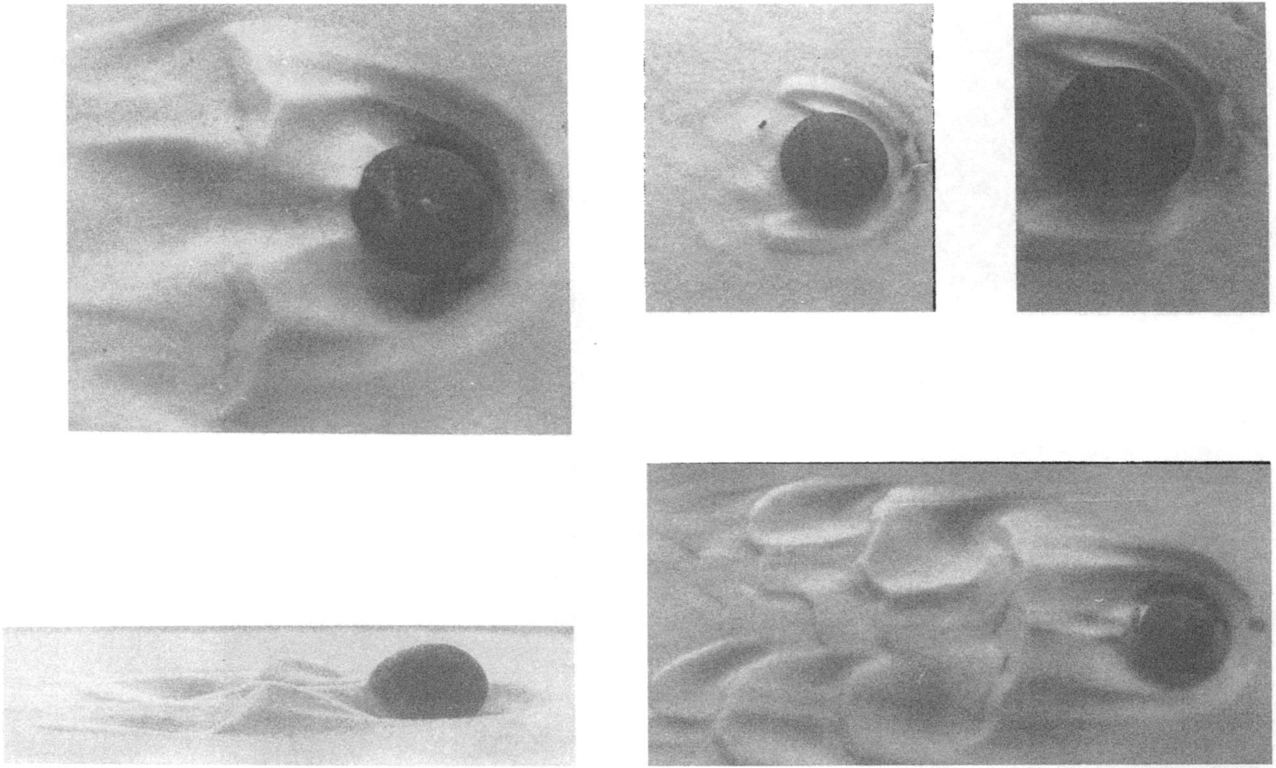

Fig.5 Gradual development of streambed deformation around an obstacle, leading to appearance of ripples along a plane bed

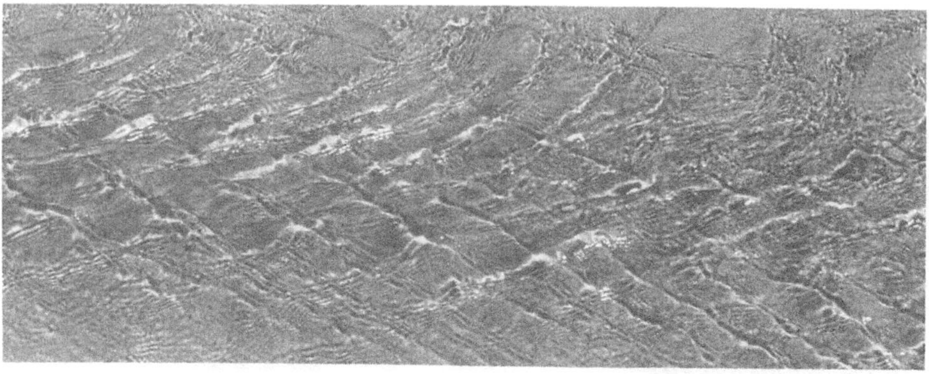

Fig.6 Oblique shear wave pattern

Fig.7 Deformation of a bar by the oblique wave pattern

of a tipping vortex at the rear top part of the obstacle, which would produce a linear scour in the obstacle lee.

The occurrence and origin of obstalce-induced bed forms were reviewed recently by Allen (1965), Richardson (1968) and Karcz (1968), and only brief additional comments would be made here.

a. High Re-number wakes rarely develop on pebbles, because at the high velocities required, the pebbles are carried away by the flow, or are buried tipping-down into the upstream scour (Johannsson, 1963). However, in larger scale obstacle-induced deformation of flow (such as reported by meteorologists and oceanographers from around mountains and islands), the wake structure may be imprinted on morphology and drainage. In most cases, elongate, oval or irregular scour-trail will develop in the rear of an obstruction or at a sharp convergence (Tison, 1940; Shen et al. 1966; Karcz, 1968). It is not clear however, whether the von Karman vortex street configuration expresses itself in the streambed geometry. The effect of a quasi-periodic system of eddies in offset alignment generated in a strained turbulent wake, was briefly mentioned amongst other possible mechanisms instrumental in development of channel sinuosity (Karcz, 1971). However, until concrete evidence is obtained, one should be wary of the lethal lure of geometrical similarity (e.g. Figs. 2 and 3).

b. In addition to structures developed around individual obstacles, obstacle-assemblages may be instrumental in initiation of other bed forms. Allen (1971) indicated that development of many streambed geometries is interpreted in either of two ways. One assumes that the bed is a passive, pliable medium, which faithfully records the flow configuration at the boundary ("passive-bed" theory). The other attributes the start of deformation to the effect of surface and bulk irregularities and inhomogeneities ("defect" theory). These two hypotheses are not mutually exclusive, but rather represent complementary or alternate ways of generating a geometrically similar pattern; their relative importance depends on conditions of flow and bed material. Often the recognition of obstacle-induced effects is hampered by observational difficulties especially over rough surfaces. Schlichting (1968) reported that behind rows of spheres a secondary current pattern developes, leading to a "negative wake effect", i.e. appearance of high velocities behind the obstacles, and of low velocity strips in between (Fig. 4) and Davies (1972) reported three dimensional, longitudinal motions in the wake interference flow developed in flow of thin layers over rugose boundaries. It is not clear however, to what extent, if at all, such phenomena may have some bearing on the problem of development of features such as rills generated in sheet flow over sloping rough surfaces. Several clearer and better documented examples will be cited below.

1. Material scoured from around obstacles on a flat bed often migrates downstream in ripples driven by the obstacle-modified velocity pattern (Fig. 5), hastening the transition from flat bed to ripple-configuration (Southard and Dingler, 1971). Moreover, numerous studies maintain that ripples are initiated at chance pile-ups of sand grains, where flow separates, and the pattern spreads along the bed (e.g. Williams and Kemp, 1971; Grass, 1970).

2. Rhomboidal marks and ripples (Hoyt, and Henry, 1963; Otvos, 1965) are attributed to obstacle-induced oblique hydraulic jumps, and to interference of wave-trains refracted by obstacles across the streampath. However, the hydraulic experience (Chow, 1959), recent studies of diagonal free-surface shear-waves (Kennedy and Robillard, 1967; Kennedy and Iwasa,

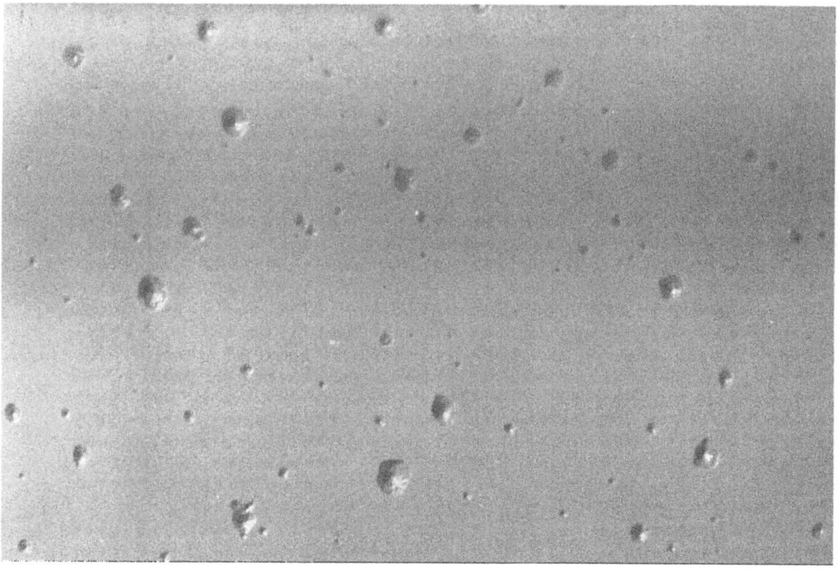

Fig.8 Pattern of pit-like defects developed in prolonged fluid stressing of clay

Fig.9 Transformation of pits into flow-aligned grooves

1968), and a study of Chang and Simons (1970), indicate that at Fr numbers above 1, rhomboidal dune patterns may dominate the bed configuration even though obstacles are absent. It appears that patterns of oblique waves (Fig. 6) imprint themselves on the bed and alter the mode of sediment transport. For example, diamond-shaped elements of streambed alignment are frequently observed in dune and bar-dominated low flows in natural channels (Fig. 7). In these relatively shallow flows over an undulating bed, centrifugal components develop and distort the hydrostatic pressure distribution, with a resulting deficit over the crests and excess over troughs. The pressure deviations increase away from the walls, and thus near the bed a three dimensional pattern of secondary motions developes transporting fluid from the walls towards the crests, and from the troughs towards the walls. Further, centrifugal pressure gradients lead to a longitudinal acceleration of the fluid which decreases away from the wall, and the pattern becomes diagonal.

3. In a discussion of erosional features of muddy beds, Allen (1971) suggested that scour is induced by bed defects, which exceed the critical size for the given flow. Flow pattern starts changing, flow separation and instability phenomena appear, and deformation of the bed extends in space and in time, until the individual scour elements reach the maximum limiting size. The original defects are attributed to depressions upon the bed surface (burrows, trails and tracks, cavities and imprints of pebbles and shells), steps—up and down the surface (cracks, joints, fracture, etc.), and objects on the bed surface (organisms, faecal pellets, pebbles, etc.).

Widespread occurrence of features of soft rock deformation, liquefaction and piping in the geological record, suggest that processes of early primary consolidation may represent an additional source of bed-defects. To examine this possibility, beds of montmorillonite soil clay were deposited from suspension along a plexiglass flume, 2 m long, 0.1 m wide and 0.15 m deep, and where subjected to prolonged flow. Velocities employed ranged from 15 to 55 cm/sec, and tractive shear from 10 to 25 dyn/dm^2. Following some short-lived erosion near the flow-entry, the bed became stable and the flow proceeded without scour, until 3 to 24 hours later when minute pits and mounds appeared at the bed surface. As the flow continued, the pits increased in number and grew up to 8 - 9 mm diameter (Fig. 8). At first the pit outlines remained circular, but with time became asymmetrical towards the downstream. At higher velocities (usually 35 to 55 cm/sec) pits became drop-shaped and finally transformed into flow-aligned straight grooves, up to 60-70 mm long and 8-9 mm wide (Fig. 9). Further growth of the grooves led to removal of the entire upper part of the bed. The mounds, in most cases, developed cracks and the surface crust was eroded, revealing pit-like craters underneath.

It appears that under conditions of prolonged fluid-stressing some internal rearrangement of bed particles occurs shortly after deposition. It is known that clays settle from suspension in flocs and aggregates. When flow-induced consolidation sets in, the aggregates are packed closer, and are progressively deformed and destroyed, with an accompanying decrease in voidage and possible realignment of the clay particles. Reduction in void-space, causes momentary local increases in pore pressure with the ultimate expulsion of the excess fluid, leading to a weakening of the bed and to appearance of surface and bulk defects. These defects later become foci of intensified scour.

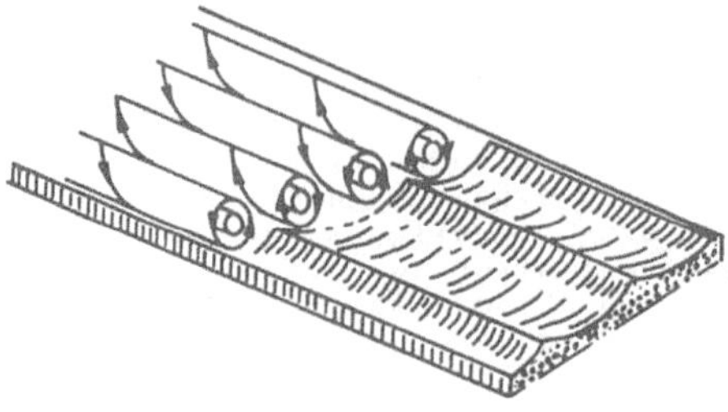

Fig.10 Longitudinal vortex rolls and their probable effect on the streambed

Fig.11 a, Sand stringers along a rigid bed

Fig.11 b, A furrowed pattern produced by sand stringers along a weakly cohesive bed

CURRENT LINEATIONS, SAND RIBBONS AND TROUGHS

Patterns of flow-aligned ridges and troughs occur in various sedimentary environments and range in size from current lineations, one to several grains in width and height, (e.g. Allen, 1964) to longitudinal aeolian dunes (e.g. Folk, 1971) and extensive sand ribbons along current-swept sea-floors. Most investigators attribute these bed-patterns, to a kinematic flow-configuration of series of longitudinal vortex rolls with alternate sense of rotation, in which the fluid particles advance helically (Fig. 10). The resulting system of spiral threads is assumed to induce scour of the granular boundary along the paths of diving motions, and deposition along paths of rising motions.

Kinematic flow patterns of this type were widely reported in meteorology, oceanography, physics and engineering and were attributed to instability mechanisms generated in stratified flows and in turbulent shear flows. Instability in fluids does not express itself in a single catastrophic event, but leads to a chain of successive instabilities and transformations, with a specific kinematic flow configuration at each stage. Despite some physical differences, development of instability in various flows follows a similar path and is governed by closely analogous equations. In sedimentology, concept of instability transformations driven by inverse density gradients was employed in interpretation of structures in turbidite sequences (Dzulynski and Walton, 1965; Dzulynski, 1966), whereas concepts of instability of parallel flows, structure of shear flow at the boundary, and effects of secondary currents were emphasized in interpretation of structures in other environments (e.g. Allen, 1968, 1969, 1971; Karcz, 1967, 1970).

Current lineation, reported from numerous sedimentary environments (for references: Allen, 1964, 1968; Potter and Pettijohn, 1964) is smallest in scale amongst longitudinal structures. It consists of a fairly regular pattern of stringers and streaks of sand grains, which form under a wide range of flow conditions, over smooth and semi-rough, flat and undulating beds. This widespread occurrence suggests that these lineations reflect a fundamental mechanism of flow and sediment transport at the bed boundary. The recent studies of the innermost zone of turbulent flow (Kline et al., 1967; Kim et al., 1971; Corino and Brodkey, 1969; Nychas et al., 1973) show that the sublayer consists of a pattern of longitudinal low-and high-speed streaks migrating downstream side by side. The process of generation and transfer of turbulent energy within the flow is due to the intermittent lifting and ejection of the low-speed streaks into the outer flow. This process occurs at random, but always in an orderly sequence of events. The higher velocity streaks are related to inflow of higher-speed fluid towards the boundary, where it spreads, displacing low-momentum fluid, lifting it up in a low-velocity streak and triggering its ejection away from the boundary into the outer flow. This ejection of sluggish fluid modifies the flow and induces further inflows towards the boundary. Some details of this process are still not clear, especially the relative significance of the streamwise and transverse vortexes reported by Kim et al., (1971) and Nychas et al., (1973). The transverse vortex is believed to form at the interface between the accelerated and decelerated fluid and preceeds the ejection. The streamwise vortex is assumed to form in ejection where outflow of low-momentum fluid generates an inflexion in the velocity profile. Instability sets in, leading to an oscillatory vortex motion, which grows and intensifies downstream.

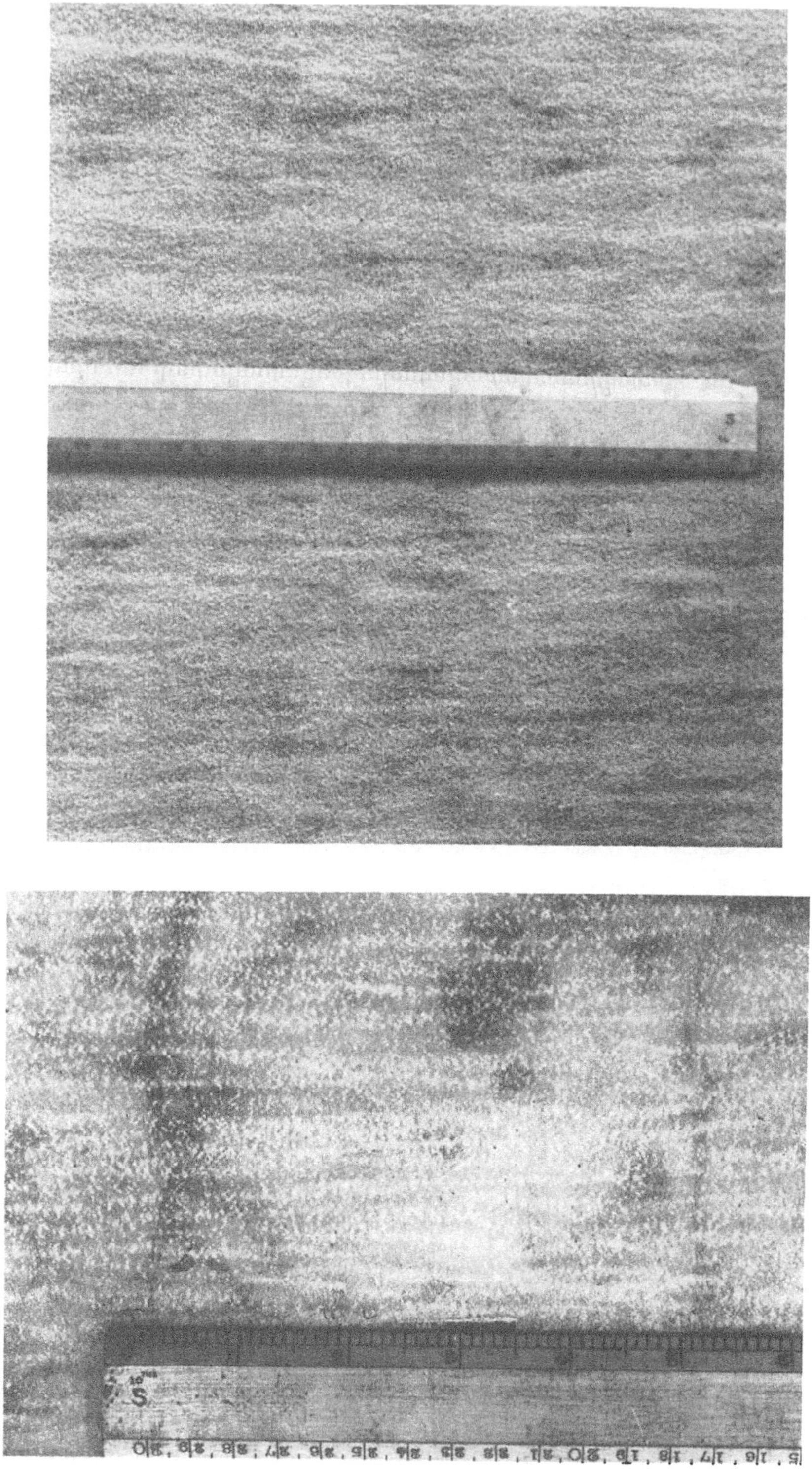

Fig.12 Anastamosing longitudinal sand streaks along a loose granular boundary

This random process, in which jets of low-momentum fluid burst away from the boundary, generates over 70% of the Reynolds stresses and it represents the principal mechanism of generation and transfer of turbulent energy.

To determine the applicability of these concepts in interpretation of mechanics of origin of current lineations, a series of experiments was carried out in the SUNY Binghamton recirculating glass channel, 10m long, 0.6m wide and 0.45 deep, driven by a 4 cusec pump regulated by a varidrive motor. The transparent glass bottom allowed observation of grain movement at the rigid boundary. The information summarized below was obtained by direct visual observation and videotape photography, and yielded a general picture of the processes associated with the current lineations and akin features. The temporal and spatial scale of the recorded phenomena is small, and thus a high-speed photography study is now being undertaken in an attempt to derive more detailed quantitative relationships.

Although the mechamism and principal features of lineations which appear along rigid, smooth and pliable, and semi-rough boundaries appears to be the same, some morphological differences exist. For example, three runs under the same conditions (10 cm/sec velocity, and extremely low rate of sand feed) along different beds produced the following results. Over the rigid glass bottom, the grains organize themselves in remarkably continuous and straight stringers and migrate downstream representing the first manifestation of an orderly geometrical pattern of sediment motion. Over a smooth, but pliable cohesive boundary of fresh-deposited clay, the same alignment appears, but the main morphological effect is formation of scour threads along the path followed by the sand grain, transforming the bed surface into a deeply-harrowed pattern (Fig. 11). Along a loose granular boundary such as quartz sand, the pattern was less continuous and consisted of braided, converging and diverging streaks up to 5-6 cm long (Fig. 12).

At low velocities (10 to 15 cm), the grains move in a somewhat intermittent motion, accelerating,coming to a halt, and moving again. This step-like motion affects groups of adjoining grains, but due to the random nature of bursting, the particles never stay long together. Moreover, the movement depends on the grain diamater, (larger grains advancing ahead of smaller) and thus a faint size-segregation appears. The grains roll or slide, changing from one mode of transport to another. Particles follow a streamwise path, with occasional arcuate or diagonal spanwise shifts, that rarely exceed one tenth of the streamwise shift. A fractional lift component is evident, so that it is tempting to relate this type of motion to saltation, even though the mechanism envisaged differs from that suggested in the literature for saltation. Length of distance travelled in each step and time-lag between consecutive jumps, depend mainly on the flow energy, and to a lesser extent on the nature of the boundary. Over a semi-rough boundary, the lift components appear to be stronger and the streamwise component somewhat weaker than over a smooth boundary. With increase in velocity, the time-lag between the jumps decreases, and the lift and lateral shifts are lengthened. Pattern of motion becomes clearer, when the sand-feed rate is increased and the observations are extended to the entire resulting pattern. At velocities of 10 to 15 cm/sec, the material advances downstream in rectilinear stringers, one to several grains wide with intervening zones of faster, clear fluid. The stringers appear to fluctuate slightly in space and time, with fluctuations corresponding to weakly intermittent motions of groups of grains, as described above. With increase in velocity, spacing between the stringers decreases, they widen, and occasionally join together into narrow bands.

Fig.13 Appearance of larger-scale banding along a sand streak pattern (flow velocity: 15 cm/sec)

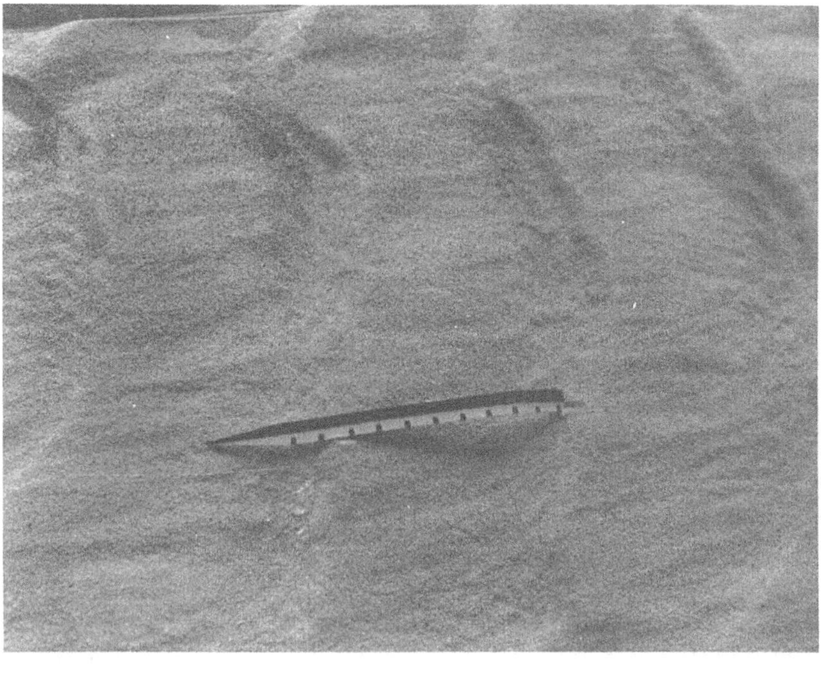

Fig.14 Current lineations in coexistence with small-scale current ripples

(Fig. 13). The most conspicuous effect however, of the increase in flow energy, is the change in the fluctuation regime. The fluctuations increase in frequency and incidence and motion becomes almost spasmodic. At any instant, each stringer or band is affected by a series of burst-induced step-like motions, which impart a streaky or patchy appearance to the entire pattern. The incidence and frequency of bursts increase with velocity of flow. Lateral fluctuations become stronger, (up to one fourth of the streamwise jump), and in plan view convey an impression of oscillating, wavy patches, alalogous to the "meandering grooves" produced by Allen (1969) at somewhat higher velocities along a weakly cohesive bed.

Relationship between current lineations and other bed forms was also noted. It is known that current lineations are often superimposed on the backs of ripples and dunes. This does not imply any drastic change in the conditions of flow because both types of configuration may coexist (Allen, 1966), the lineation representing the structure of the sublayer only (Fig. 14). Coexistence of the two forms of transport was particularly well seen, when the sediment supply to a longitudinal streak pattern was temporarily increased. This increase triggered an immediate appearance of small, usually lunate ripples, which slowly migrated downstream. The ripple form was preserved, but the sediment supply to the ripple, and the sediment leaving the ripple continued to maintain a longitudinal structure. Also across better developed sinuous ripple-patterns, the grains migrated from one ridge to another along pronounced straight paths (Fig. 15).

Increase in flow velocity to 25-30 cm initiated a replacement of streak-patterns by sinuous or lunate ripples. Commonly initiation of ripples is attributed to random appearance of pile-ups of sand grains which induce separation of flow (Allen, 1968, 1969; Williams and Kemp (1971). Present observations suggest that such pile-ups, and the corresponding scars from which the material was removed, are a result of the intensified bursts along the sand-streaks, with transverse alignment probably induced by the transverse vortex motions, described by Kim et al. (1971) and Nychas et al. (1973). It is known that bursting in the sublayer, is accompanied by larger-scale flow configuration in the outer zone of the boundary layer, suggestive of presence of thatched hairpin-vorticity systems across the flow. The secondary motions along the hairpin elements, towards and away from the boundary, may be capable of inducing the appearance of seminal hollow and mounds patterns from which ripple configurations grow. This is discussed in more detail in the section dealing with the origin of elliptical scours.

Sorby (1908) and many others reported current lineations at velocities higher as well as lower, than those required for generation of small ripples.

As shown above, current lineations may ride on the backs of well developed bed undulations and an apparent equilibrium exists between the two bedforms. However, when velocity is increased to a value corresponding to transition to the upper plane bed phase of sediment transport, longitudinal alignment intensifies, and gradually obliterates and replaces the ripples. The high-velocity streak-pattern is quite similar to that recorded before the onset of ripples, but individual streaks are wider ("patchy" appearance), show a strong lateral component due to the increased incidence and frequency of bursting, and bear evidence of presence of a larger-scale longitudinal configuration. The streaks show a rapid agglomeration into bands 2-8 cm wide, which grow, transforming the bed into a series of fairly wide and rounded ridges separated by narrower, cuspate-downwards troughs. (Fig. 16)

Fig. 15a **Linear paths of grain** movement across a pattern of transverse ripples

Fig. 15b A small barkhan ripple heralding the transition from **streak regime** to ripples

Deformation of a transverse ripple configuration starts with the appearance of a number of straight "dislocation" lines normal to the ripple crests, along which the ridges are sheared and displaced in a strike-slip fashion. ("lanes" of Allen, 1968; "threads of scour", Karcz, 1967). Sense of displacement alternates for the adjoining lines. With time, a thread of scour appears along the line, the sediment is removed sidewards where accumulation occurs and the remnant crests and troughs are progressively leveled and uprounded. (Fig. 17) This sequence of events, observed in the flume and in the field, (Fig. 18) leads to a very regular pattern of ridges with intervening troughs, (Fig. 19) an order of magnitude larger than current lineations which corresponds to many natural and experimentally produced bed configurations discussed in the literature (e.g. flume-sand ribbons and scours - Casey, 1935; Vanoni, 1946; mud-furrows - Hall, 1843; Culbertson, 1967; windrow ridges - Tanner, 1960; Harrow marks, Karcz, 1967; longitudinal ridges, Dzulynski and Walton, 1963, 1965. In all these cases, bed-morphology is related to a pattern of secondary motions shown in Figure 10, which are known to arise as a result of instability of inversely stratified flow with lateral shear, instability of parallel shear flow, instability of boundary layer flow over concave surfaces, instability at transition to fully developed turbulence, and appearance of secondary currents in channels and conduits. The geological applicability of some of these mechanisms was discussed by Allen (1968, 1969a, 1969b, 1971), Dzulynski (1966), Dzulynski and Walton (1963, 1965) and Karcz (1967, 1970) and needs not be discussed here. It must be emphasized however, that in all cases, transformation of ripples to a regular longitudinal pattern is morphologically analogous to the gradual replacement of two-and three dimensional wave patterns by longitudinal vortex roll-system predicted by the non-linearized stability theory, and recently observed in a water channel.

Also the experimental evidence outlined above, suggests that replacement of ripples and accompanying current lineations by a longitudinal pattern, is due to appearance of a larger-scale kinematic flow configuration. Such a structure may be due to the development of secondary currents due to shear at the channel boundaries, and to unequal distribution of turbulent velocity fluctuations. The presence of such secondary circulation patterns is amply documented in studies of velocity and sediment distribution patterns in rivers and channels (for references see: Einstein, 1972; Leliavsky, 1965). Another possibility, suggested by the close association of current lineations and sand ribbons observed at transition to upper plane bed, is that the lineations represent only footprints of analogous, larger-scale structures within the flow, the effect of which becomes more pronounced at higher flow-energies. Such a close architectural analogy between the structural elements within the sublayer, the outer zone of the boundary layer, and the main flow was postulated by Kovasznay (1967, 1970), Kovasznay et al. (1970), Townsend (1970) and many others. It appears possible, therefore, to regard the difference in scale of geometrically similar bed forms encountered in nature, as the expression of difference in scale and level of the parent kinematic flow patterns.

ELLIPTICAL SCOURS AND TROUGHS

Systems of flow-aligned elliptical, arcuate, and spindle-shaped scour hollows, are a common feature of many straight reaches of river channels devoid of meandering tendencies. At low water, they form thatched or scattered puddles, (Fig. 20) partly or fully infilled with sediment, at high water echo-sounder records patterns of crowded elongate hollows migrating downstream.

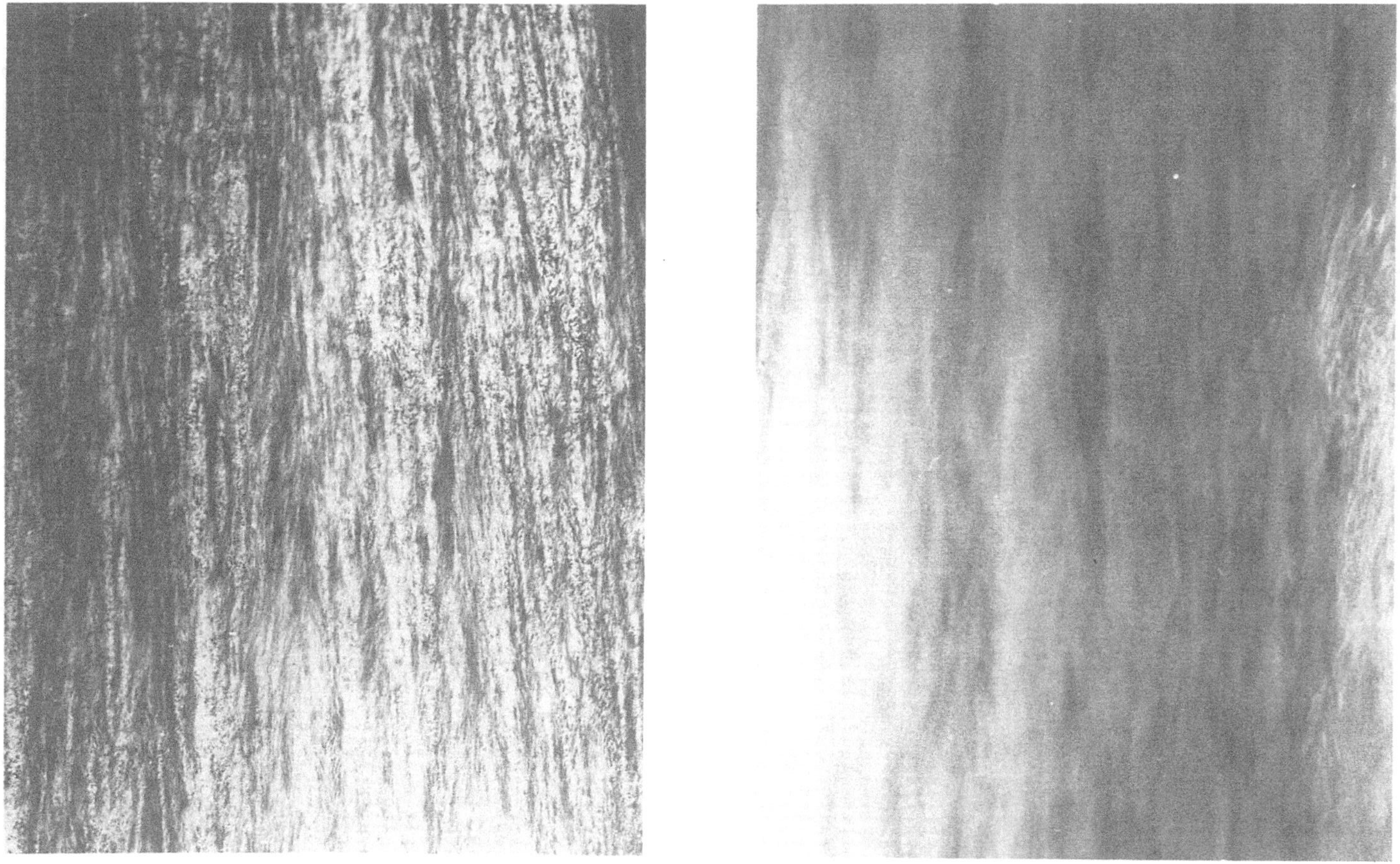

Fig. 16 The appearance of larger scale bands in a high-flow velocity.

Some of the scour elements are associated with comparable in size megaripples and sand waves, and are spread fairly evenly along and across the channel. Other, usually much larger scours are crowded or scattered along a smooth bed or appear only along some streamwise or spanwise segments of the channel, with no relation to the distribution of smaller bed forms.

Scour elements of the first type, and many of the second type, referred to as elliptical scours, reach a width of 8 m and up to 50 m in length and are preserved in the geological record as trough-type cross-stratification. Two mechanisms of origin have been proposed in the geological literature, both stressing the deposition of cross-stratified units. On explanation involves a "scour and fill" process, in which kolk-vortices, unrelated to the lee eddies to the accompanying bed-undulations, excavate hollows, subsequently infilled by advancing bed-forms (Harms and Fahnestock, 1965). Somewhat similar ideas was proposed by Wood and Smith (1959) to explain the origin of some flute marks and grooves in turbidites. The origin and infilling of elliptical scours has also been attributed to migration of ripples, whose troughs are enclosed hollows, or that contain enclosed hollows. These spoon-shaped scours are viewed as features of spanwise instability associated with skewed catenary ripples, attributed mainly to the development of flow-aligned vortex rolls (Taylor-Gortler vortices) along the concave portions of the ripple-profiles. The whole pattern (catenary ripples and scours), like other three dimensional ripple-trains, is explained in terms of flow separation patterns behind single or paired, skewed negative steps (Allen, 1968, 1969b).

Thatched scours along plane beds, and the instantaneous emergence of mound and hollow configuration all over the bed, observed in experiments pose the question as to whether or not initiation of such patterns may be triggered by boundary flow-configuration. As already mentioned, flow in the sublayer is dominated by burst-like regime, in which lumps of low-momentum fluid are ejected away from the boundary. Analogous, but larger and fewer bursts were recently found to emanate also from outer zones of the boundary layer. The largest bursts appear at the interface between the boundary layer and the flow as a series of bulges of low-momentum fluid, acting as obstacles within the faster flow. With time, these disturbed zones become elongate and merge with the surrounding fluid (Kovasznay, 1967, 1970). Similar configurations of downstream migrating bulges acting as obstacles and producing isolated or connected horseshoe-or plow-shaped turbulent zones was reported by Klebanoff et al. (1962) in study of transition from laminar to turbulent flow. Kline et al. (1967), Willmarth and Tu (1967) and others described hairpin-shaped vorticity systems aligned across the flow with ices alternately at the boundary and swept downstream. (Fig. 20) It is now known that these structures in the innermost zone of turbulent flow have larger analogues such as the "inclined double-roller eddies" of Townsend (1970), which dominate the flow structure farther away from the boundary.

It appears, therefore, that continuous spiral-like, or scattered, isolated ("holes in the carpet") vorticity systems are present within the flow. In both cases, streamwise and spanwise components are present, and secondary motions are directed towards and away from the boundary. If the presence of such configurations is assumed over an erodible boundary, scour may be expected in segments where the flow components dive towards the boundary, whereas no change or deposition will occur where they rise. Such a mechanism would account for initiation of germinal patterns of three dimensional ripples, as well as for the initiation of elliptical scours along a fairly smooth bed.

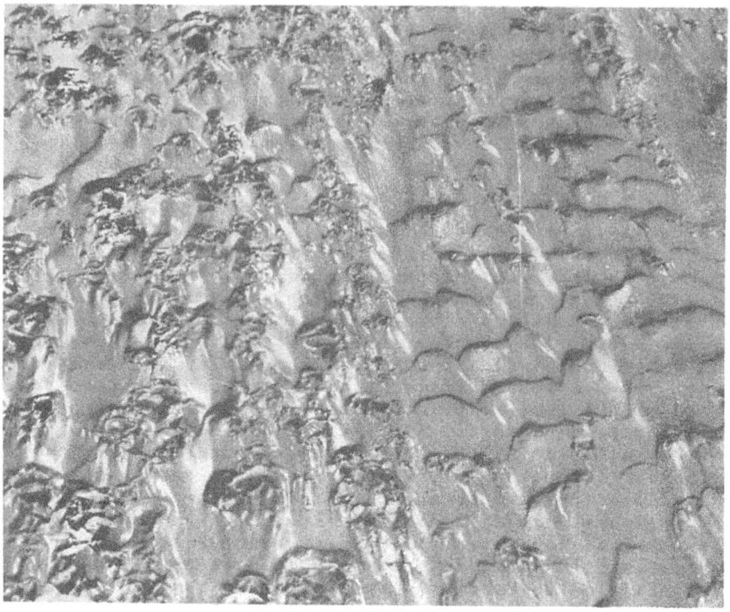

Fig.17 Replacement of transverse ripples by larger scale flow-aligned structures

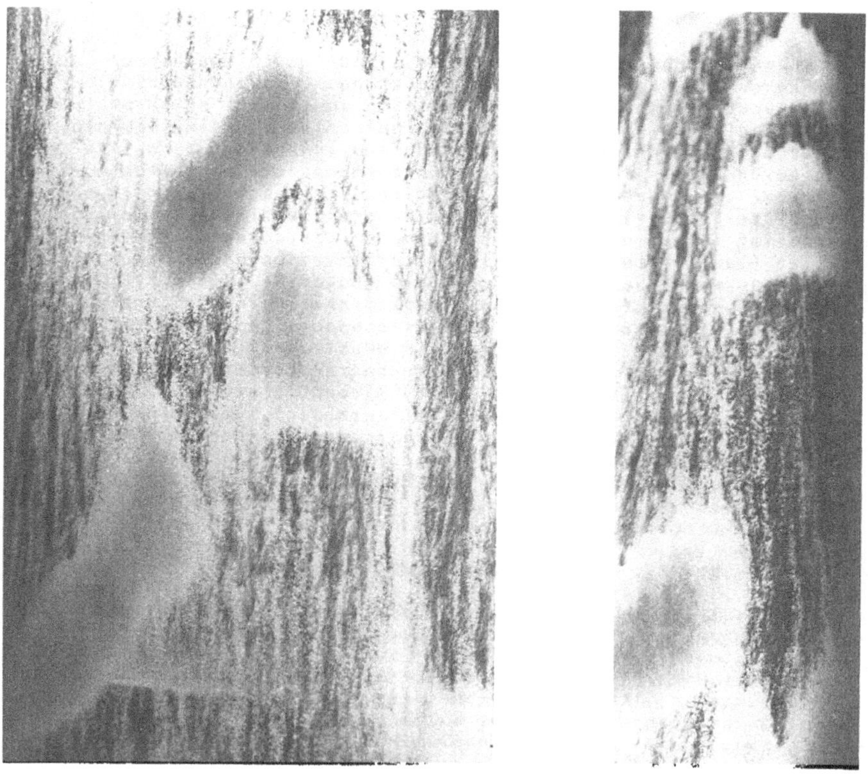

Fig.18 Gradual obliteration and displacement of ripples along sharp lines ("lanes")

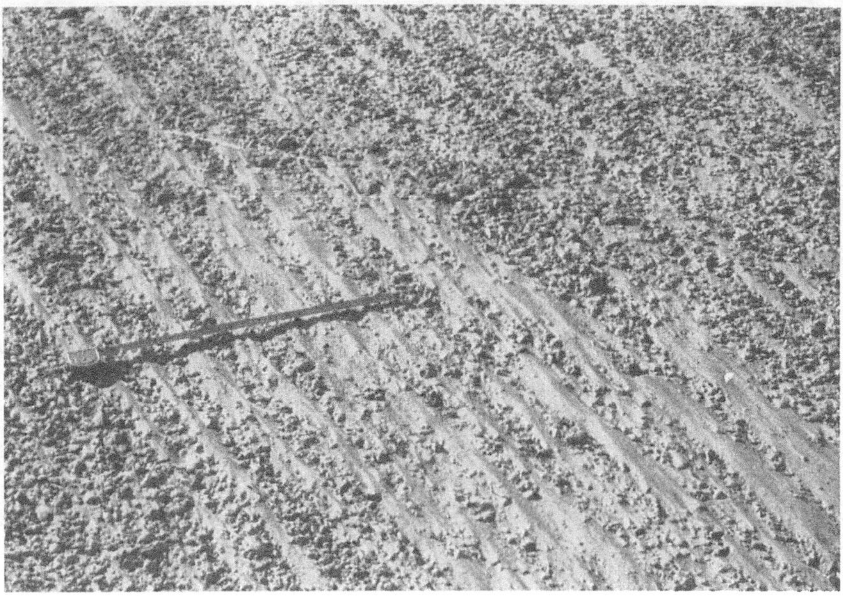

Fig.19 A fully developed pattern of flow-aligned ridges and troughs

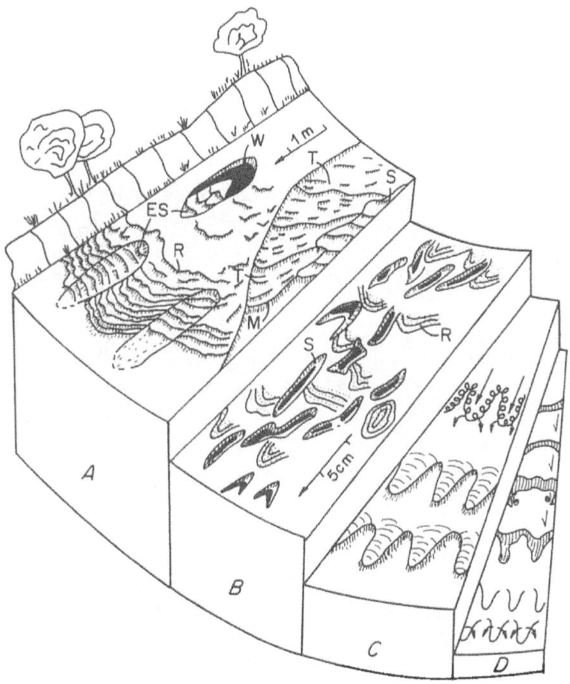

Fig.20 A. An ephemeral streambed after a flash flood. Elliptical scours (ES), partly drained (W) accompanied by ripples (R) and bars(M) developed along side shoals (S).

B. Experimentally produced pattern of scours (S) and mounds(R), along a pumice-paved laboratory channel

C. Geometry of the flow pattern instrumental in generation of elliptical scours

D. Development of flow configurations in a natural transition

Coleman (1969) has recently described migrating troughs, up to 100 m wide and well over 2000 m long from peak flows along the Brahmaputra River, when the dominant phase of transport appeared to correspond to the upper plane bed stage. Such troughs may represent gigantic and ribbons and troughs, discussed in the previous section. Equally however, they may represent the result of secondary flows induced by transverse water slopes associated with the rising stage, and unequal turbulent velocity distribution across the channel. Somewhat similar, migrating narrow deeps have been widely described in river navigation literature. Unfortunately in most cases, it is difficult to determine, whether the troughs are affected by secondary circulation patterns related to development of channel sinuosity and meandering thalwegs.

SUMMARY

Flow-aligned elements of streambed-scour may be regarded as a product of secondary, helicoidal motions induced by flow instability. In spite of the implied uniformity of origin, they vary greatly in size, shape, and distribution-density within the channel. Field evidence and experiments suggest that this variability reflects possible differences in the parent kinematic flow patterns. Three main groups of structures appear, irrespective of channel geometry and apparently unrelated to meandering tendencies:

a. Structures induced by "defects", such as obstacles, cavities etc., with features of scour and deposition emanating from the individual defects as a result of flow separation phenomena in front and rear of the obstacle.

b. Patterns of continuous longitudinal ridges and troughs, attributed to flow-configurations characterized by streamwise-persisting, rhythmic spanwise variation in velocity, suggestive of a system of longitudinal vortex rolls with alternating sense of rotation. The finest of such structures is current lineation, which appears to represent a manifestation of the fundamental structure of the flow at the boundary. Such lineations are present at velocities both lower and higher than those required to produce ripples. Larger sand ribbons are best visible at, or close to, transition to upper plane-bed phase of sediment transport.

c. Patterns of elliptical, arcuate and spindle-shaped scours, which appear in association with large bed undulations or along otherwise planar streambeds. These structures are interpreted in terms of flow separation behind skewed negative steps. It is possible however, that in some cases they are triggered by downstream-migrating, hairpin-shaped vorticity systems, with secondary flows pointing towards and away from the bed. Some larger scale scours, which appear scattered along segments of the channel, are probably the result of secondary currents induced by spatial or temporal changes in flow parameters or in hydraulic geometry.

All these structures show a considerable variation in scale. This variation may be simply an indication as to whether the structure is a manifestation of the kinematic flow pattern within the sublayer, within the boundary layer or within the main flow.

ACKNOWLEDGEMENTS

During the past few years, I profited immensely from occasional discussions and arguments with J.R.L. Allen, E. K. Walton, K. J. Hsu and J. Southard. Some of the field observations along the ephemeral streams in the southwest U.S. were funded by the Center for Solid Earth Geology, SUNY at Binghamton. Marie Morisawa tried to make the manuscript more readable.

BIBLIOGRAPHY

Allen, J.R.L., 1964. Primary current lineation in the Lower Old Red Sandstone (Devonian) Anglo-Welsh basin. Sedimentology, 3:89-108.

Allen, J.R.L., 1968. Current Ripples. North-Holland, Amsterdam, 433 pp.

Allen, J.R.L., 1969a. Erosional current marks of weakly cohesive mud beds. J. Sediment. Petrol. 39:607-623.

Allen, J.R.L., 1969b. On the geometry of current ripples in relation to stability of fluid flow. Geograf. Ann., SIA, 61-96.

Allen, J.R.L., 1971. Transverse erosional marks of mud and rock: the physical basis and geological significance. Sedim. Geology, 5, 167-385.

Berger, E., and R. Wille, 1972. Periodic Flow phenomena, Ann. Rev. Fluid Mech., 4, 313-340.

Casey, H.J., 1935. Mitt. Preussische Versuchsanst. Wasserbau Schifbau, 1,

Chandrasekhar, S., 1961. Hydrodynamic and hydromagnetic stability: Oxford, Clarendon Press, 654 p.

Chang, H.Y. and Simons, D.B., 1970. The bed configuration of straight sand bed channels when flow is nearly critical. J. Fluid Mech., 42:491-495.

Chow, V.T., 1959. Open Channel Hydraulics, McGraw Hill, New York, 608 pp.

Coleman, J.M., 1969. Brahmaputra River: channel processes and sedimentation. Sediment. Geol., 3:129-239.

Corino, E.R., and R.S. Brodkey, 1969. A visual investigation of the wall region of turbulent flow, J. Fluid Mech., 37, 1,

Culbertson, J.K., 1967. Evidence of secondary circulation in an alluvial channel, U.S. Geol. Surv. Prof. Pap. 575-D, 214.

Dzulynski, S., 1966. Sedimentary structures resulting from convection-like patterns of motion, Ann. Soc. Geol. Pologne, 36, 3,

Dzulynski, S., and Walton, E.K., 1963. Experimental production of sole markings: Edinburgh Geol. Soc. Trans., v. 19, p. 279-305.

Dzulynski, S., and E.K. Walton, 1965. Sedimentary Features of Flysch and Graywackes, 274 pp., Elsevier, Amsterdam.

Einstein, H.A., 1972. On secondary currents and sediment motion in: River Mechanics, H. W. Shen (ed). 18/1 - 18/10

Einstein, H., and Li, H., 1958. Secondary flows in straight channels: Am. Geophys. Union Trans., v. 39, p. 1085-1088.

Engelund, F., and Hansen, E., 1967. A monograph on sediment transport in alluvial channels: Copenhagen, Teknisk Forlag, 62 p.

Folk, R.L., 1971. Longitudinal dunes of the northwestern edge of the Simpson Desert Northern Territory, Australia, 1. Geomorphology and grain size relationships, Sedimentology, 16, 5-54.

Grass, 1970. The initial instability of fine sand, J. Hyd. Division ASCE, 96, 619-632.

Grant, H.L., 1958. The large eddies of turbulent motion, J. Fluid Mech., 4:149-190.

Hall, J., 1843. Geology of New York: New York, Carrol and Cook, 683 p.

Harms, J.C. and Fahnestock, R.K., 1965. Stratification,bed forms and flow phenomena (with an example from the Rio Grande). In: G. V. Middleton (Editor), Primary Sedimentary Structures and their Hydrodynamics Interpretation - Soc. Econ. Paleontol. Mineral., Spec. Publ., 12:84-115.

Hoyt, J.H. and Henry, V.J., 1963. Rhomboid ripple mark, indicator of current direction and environment. J. Sediment. Petrol., 33:604-608.

Johansson, C.E., 1963. Orientation of pebbles in running water a laboratory study. Geog. Annaler, A, 45, 85-112.

Karcz, I., 1967. Harrow marks, current aligned sedimentary structures, J. Geol. 75:113-121.

Karcz, I., 1968. Fluviatile obstacle marks from the wadis of the negev (southern Israel). J. Sediment. Petrol. 38:1000-1012.

Karcz, I., 1970. Possible significance of transition flow patterns in interpretation of some natural bed forms. J. Geophys. Res. 75:2869-2873.

Karcz, I., 1971. Development of a meandering thalweg in a straight erodible laboratory channel, J. Geol. 79, 234-240.

Kennedy, J.F. and Iwasa, Y., 1968. Free surface shear flow over a wavy bed. J. Hydraul. Div. A.S.C.E., 94:431-454.

Kennedy, J.F. and Roubillard, L., 1967. Some experimental observations on free surface shear flow over a wavy boundary. Proc. 12th Congr. Int. Assoc. Hydraul. Res., 1:41-48.

Kim, H.T., Kline, S.J., and W. C. Reynolds, 1971. Production of turbulence near a smooth wall in a turbulent boundary layer, J. Fluid Mech., 50, 133-160.

Klebanoff, P.S., Tidstrom, K.D., and Sargent, L.M., 1962. The three-dimensional nature of boundary layer instability: Jour. Fluid Mech., v. 12, p. 1-34.

Kline, S.J., Reynolds, W.C., Schraub, F.A., and Runstadtler, P.W., 1967. The structure of turbulent boundary layers. J. Fluid Mech., 30:741-775.

Kovasznay, L.S.G., 1967. Structure of turbulent layer, Physics of Fluids, S25-S30.

Kovasznay, L.S.G., 1970. The turbulent boundary layer Ann. Rev. Fluid Mech., 2, 95-112.

Kovasznay, L.S.G., Kibbens, V., and Blackwelder, R.F., 1970. Large scale motion in the intermittent region of a turbulent boundary layer, J. Fluid Mech., 41:283-325.

Kuo, H.L., 1963. Perturbations of plane Couette flow and origin of cloud streets: Physics Fluids, v. 6, p. 195-211.

Leliavsky, S., 1965. Design textbooks in Civil Engineering, v.4, River Mechanics; Chapman and Hall, London, 228 pp.

Nycha S., Hershey, H., and R. S. Brodkey, 1973. A visual study of turbulent flow, J. Fluid Mech., 61, 513-540.

Otvos, E.G., 1965. Types of rhomboid beach surface patterns. Am. J. Sci. 263:271-276.

Potter, P.E., and F. J. Pettijohn, 1963. Paleocurrents and basin analysis, Springer Verlag, Berlin, 296 pp.

Richardson, P.D., 1968. The generation of scour marks near obstacles. J. Sediment. Petrol., 38:965-970.

Roshko, 1961. Experiments on the flow past a circular cylinder at very high Reynolds numbers, J. Fluid Mech., 10, 345-356.

Schlichting, 1968. Boundary layer theory, McGraw Hill, New York, 747 pp.

Shen, H.W., Schneider, V.R., and Karaki, S., 1969. Local scour around bridge piers, J. Hyd. Division ASCE, 95, 6891 p.

Sorby, H.C., 1908. On the application of quantitative methods to the study of the structure and history of rocks. Q. J. Geol. Soc. London, 64, 171-232.

Southard, J. B., and Dingler, J.R., 1971. Flume study of ripple propagation behind mounds on flat sand beds, Sedimentology 16,

Stuart, J.T., 1965. Hydrodynamic stability. Appl. Mech. Rev., 18:523-531.

Willmarth, W.W., and Tu, B.J., 1967. Structure of turbulence in the boundary layer near the wall: Physics of fluids, v. 10, p. S134-S137.

Wood, A., and Smith, A.J., 1959. The sedimentation and sedimentary history of the Aberystwyth grits (upper Llandoverian): Geol. Soc. London Quart. Jour., v. 114, p. 163-189.

Williams, P.B., and P.H. Kemp, 1971. Initiation of ripples on flat sediment beds, J. Hyd. Division, ASCE, 97, 505-522.

Wu, T.Y., 1972. Cavity and wake flows, Ann. Rev. Fluid Mech., 4, 243-284.

Vanoni, Vito, 1946. Transportation of suspended sediment by water: Am. Soc. Civil Engineers Trans., v. 111, p. 67-102.

Tani, I., 1969. Boundary layer transition. In: W. R. Sears and M. van Dyke (Editors), Annual Reviews of Fluid Mechanics, I. Annual Reviews, Palo Alto, Calif., pp. 169-196.

Tanner, W.F., 1960. Shallow ripple mark varieties: Jour. Sed. Petrology, v. 30, p. 481-485.

Townsend, A.A., 1970. Entrainment and structure of turbulent flow, J. Fluid Mech., 41:13-46.

PART III

The third group of papers are concerned with alluvial landforms of river valleys. These include two kinds of patterns created by rivers flowing through and around sediments they have deposited, resulting in meandering or in braided configurations.

A river meandering on its floodplain creates a landscape with many features such as meander scrolls, abandoned channels, point bars, swamps, lakes, sloughs, cutoffs, levees and flood basins. These forms are a result of the magnitude and dynamics of river discharge and, hence, statistical relationships have been described between discharge and meander length and between discharge and meander curvature. River bends are destined to change, with erosion on the outside banks and deposition on the inner bank causing lateral and downstream migration of the curve.

Brice traces alterations in the meandering pattern of the White River system utilizing air photographs. He measures morphometric factors such as loop curvature, loop size and loop orientation in his analysis. According to his determination, meander loops do not evolve sequentially along a reach, but rather, intermittently with a tendency toward elongation through time. Loops oriented in a downstream direction seem to have the greatest potential for erosion. The fit of circular arcs to the meander pattern shows that arcs are a basic element of the shape of river bends.

The other kind of river valley landscape examined in Part III is braiding. The alluvial pattern of a braided river consists of channelways which divide and reunite, wandering around and sometimes over, alluvial islands and bars. Bars may be transverse, oblique or longitudinal to the general downstream flow direction. They are usually composed of pebbles, gravel, or sand with very little or no silt and clay. Braided reaches occur under highly variable discharge and abundant bed load. The bars and divided channels are considered effective hydraulic devices to enable the river to transport its load.

Smith's paper is a study of the difference in braiding of two rivers in Canada. Using Mackin's concept of upstream and downstream controls on aggradation he examines braided deposits. For the river with upstream control of aggradation, a thick wedge of deposits thins downstream. The sediments are composed of coarse outwash gravel with large cobble lenses representing scour pools. Reaches of such a river have steeper gradients to increase the ability to transport the load. This type of channel deposit and pattern represent true braiding, with many unstable channels threading through bars. Changing channelways result from erosion of non-cohesive banks.

On the other hand, downstream control of aggradation is indicated by a wedge of deposits which thins upstream. The beds are laid down in backwater reaches and thus contain silts. This kind of aggrading channel Smith calls anastomosing - to be differentiated from braiding by a lower gradient, finer sediments and more stable channelways.

In the last paper of Part III, Fahnestock contrasts two braided rivers in Alaska. He finds that they are different in regard to slope, intricity of deposits and in stability. The morphologic difference are due to differences in flow pattern and bed load.

On the Knik, repeated catastrophic floods from annual spring breakouts result in a channel pattern and morphology which is stable. The lower reaches have a gentle gradient over which sands and fine gravel are transported. On the Matanuska, lower peak flows but widely varying discharges, as well as a coarser bed load, cause an unstable channel pattern and changeable morphology. The river gradient is steep in order to aid in the transport of coarse sediments with available discharge.

In addition, Fahnestock discusses the sedimentary structures found in the banks of the Knik River. Cut and fill structures he relates to shifting channelways or to movement of dunes along the bed. Tabular cross bedding results from avalanching at the steep front of gravel waves moving into deep, still water. Scour pockets are formed as water moves around isolated blocks of ice which cause separation of flow (see Part II, this volume).

8

MEANDERING PATTERN OF THE WHITE RIVER IN INDIANA–AN ANALYSIS

James Brice

ABSTRACT

The meandering patterns of three reaches of the White River system in Indiana, ranging in length from 57 to 98 km, are analyzed. The patterns of these reaches are dominated by segments that approach constant curvature; and these are either tangent with adjacent curved segments or connected to them by nearly straight segments. A simple meander loop has only one segment of constant curvature whose length exceeds its radius. A compound loop includes two or more simple loops, whose curvature is commonly directed toward the same side of the river. Simple loops are the basic unit for analysis and they consistently have a unimodal log normal size frequency distribution for both length and radius. Some subjectivity is involved in the delineation of simple loops, but radius is less subject to operator variance than length and height; moreover, radius is probably the fundamental property of a meander loop.

For the East Fork White River reach, the river centerlines for 1937 and 1968 (as obtained by simultaneous projection of aerial photographs) were plotted. Radius, length, and orientation of each loop (about 100 on each centerline) were measured and the direction of migration was determined from displacement of the loop centroid. The area eroded by migration of each loop, which ranged from near zero to 14 acres, was measured. No correlation was found between the size of a loop and the area eroded, and the potential for erosion is apparently the same for loops ranging in radius from 100 to 300 meters. The meander loops along the reach have not evolved either sequentially or simultaneously (as implied by laboratory experiments) but intermittently, first in one part of the reach and then in another. No specific loop size seems uniquely characteristic of the river or more stable than the rest. In the evolution through time of individual loops, a general tendency toward elongation (decrease in L/H ratio) was discerned, as was a weak tendency toward decrease in radius.

According to the frequency distribution of loop orientations, the axes of most loops on East Fork White River tend to point approximately at a right angle to mean downstream direction, but orientations are diverse and many loops point upstream. The direction of loop centroid movement for the period 1937-1968 lies in a quadrant between the azimuth of the loop axis and a perpendicular to the axis that points in a downstream direction. Flat loops tend to migrate perpendicular to their axes, whereas more elongate loops tend to grow in the direction of their axes.

INTRODUCTION

Three reaches of the White River system in Indiana (Figure 1) have been chosen as particularly suitable for analysis of the meandering channel pattern because they seem to be typical of meandering rivers. In addition, the meanders of these reaches are well developed and diverse in form, the rivers flow in broad valleys of uniform slope and are underlain by reasonably homogeneous alluvium, and full coverage by recent topographic maps and sequential aerial photographs is available. A U.S. Geological Survey gaging station having a satisfactory period of record is located on each reach. The White River system is unusual in that no large reservoirs have been constructed on it. Also, in the reaches chosen there has been a minimum amount of interference with the natural behavior of the rivers, although a few artificial meander cutoffs have been made, and riprap or other measures intended to halt bank erosion have been applied from place to place by private landholders. In a sample of 50 meandering alluvial rivers of comparable size in eastern United States, for which sequential aerial photographs have been obtained, the White River reaches are among the most rapid in rate of lateral migration and meander development. Finally, an analysis of the meandering pattern of the White River by the method of power spectral analysis has been made by Chang and Toebes (1970) and is available for comparison with the method used here.

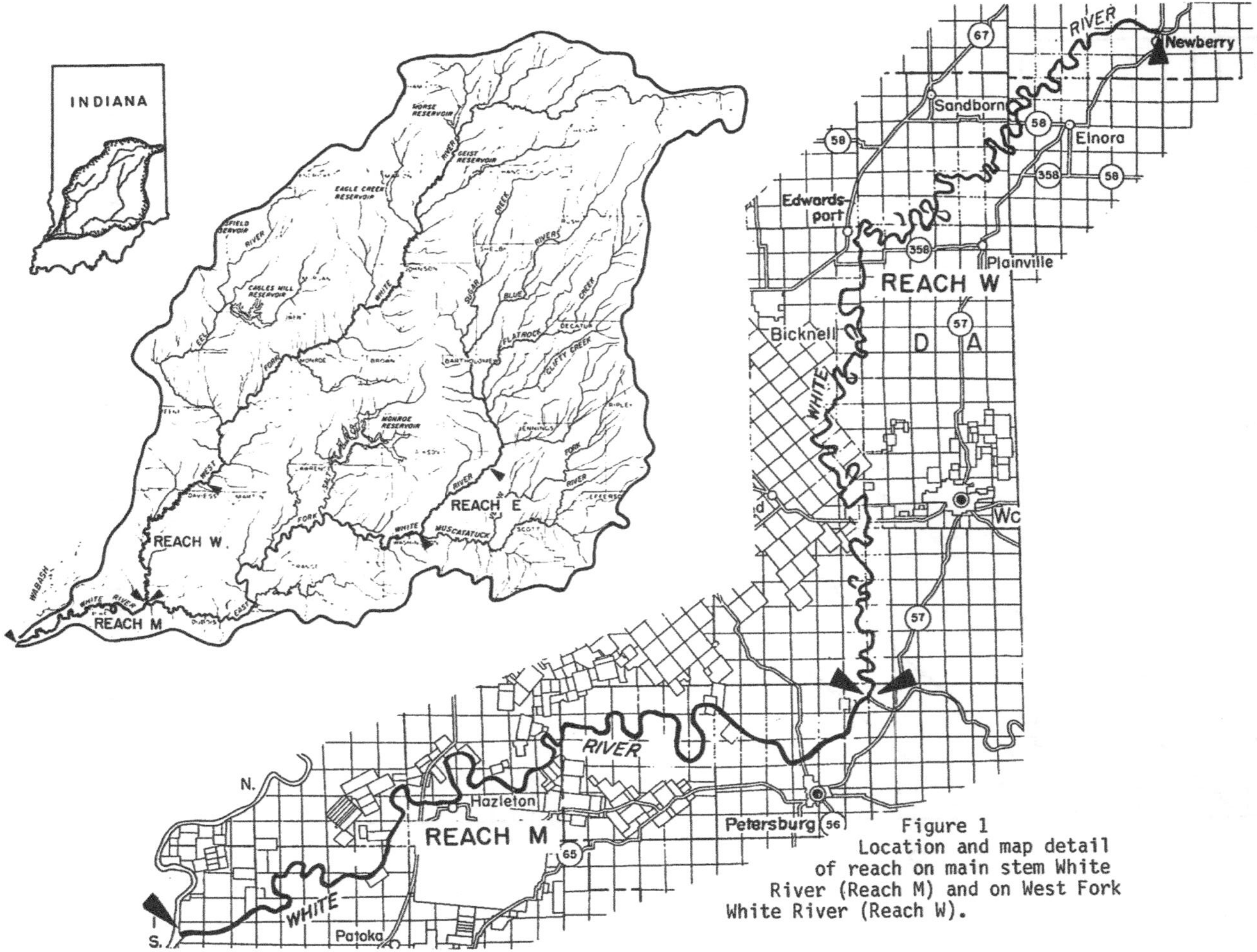

Figure 1
Location and map detail
of reach on main stem White
River (Reach M) and on West Fork
White River (Reach W).

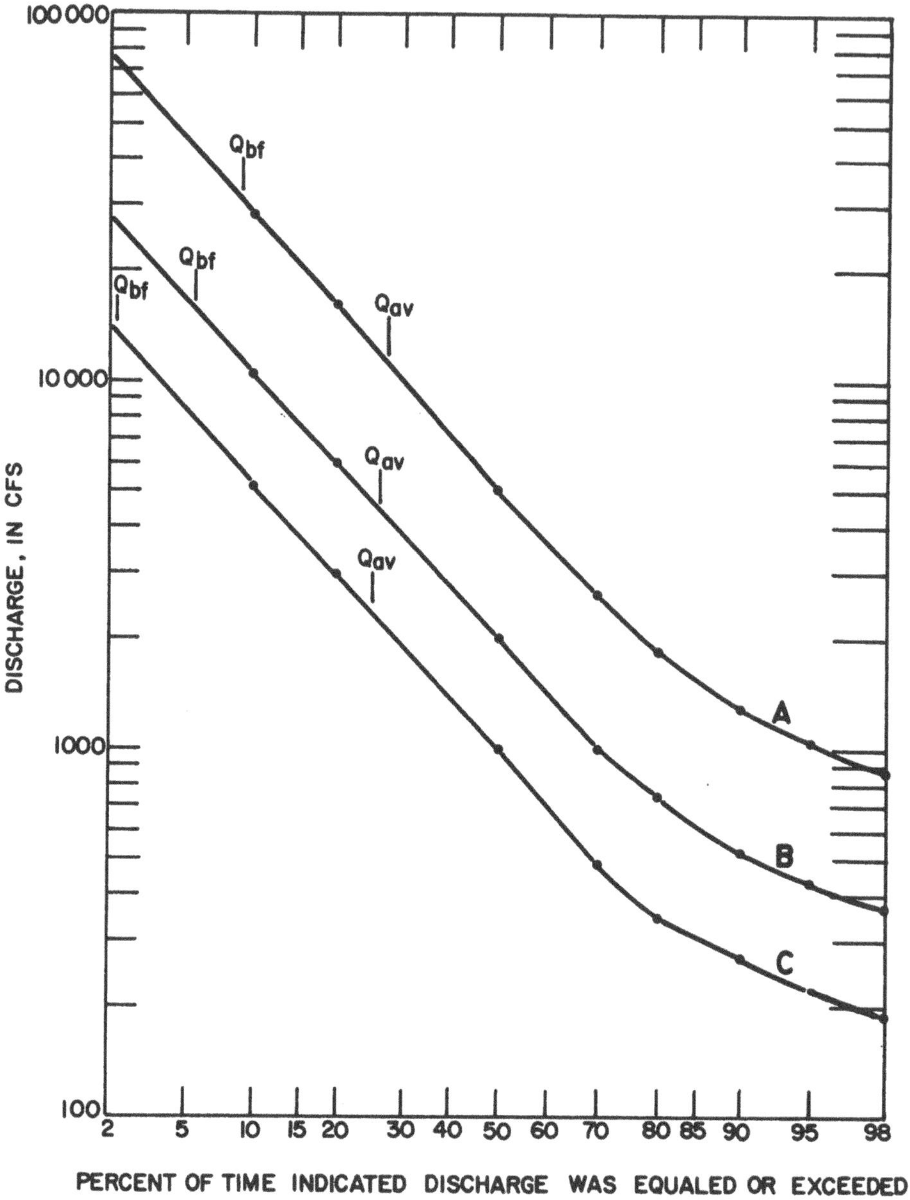

Figure 2 Flow duration curves. A. White River at Petersburg (Reach M). B. White River at Newberry (Reach W), 1928-1965. C. East Fork White River at Seymour (Reach E), 1927-1965.

A small-scale but accurate representation of the pattern of Reaches M and W is given in Figure 1, and a more detailed representation of the pattern of Reach E is given in a later section. Reach M, on the main stem White River, begins at the confluence of the East and West Forks and ends at the confluence with the Wabash River. Reach W, on West Fork White River, begins at the gaging station at Newberry and ends at the confluence with East Fork White. Reach E, on East Fork White River, begins at the gaging station at Seymour and ends at the confluence with the Muscatatuck River. Some basic data on the reaches are given in Table 1.

Table 1 General morphologic and hydrologic data for Reaches M, W, and E

	Reach M (1958)	Reach W (1958)	Reach E (1968)
Reach length along river, km	78.8	98.4	56.9
Reach length along valley, km	52.3	59.7	33.4
Total sinuosity	1.48	1.65	1.7
River slope, m/m	0.000144	0.000192	0.00028
Valley slope, m/m	0.00026	0.000326	0.000485
Median river width, m	140.	85.	60.
Bank height, m	5.5-7.5	4.-5.	2.7-3.3
Drainage area at gage, mi^2	11,139	4,696	2,333
Average discharge, cfs	11,310	4,521	2,365
Bankfull discharge, cfs	30,000	15,500	13,500
Bankfull stage, ft	16.	13.	14.

GEOMORPHIC HISTORY

Southern Indiana is crossed by north-south trending belts of Paleozoic strata that crop out on the eastern side of the Illinois Basin. Reaches M and W are within the Mitchell Plain and are underlain by rocks of Pennsylvanian age. Reach E lies within the Scottsburg Lowland, which is an erosional plain of slight relief developed in most places on the easily eroded New Providence Shale and similar rocks of Mississippian age. The White River basin does not appear to have been very much extended or diminished by the Pleistocene glaciations. The Illinoian ice sheet covered most of the basin, including the area of all three reaches described herein, but the Wisconsin glacial border crosses the central part of the basin and lies to the north of the reaches. According to a map by Thornbury (1950, p. 10), the valleys now occupied by the reaches served as sluiceways during the Wisconsin glaciation and were filled with valley train deposits.

The stratigraphy and composition of the valley train deposits and of the overlying Holocene alluvium have been usefully summarized by Wayne (1963). The valley train deposits are regarded as the outwash facies of the Atherton Formation and consist mainly of sand and gravel. In the valleys of major rivers, including the reaches described here, the Atherton Formation is overlain by the alluvial facies of the Martinsville Formation, which range in thickness from about 1 meter along small streams to 10 meters or more along large rivers. All cut banks that were examined along the reaches are identified as exposures of the Martinsville Formation, the base of which is about 7-8,000 years old, according to radiocarbon dates cited by Wayne and Zumberge (1965, p. 73).

The development of the present river courses could be traced back no further than about 1880, the date of the earliest adequate maps that are available at the Indiana State Library.

WATER DISCHARGE

Water discharge characteristics as measured at gaging stations on each of the study reaches are summarized in Table 1 and Figure 2. Each gage is located at or near the upstream end of a reach and such values as drainage area and discharge will increase in a downstream direction through a reach. However, no tributaries of substantial size enter any of the reaches.

The flow duration curves in Figure 2 are drawn from data given in Rohne (1972), which emphasize low flow characteristics and extend from 10 percent to 98 percent on the probability scale. It has been considered safe to extrapolate the curves to the 2 percent point on the scale. The bankfull discharges indicated on the curves are based on the discharges corresponding to bankfull stage as listed in U.S. Geological Survey Water-Supply Paper 1675. The values for percentage duration of bankfull stage are similar to those obtained for the White River by Dury (1961, p. 53), but the return period for bankfull stage seems to be much lower (more frequent) than indicated by Leopold and others (1964, p. 319-320) for streams in general. On the West Fork White at Newberry during the period 1929-1961, 28 floods, or an average of about 1 each year, had a peak stage of 20 feet or more and definitely submerged a substantial part of the floodplain. On the East Fork White at Seymour during the period 1923-1961, 70 floods, or an average of almost 2 a year, had a peak stage of 16 feet or more and submerged at least part of the floodplain. It seems apparent that the floodplain is rather frequently submerged on these reaches of the White River. The floodplains, particularly along Reach E, bear many small channels doubtless formed by scour during overbank flow, and the frequency of overbank flow is probably responsible for the numerous cutoffs on Reaches E and W.

SUSPENDED SEDIMENT, BED MATERIAL, AND VALLEY ALLUVIUM

No data on suspended sediment have been published for Reaches M and W, but some data are available for Reach E. For water year 1967, daily mean concentrations of suspended sediment ranged from 6 to 451 ppm, and the average discharge weighted concentration was 117 ppm. In a map prepared by Rainwater (1962), streams of eastern U.S. are placed into three sediment concentration classes: 0-270 ppm, 270-1900 ppm, and 1900-6300 ppm. Streams of the White River drainage basin fall with the class of 0-270 ppm and are hence among the less muddy in eastern U.S. Particle size analyses of suspended sediment at Seymour (U.S. Geological Survey Water-Supply Paper 2012, p. 260) indicate that the sediment tends to be rather fine, having an average of about 55 percent in the clay fraction (less than 4 microns) and about 85 percent in the silt-clay fraction (less than 62 microns).

The bed material of all three reaches can be adequately described for present purposes as consisting mainly of sand but with local concentrations of fine gravel. The gravel of Reach M is coarsest; a representative sample observed on a lateral bar had a median diameter of about 3 cm and a maximum diameter of about 10 cm.

In general, the banks of all three reaches consist of sand and fine gravel at the base, grading upwards to loess-like silt toward the top. The silt tends to be mixed with fine sand and is not rich in clay. Pipette analysis of a representative silt sample from the bank of Reach W near the town of Newberry indicated a clay content of 8 percent and a silt-clay content of 61 percent. But the silt-clay content of the bank materials can scarcely be taken as a measure of bank cohesiveness, because the banks are inhomogeneous both vertically and laterally. In some places, undercutting of a stratum of clean sand at the base of a bank was observed to cause slumping of more cohesive materials at the top. In other places, the normal sequence of bank materials grades laterally into sand, which forms almost the entire height of the bank for a distance of several meters and is easily eroded. Nevertheless, the cut banks appear to be about average in homogeneity for rivers generally, and the resistance of the banks is also judged to be in the average range. Outcrops of bedrock occur in a few places where the rivers are near the valley sides (or near one of the few outliers of bedrock that form "islands" in the valley alluvium), but on the whole the channels are cut entirely in alluvium.

Fisk (1944, p. 51) has demonstrated that resistant "clay plugs", formed by accumulation of fine sediment in the oxbow lakes, occur in the valley alluvium of the lower Mississippi River and account for cutoffs and distortion of meander pattern. No clay plugs were observed in the cut banks along the White River reaches. In addition, intersection of sediment-filled oxbow lakes by active meanders can be observed on aerial photographs at many localities, and the former lakes have no apparent effect on meander configuration.

Similarly, natural levees are an important element in the morphology of the lower Mississippi and certain other large rivers, especially in the very lowermost part of their courses. At most places along the White River, no natural levees could be detected in the field nor from careful study of aerial photographs and topographic maps. However, Ulrich and others (1943, p. 45) report that a particular soil type (Genesee loam) occupies natural levee positions and areas in the bends of the White River (Reaches M and W of this report). The distribution of this soil type, as shown on their accompanying soils map of Knox County, is discontinuous and is mainly on the inside of meander loops. There may be some confusion in the soils report between natural levees and meander scrolls; and indeed the distinction between the two is not always clear. Features identified as discontinuous natural levees, having a height of about 3 feet, are present along Reach M in the vicinity of Petersburg. In most places, natural levees are either absent or else are insignificant in the morphology of the reaches.

MORPHOMETRIC ANALYSIS

Measurements were made on the meandering pattern as represented by the river centerline, which was traced from U.S. Geological Survey 7 1/2' topographic maps for Reaches M and W and from vertical aerial photographs for Reach E. The centerline applies to a river stage at or near the median point on the flow duration curve, an approximation to the modal stage and to the "normal" stage at which the Topographic Division of the U.S. Geological Survey represents rivers on topographic maps. Because the river centerline shifts with river stage, and the channel centerline is indefinite at bends, no great precision can be attained in tracing a centerline. In this paper, the centerline at bends is drawn halfway between the shoreline of the point bar and the cut bank on the outside of the bend.

This analysis of the meandering pattern is based on the observation that the pattern consists dominantly of circular arcs that are either tangent with adjacent arcs or connected to them by nearly straight lines. Many persons have observed the circular aspect of river meanders, and Langbein and Leopold (1966, p. H4) report that the radius of curvature is nearly constant for a full third of the length of a symmetrical meander loop. Sooky (1949) has applied a model consisting of circular arcs joined by straight reaches in his analysis of meander geometry, and Chitale (1970) has proposed that meanders can be suitably described by the use of circles. What seems to have escaped general attention is the fact that two successive circular arcs, curved toward the same side of the stream, are a characteristic feature of the natural meandering pattern and must be taken into account in an analysis of the pattern. The only reference in the literature to the occurrence of such arcs seems to be a brief statement by Winkley (1972, chapt. 19, p. 68).

The fit of circular arcs to a subreach (E3, 1937 course) of Reach E is demonstrated at top in Figure 3. The centerline was traced from a vertical aerial photograph, which is reproduced in a later section of this paper. Each circular arc was fitted to a meander loop by superposing a series of closely spaced circles inscribed on transparent acetate film, then marking the center of radius, and drawing the circle with a compass. The fit of the circles is more striking for this particular reach than for most, but it is sufficiently good for all the study reaches to suggest that the circular arc is the basic element of the meandering pattern.

A useful and more general method for analyzing the meandering pattern has been devised by Langbein and Leopold (1966). The angular deviation, from mean downstream direction, of each small segment of a channel is plotted against the channel distance of the segment from the inflection point of a meander or from an arbitrary starting point. In Figure 3 at bottom, a subreach of Reach E is analyzed according to this method, except that the azimuth, from true north, of each segment is plotted rather than its deviation from mean downstream direction. A mean downstream direction line can be drawn on the plot along any

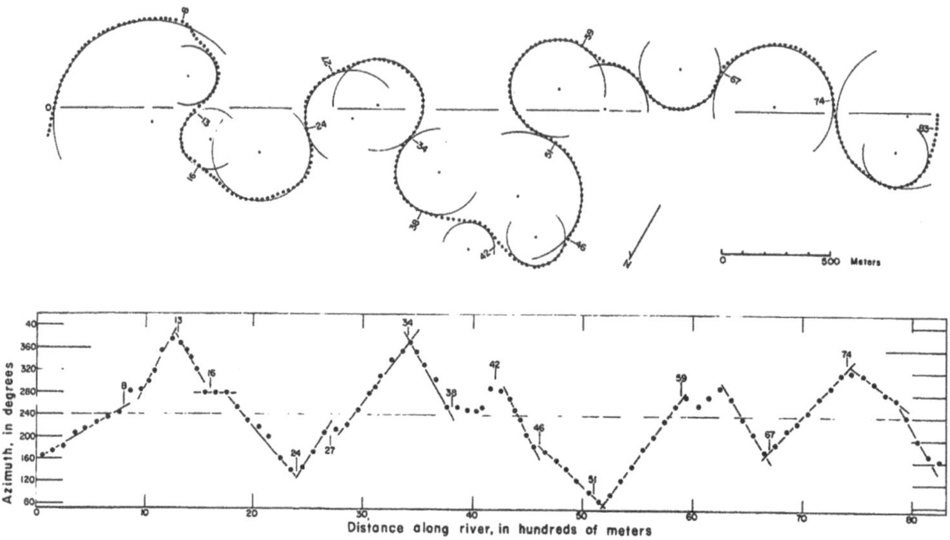

Figure 3 Top, in dotted line, centerline of Subreach E3, East Fork White River. Bottom, plot of river distance vs. azimuth of river centerline as measured from true north.

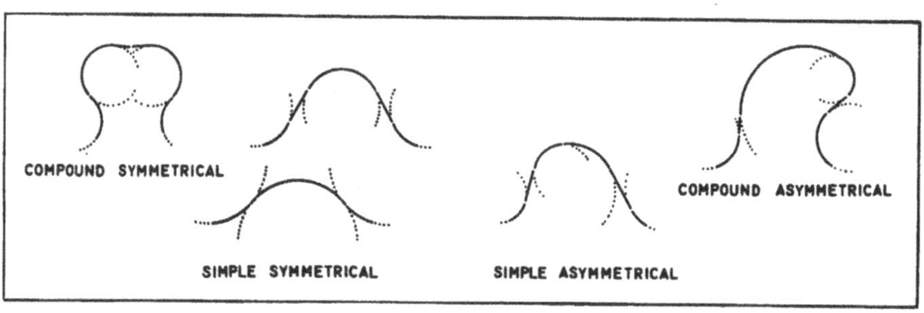

Figure 4 A classification of meander loops according to form.

azimuth that seems appropriate (in this case, at 240°). Azimuth is more convenient to measure than deviation from mean downstream direction, to which either a left or right designation must be assigned. River distances have been measured with bow dividers, which are generally more accurate and satisfactory than other devices (Water Resources Council, 1968). Corresponding distances along the river are marked on the pattern and on the plot in order that the two may be compared more readily. An arc of uniform curvature on the pattern plots as a straight line of uniform slope; straight segments of the pattern plot as straight lines parallel to the x-axis; and channel segments of varying curvature plot as curved lines.

Langbein and Leopold (1966) propose that a symmetrical stream meander (two matched opposing loops) plots as a sine curve, and that the meander thus has the form of a "sine generated" curve. The straight segments that form the limbs of the sine curve represent circular arcs at the apex of meander loops, and the curved segments at the apex of the sine curve represent lines of varying curvature on the limbs of meander loops. This hypothesis is difficult to test for the White River because of the rarity of matched pairs of opposing loops. Its application to individual meander loops is not very convincing, although it has been applied by Daniel (1971) to selected meander loops on Indiana streams.

Plots of the White River pattern all have the general configuration of the example given at bottom in Figure 3. Straight sloping lines that represent arcs of uniform curvature either intersect sharply (are tangent) or else are connected by horizontal segments (straight lines). River segments that deviate from this rule occur in many places, but these do not constitute a significant portion of the total stream length.

According to the hypothesis proposed here, the meandering pattern of the White River reaches consists dominantly of segments of nearly constant curvature that are either tangent with adjacent curved segments or connected to them by nearly straight lines. When the chord of such a segment exceeds its radius and is not greater than about 7 times its height, it is arbitrarily regarded as a meander loop. The curved segments occur either singly or in combination to form individual loops. Tangent loops may be either opposing or non-opposing; if opposing, the centers of curvature are so spaced relative to radius that neither loop is more than a half circle. Connected loops may also be either opposing or non-opposing; if opposing, the centers of curvature are so spaced that no sharp inflection point occurs.

As shown in Figure 4, symmetrical meander loops that have only one segment of constant curvature are classified as simple symmetrical. Simple asymmetrical loops have two segments of constant curvature, but the chord length of one of these is less than its radius; that is, it does not meet the arbitrary criterion for a loop. Compound loops, which demonstrably evolve from simple loops, include two or more simple loops whose curvature is commonly (but not necessarily) directed toward the same side of the stream. Compound loops may be either symmetrical or asymmetrical. A wide range of forms may, of course, be assumed by the loops in each category.

The method used to delineate simple meander loops for measurement, as applied to a part of Reach M, is illustrated in Figure 5. The stream centerline is traced directly from the East Mount Carmel (Ind.--Ill.) 7 1/2' map, on which it forms the boundary between two counties. The map is compiled from aerial photographs taken in 1958. Simple loops are marked off by chords, which are positioned in continuity from loop to loop except where this would cause a loop to have limbs of unequal length about its axis. Loop length, which is measured along the chord, is equivalent to half the wavelength of a meander as defined by Leopold and others (1964, p. 295). The axis of a loop divides it symmetrically, and loop height(amplitude)is measured along the axis. Radius is regarded as a more primary form property than length because loops of the same radius can have different lengths, depending on their stage of development. As a measure of loop size, radius is also preferred because the selection of chords for measurement is more subjective than the selection of arcs. Only simple loops are treated statistically, because no consistent means of expressing the properties of compound loops is apparent and their physical significance is doubtful. The sinuosity due to simple loops is measured

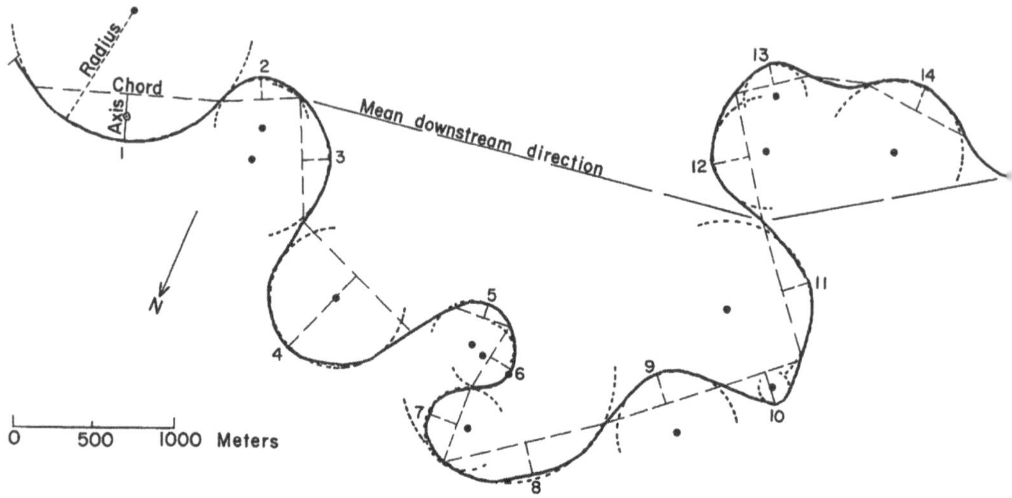

Figure 5 An example of the method of delineating meander loops, main stem, White River.

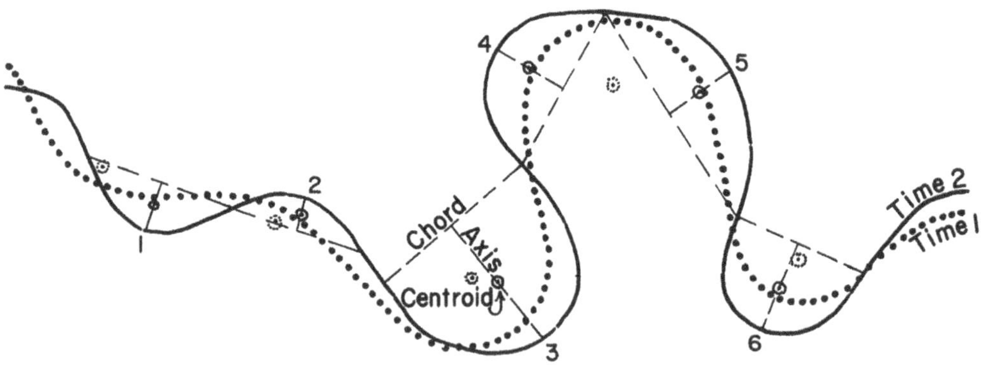

Figure 6 Illustration on a hypothetical reach, of the movement of loop centroids from Time 1 to Time 2.

relative to distance along the chords, whereas total sinuosity is measured relative to the mean downstream direction line, which is shown by long dashes in Figure 5.

Relative positions of the stream centerline on different dates, as illustrated by a hypothetical reach in Figure 6, were obtained by simultaneous projection and matching of sequential vertical aerial photographs. Earlier photographs were obtained from the National Archives and Records Center, and later photographs from the U.S. Dept. of Agriculture. The earliest and the latest photographs of a particular part of the river were photocopied with Kodachrome II film to produce 35-mm transparencies. Two Leitz Prado Universal projectors, mounted side by side on a sliding track, were used to project the 35-mm slides simultaneously onto a rigid screen. By adjustments of distance, tilt, and rotation, fixed reference points on the two images were matched for scale and position, and the river courses (centerlines and both banks) of different dates were traced on the same sheet of paper at a scale of 1:10,000. When all sources of error are considered, the earlier banks and centerline are estimated to be within ± 15 meters of their correct position relative to the later banks and centerline; and most of the results are somewhat more accurate than this. At a scale of 1:10,000, reference points on the images must be matched within ± 1.5 mm to achieve the estimated accuracy. This procedure was carried out for all of Reach E, part of Reach W, and none of Reach M.

By delineating the loops on each centerline, according to the procedure previously described, the change in shape, position, and orientation of a loop can be measured. The shape of a loop at a particular time can be described by its length (L), height (H), and radius (R). The flatness or elongation of a loop is approximated by its L/H ratio. The position of a loop is approximated by the position of its centroid, which is the point midway on its axis; and the movement of the loop can be approximated by the distance and direction that its centroid moves from Time 1 to Time 2. In Figure 6, the chords and axes of Time 1 loops are not shown, but the centroids are shown in dotted circles. The orientation of a loop is taken to be the azimuth of its axis, as measured clockwise from mean downstream direction. The loops numbered 4 and 5 in Figure 6 have evolved from a single simple loop and have no counterparts in Time 1.

Daniel (1971) has developed a method for the analysis of the growth and movement of meander loops, which he applied to one meander on each of 6 streams in Indiana. The method is based on the assumption that a meander loop can be described as a sine-generated curve, according to the hypothesis of Langbein and Leopold (1966), and it would be time consuming to apply to a statistically significant number of loops. The method proposed here, although less refined, is much more rapid and is believed to be adequate for the purpose.

SIZE FREQUENCY DISTRIBUTION OF MEANDER LOOPS

The cumulative frequency distribution of loop lengths and radii for each reach, plotted on log probability paper, is shown in Figure 7. The curves are fitted by eye, but the distributions are definitely log normal. Some scatter is to be expected at the extremes of the probability curve, where the size is based on very few loops. Because of the symmetry of the distributions, the median can be taken as an approximation to both the mode and the mean. In most studies involving the size of meanders (for example, Leopold and Wolman, 1960) a representative symmetrical pair of loops has apparently been selected for measurement; or else size has been based on the mean value of a specified number of loops, such as the 6 or 8 used by Carlston (1965). It seems likely that an experienced investigator can subjectively select, with some degree of reliability, the modal loop on most river reaches. However, the tacit assumption, doubtless based on experiments such as those of Friedkin (1945), has been that all meanders loops on a reach should be of the same size; and that the larger and smaller are aberrant and of no particular geomorphic or hydraulic significance. The viewpoint taken here is that all meander loops on a natural stream are a valid part of the loop population and have some significance, even if this is not known at the present. The size frequency distribution of meander loops on about 25 streams in eastern U.S. has been determined by the method described here and all have proved to be log normal, although a few are bimodal. A log normal size distribution of meander loops is probably characteristic of most natural streams.

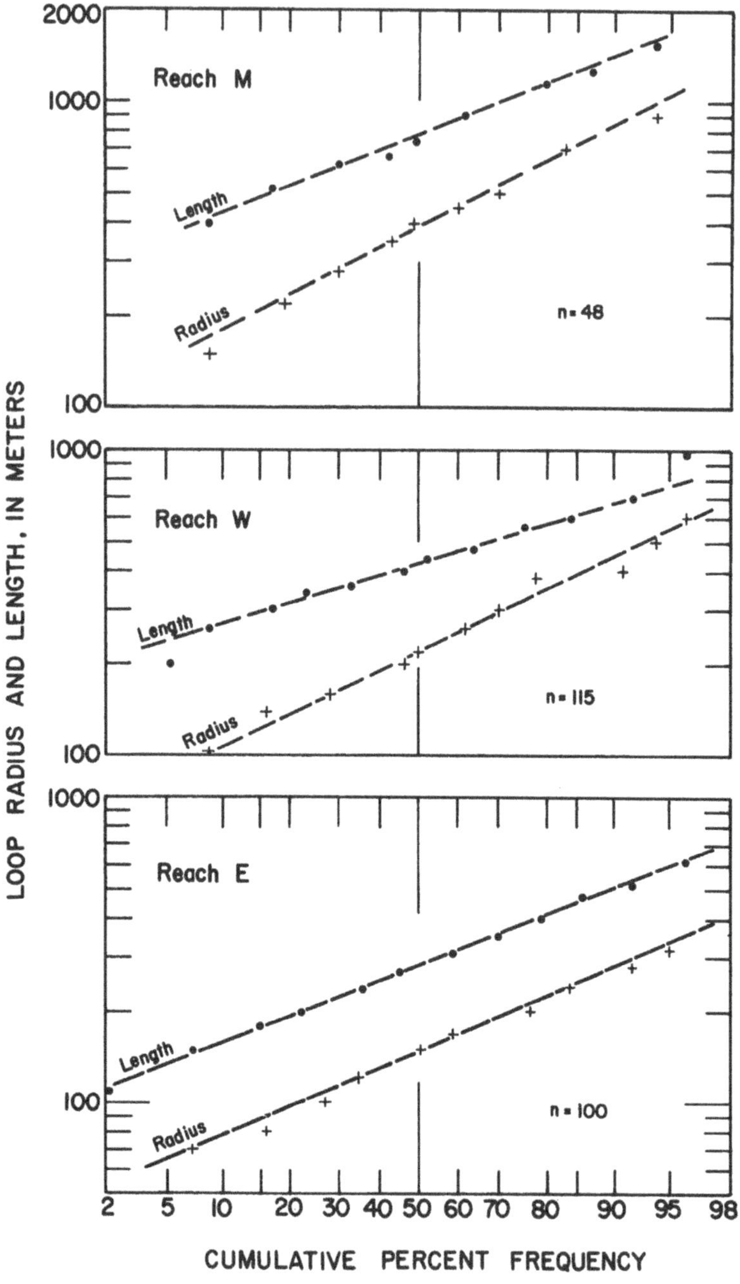

Figure 7 Cumulative frequency distribution of lengths and radii of simple loops, plotted on log probability paper.

The statistical treatment of log normal size distribution curves has been thoroughly explored by sedimentary petrologists, mostly with the hope of applying them toward genetic interpretation of the grain size distribution of sediments. The statistical aspects of this work are applicable to the size distribution curves of meander loops, and among the more useful references are Krumbein and Pettijohn (1938) and Friedman (1962). Statistical measures that have been shown to have some genetic significance for sediments (if not for sedimentary rocks) are standard deviation, skewness, and kurtosis. Preliminary investigation of the size distribution curves for meander loops indicates that skewness is typically near zero and that kurtosis values fall within a narrow range. Measures of dispersion are useful for descriptive purposes and may prove to have some genetic significance.

Of the many measures of dispersion that might be applied, an adaptation of the graphical measure proposed by Inman (1952, p. 130) seems most suitable because of its simplicity and its good correlation, for distributions of low dispersion, with log standard deviation. Although cast in the form of logs to the base 2 (Phi units), it has the following form for common logs:

$$\text{Log } D_g = 1/2 \ (\log P_{84} - \log P_{16}),$$

Where D_g is a graphical deviation measure that approximates the log standard deviation, and P_{84} and P_{16} are the percentiles one standard deviation on either side of the mean for a normal distribution. A simpler measure of dispersion that may be adequate is simply the ratio P_{84}/P_{16}, where the percentile measures are expressed in ordinary units rather than logs. Values for these measures as applied to radius are given in Table 2. Log D_g values indicate that the spread or dispersion between P_{84} and P_{16} for radius is about 1.5 X greater for Reach M than for Reach E. The P_{84}/P_{16} values indicate that the radius of larger meander loops is about 2.7 times greater than the radius for the smaller loops for Reach E and about 3.8 times greater for Reach M. For a sample of 25 meandering streams in eastern U.S., values of D_g for loop radius range approximately from 0.2 to 0.3 and values for P_{84}/P_{16} range from 2.5 to 3.9. These same measures can be applied to loop lengths, although length is probably a less reliable indicator of meander size than is radius.

The method of power spectral analysis, based on change in stream direction with stream distance, has been applied to rivers in New Guinea and Australia by Speight (for example, 1965) and to the White River by Chang and Toebes (1970). Of the many White River reaches analyzed by Chang and Toebes, three correspond approximately to the reaches described herein. The results may not be strictly comparable because the methods are quite different, but both methods produce a mean value of meander radius and meander wavelength.

Table 2 Meander dimensions obtained in this study as compared with those reported by Chang and Toebes (1970).

	This study			Chang and Toebes, 1970		
	Reach M	Reach W	Reach E	Reach M1	Reach W2	Reach E10
P84 radius, m	730	390	240	---	---	---
P16 radius, m	190	125	90	---	---	---
Log Dg	0.29	0.24	0.21	---	---	---
P84/P16	3.84	3.12	2.66	---	---	---
Mean radius (Rm), m	380	220	150	753	440	336
Mean wavelength (Lm), m	1520	840	580	3100	2460	2030
Lm/Rm	4.2	3.8	3.9	4.1	5.6	6.0
Mean wavelength of selected loops, m	1450	1140	920	---	---	---

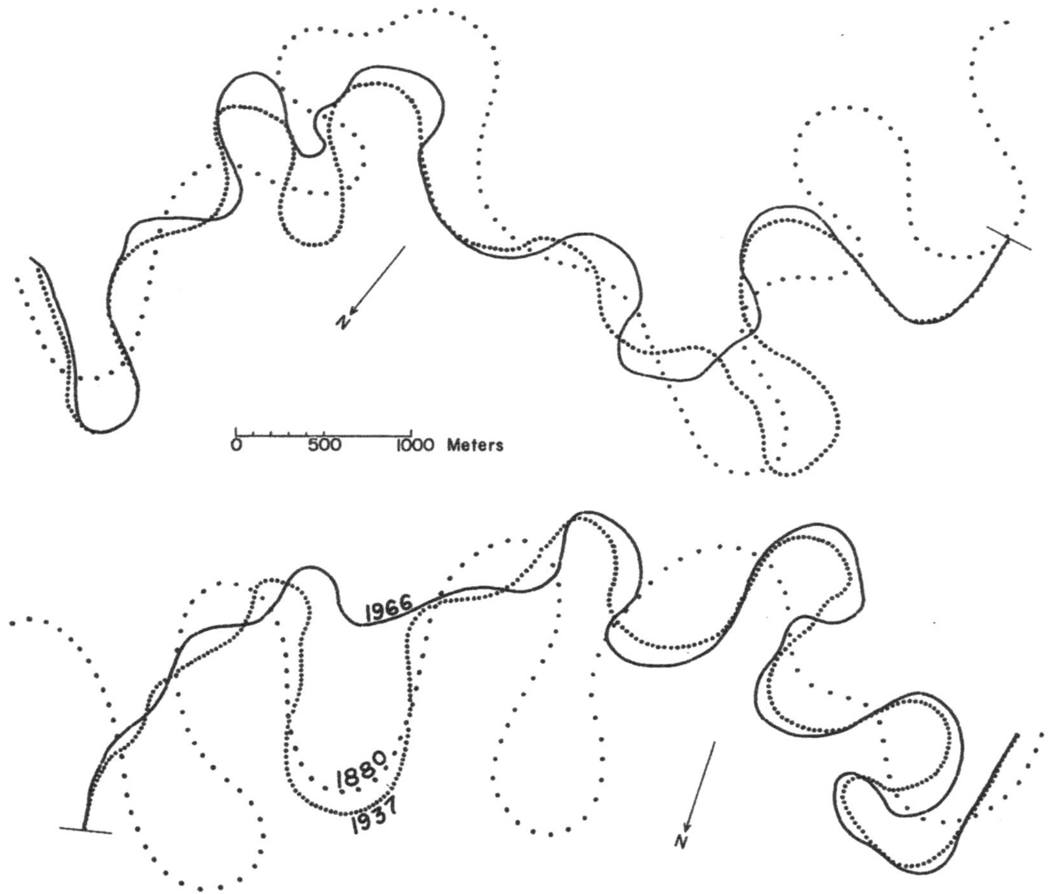

Figure 8 River centerlines for 1880, 1937, and 1966, West Fork White River upstream from Edwardsport.

In Table 2, results of the two methods are compared and also given for comparison are mean values based on four subjectively selected loop pairs that look representative and are fairly symmetrical. The mean values obtained by Chang and Toebes are much the higher, probably for the following reasons: 1) The spacing of stream distance points that was used by Chang and Toebes is 528 ft or 158.5 m, and this is too great to detect curves of small radius. About 30 percent of the loops on Reach E have radii less than 100 meters. 2) Compound meander loops are measured, as well as large scale inflections in the river course like the one in Figure 5. If these have any physical significance, it is not the same as for simple meander loops, which are characterized by a comparatively regular geometry and which function as units in the lateral migration of the river. Also, the multi-modal distribution of meander wavelengths reported both by Speight and by Chang and Toebes are probably due in part to compound meanders.

LATERAL EROSION AND THE MIGRATION OF MEANDER LOOPS

Reach W upstream from Edwardsport

A section of Reach W upstream from the town of Edwardsport is of particular interest because of the complexity of its pattern and the rapidity with which meanders have evolved and been cut off. In Figure 8, the river centerline dated about 1880 has been plotted from old maps (Illustrated Historical Atlas of Knox Co., Ind., 1880, D. J. Lake and Co., Philadelphia) and from oxbow lakes. The maps are not accurate planimetrically but they do help to establish the time that cutoffs occurred and they give the general configuration of the pattern. As shown by the 1880 centerline (and more strikingly, on aerial photographs by the configuration of oxbow lakes that were cut off prior to, and just after, 1880) the pattern of 1880 was more regular and more sinuous than subsequent patterns. Also, the earlier meander scrolls, which are particularly well shown on the 1937 photographs, are more closely and uniformly spaced than subsequent scrolls.

Clearing of the forests on the floodplain has probably resulted in a more rapid rate of meander growth and a less regular configuration of individual meanders. According to Ulrich and others (1943) most of the land in Knox County (whose boundaries are formed by Reaches W and M of this report) originally supported a very fine stand of timber, and it can be safely assumed that the bottomlands were completely forested. About 1815, the county was laid out with its present boundaries; and by 1880, 80 percent of the county was in farms. The bottomlands, including much of the river banks, were probably cleared of forests between 1815 and 1880. A similar chronology will apply to Jackson County, in which Reach E is located. At the date of the latest available aerial photographs (1966, 1968) forested land along the rivers is confined mainly to the inside of meander loops and the outside of loops is typically cleared to the very edge of the banks.

Although the clearing of forests has doubtless affected the changes in pattern with time, some inferences that may have general application to meander evolution can be made from Figure 8. 1) Very elongate loops, such as the one at top left, do not migrate very much in a direction perpendicular to their axes, and such migration probably occurs mostly in an earlier stage when a loop is less elongate. 2) A cutoff, such as the two in the diagram at top that occurred between 1937 and 1966, may result in a rapid rate of growth of adjoining meanders; but this effect does not extend very far upstream or downstream. 3) Compound meanders evolve from simple meanders, as indicated by several examples at the downstream end of the lower diagram.

Lateral erosion and loop evolution on Reach E

In Figure 9, the river centerlines of 1937 and 1968 for Reach E (divided into 9 subreaches) have been plotted from aerial photographs, and the relative positions of the two centerlines is estimated to be accurate within $\pm$ 15 meters. The area affected by the growth of each loop on Reach E was determined by tracing the area between the 1937 and 1968 banklines (not channel centerlines) on paper of uniform thickness, cutting out the traced areas, and weighing them on an analytical balance. Erosion along straight segments of the channel was negligible and, of a total of 100 loops, about 50 grew by a measurable amount.

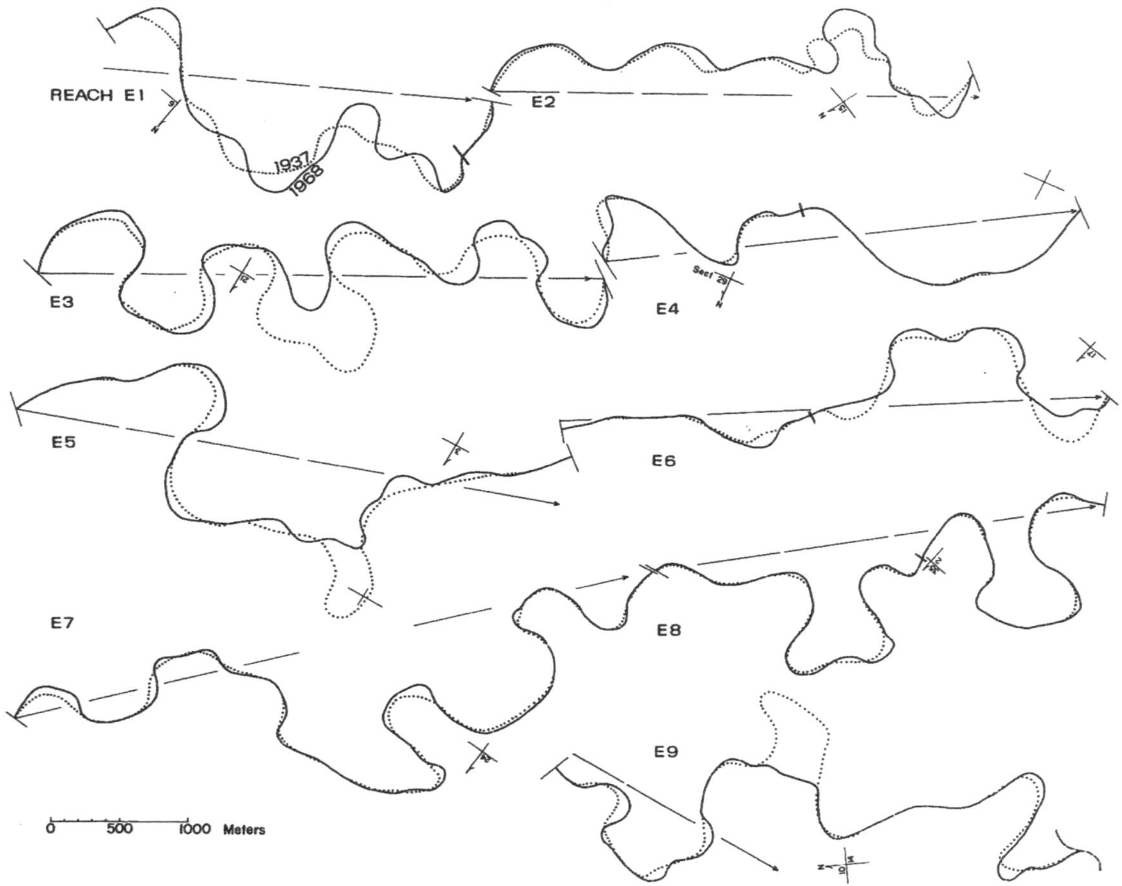

Figure 9 River centerlines for 1937 and 1968, Reach E on East Fork White River.

Total areas eroded and other pertinent data on the subreaches are given in Table 3. Since channel width remained constant, the eroded areas on the outside of loops were replaced by equivalent areas on point bars, but this may involve a change in land ownership and the land on point bars is of little immediate agricultural value. The sinuosity of some subreaches changed substantially during the period, but the total sinuosity of Reach E remained about constant. Erosion rates were much higher at loops in the upstream subreaches and no convincing reason for this could be assigned, either by field study or study of aerial photographs. Moreover, quite different erosion rates were found for loops that seem essentially similar in size, orientation, and bank resistance. Some tendency for rapid erosion at loops upstream and downstream from cutoffs (of which three occurred during 1937-1968) can be discerned, but many rapidly eroding loops are far from any cutoff. It is significant that the meander loops along this reach are not evolving either sequentially or simultaneously (as is implied by laboratory experiments on streams) but that growth has apparently occurred intermittently, first in one part of the reach and then in another.

Table 3 Sinuosity and areas of lateral erosion, subreaches of Reach E, 1937-1968

Subreach	Centerline length in 1968 (meters)	Valley length (meters)	Sinuosity 1937	Sinuosity 1968	Lateral erosion, 1937-1968 (acres)
E1	5,070	2,820	1.52	1.80	62.1
E2	5,020	3,470	1.30	1.43	49.4
E3	7,880	4,100	2.15	1.92	68.3
E4	4,770	3,460	1.35	1.37	7.0
E5	6,140	4,000	1.6	1.53	23.0
E6	5,020	3,970	1.29	1.27	17.9
E7	8,700	4,750	1.70	1.83	27.3
E8	7,380	3,280	2.14	2.25	14.8
E9	6,930	3,600	2.11	1.92	27.5
Total reach	56,910	33,450	1.68	1.70	297.3

Vertical aerial photographs, taken in 1937 and 1968, of Subreach E3 and part of E2 are reproduced in Figure 10. Discharge at the time of the 1937 photograph was about 1,240 cfs and at the time of the 1968 photograph, about 500 cfs. Distribution of vegetation shown in the 1968 photograph is typical of all three study reaches described herein.

The valley profile and river-surface profile of Reach E, based on 7 1/2' quadrangle maps having a contour interval of 10 feet, is shown in Figure 11. Each dashed line between the profiles connects a point on the valley flat to an immediately adjacent point on the river surface, such that the vertical interval between a pair of points represents bank height. Inasmuch as the slope of the valley flat is nearly constant, variations in slope of the river are due either to variations in sinuosity or to changes in bank height. The bank height at the downstream end of the reach is about 2.7 m (9 ft) as compared with 3.3 (11 ft) at the upstream end. Local variations in river sinuosity or behavior are sometimes attributable to corresponding variations in valley slope (Schumm and others, 1972), but this does not apply to Reach E.

To test the hypothesis that loops of a particular size are more stable than the rest, the area eroded by each loop between 1937 and 1968 on Reach E was plotted against loop radius (Figure 12). No correlation is apparent, and it is therefore concluded that the potential for erosion is about the same for loop ranging in radius from 100 to 300 meters. However, loops on the 1937 course that had a radius of 80 m or less were apparently unstable and were obliterated with time. See, for example, the loops downstream from "a" on the 1937 photograph, Figure 10.

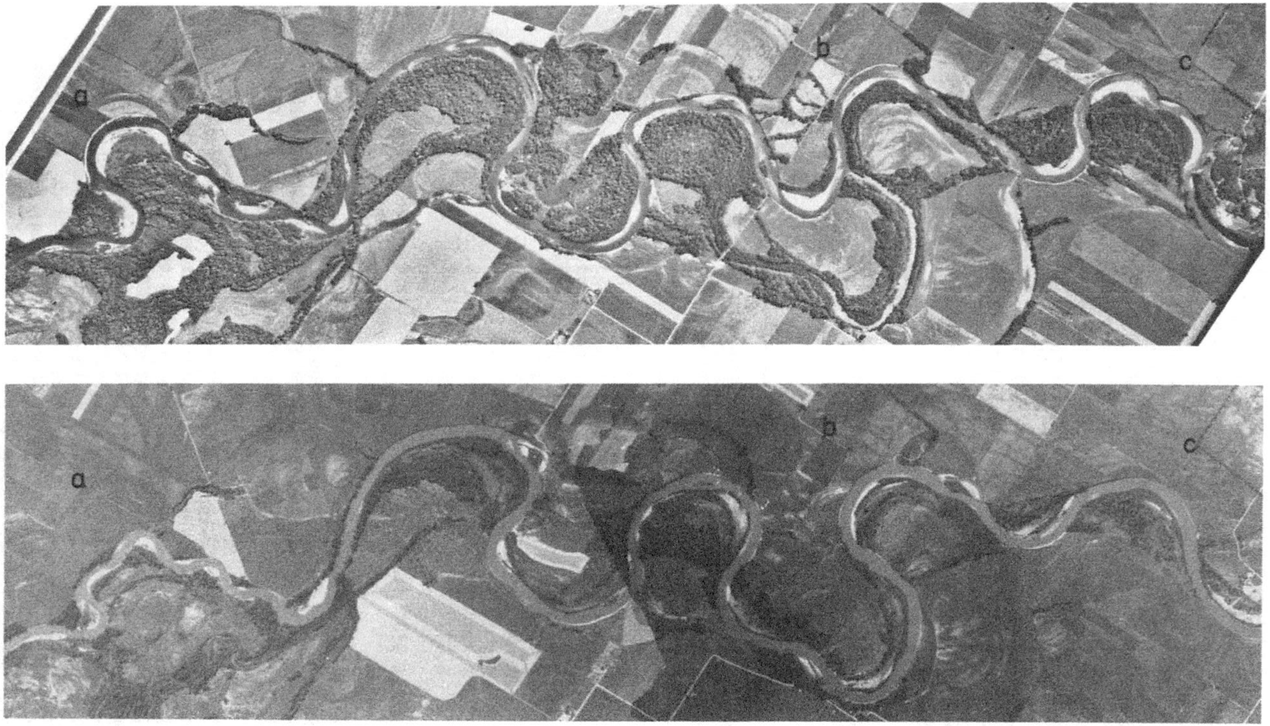

Figure 10 Evolution of meander loops on East Fork White River (Subreaches E2 and E3) as shown by comparison of vertical aerial photographs flown on 26 September 1968 (top) and 19 August 1937 (bottom). Downstream direction to right, north toward bottom. Corresponding letters are at identical ground positions, to serve as reference points. At point "a", a loop has migrated laterally about 120 meters. At "b", a simple loop has become compound.

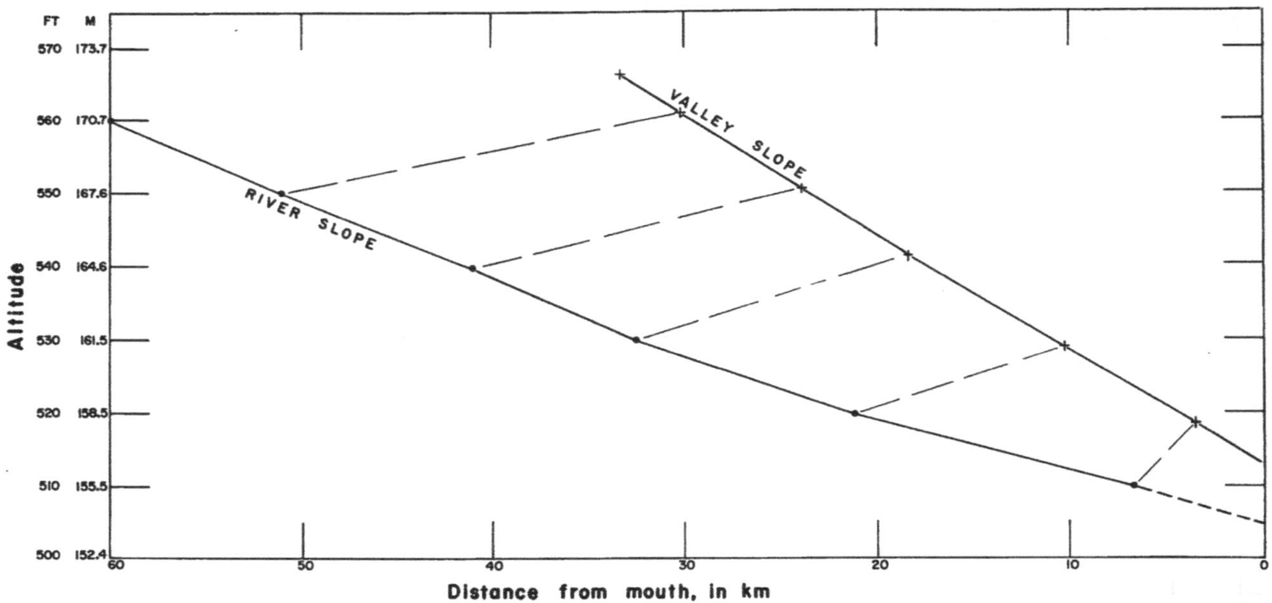

Figure 11 Longitudinal profile for valley and for river, Reach E.

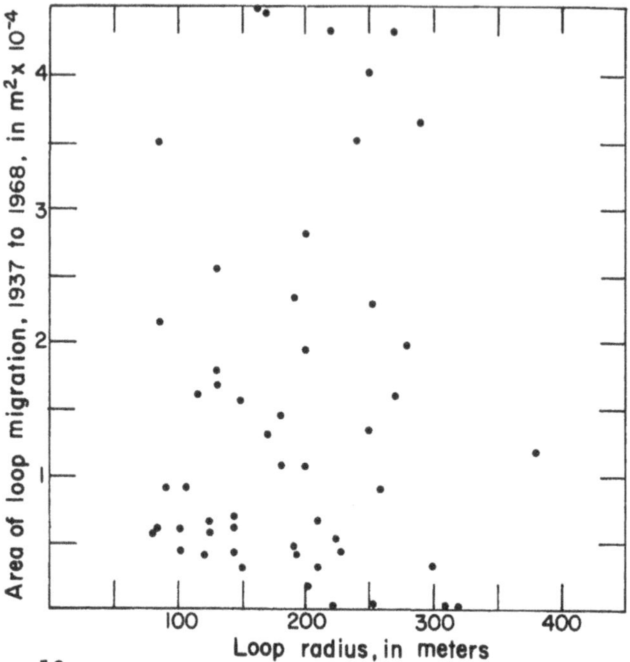

Figure 12
Scatter diagram for loop radius versus area eroded by migration of loop, 1937 - 1968, Reach E.

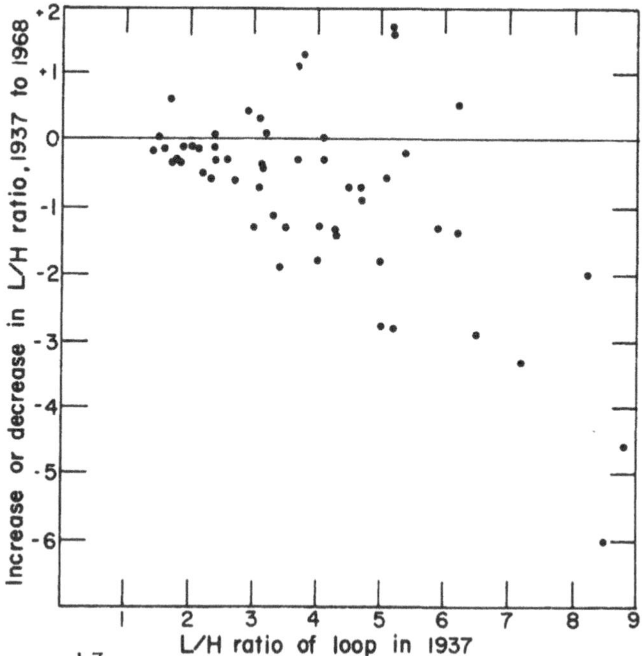

Figure 13
Scatter diagram for L/H ratio of loop versus change in ratio between 1937-1968, Reach E.

Meander loops on Reach E have grown sufficiently that a comparison of the properties of a loop in 1937 with its properties in 1968 may give some insight into loop evolution. In Figure 13, the L/H ratio of a loop in 1937 is plotted against the change in L/H ratio between 1937 and 1968. Most loops show a decrease in L/H ratio (become more elongate) and the flatter loops tend to show a greater decrease than the more elongate loops. This implies that a loop becomes more elongate as it evolves, and that the rate of elongation is more rapid in the earlier stages of evolution.

A study of meander scrolls within loops indicates that the radius of a loop tends to decrease during the earlier stages of its elongation, but this was not supported by a plot of change in radius against L/H ratio. Of 57 loops, 28 decreased in radius, 9 retained about the same radius, and 20 increased in radius during the period 1937-1968. All loops having an L/H ratio greater than 6 decreased in radius. If a large population of loops in an early stage of evolution was available, it is likely that a decrease in radius with evolution could be demonstrated.

Loop orientation and direction of loop migration on Reach E

The distribution of loop orientations according to azimuth for the 1968 centerline of Reach E is shown in Figure 14. Although most loops tend to be oriented approximately at a right angle to mean downstream direction, the orientations are diverse and many loops are oriented in an upstream direction.

The movement of loop centroids provides some insight into the migration of meander loops relative to mean downstream direction. In Figure 15, the direction of centroid movement for the period 1937-1968 is shown by arrows. Centroid movement has a downstream component for most loops but it is commonly a minor component. In general, the direction of centroid movement lies in a quadrant between the azimuth of the loop axis and a perpendicular to the loop axis that points in a downstream direction. The direction of centroid movement within this quadrant depends on the orientation of the loop axis, the configuration of adjoining loops, and on other factors that are difficult to anticipate, such as a lack of uniformity in bank resistance.

A comparison of centerlines of different dates, as in Figures 8 and 9, suggests that flat loops (high L/H ratio) tend toward downstream translation, or migration perpendicular to the loop axis; whereas more elongate loops tend to grow in the same direction as the loop axis. To test this hypothesis, L/H ratio was plotted against the angle of centroid movement relative to the loop axis (Figure 16). Although the correlation is not good, a general tendency for flatter loops to migrate at higher angles to the loop axis can be discerned.

Migration of elongate loops along the loop axis is supported by the observation that the later meander scrolls within elongate loops are concentric both on the upstream and downstream side of the loop, whereas the earlier scrolls have been truncated on the upstream side.

To examine the possible effect of loop orientation on migration rate, the azimuth of each migrating loop on Reach E was plotted against the area eroded between 1937 and 1968 (Figure 17). The results indicate that loops oriented in a downstream direction have the greatest potential for erosion.

ACKNOWLEDGMENTS

The work reported here was funded by a grant from the U.S. Army Research Office-Durham. Warren E. Grabau, of the U.S. Army Waterways Experiment Station in Vicksburg, suggested the use of loop centroid as a means of measuring the migration of meander loops and also provided much helpful discussion on the problems of measuring meanders.

REFERENCES

Carlston, C.W., 1965, The relation of free meander geometry to stream discharge and its geomorphic implications: Amer. Jour. Sci., v. 263, p. 864-885.

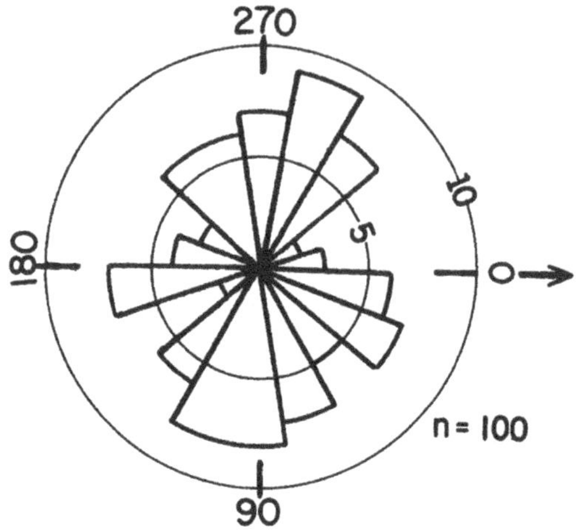

Figure 14
Distribution of loop orientations for the 1968 course of Reach E.

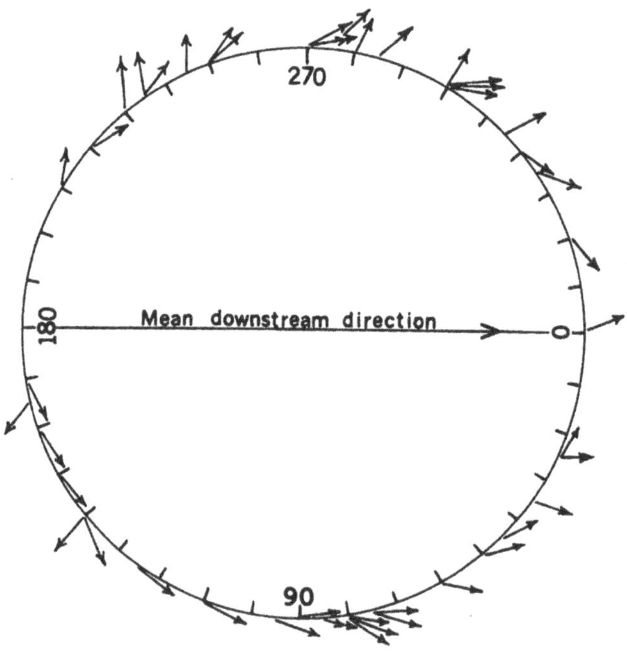

Figure 15
Direction of movement of loop centroids, 1937-1968, relative to mean downstream direction, Reach E.

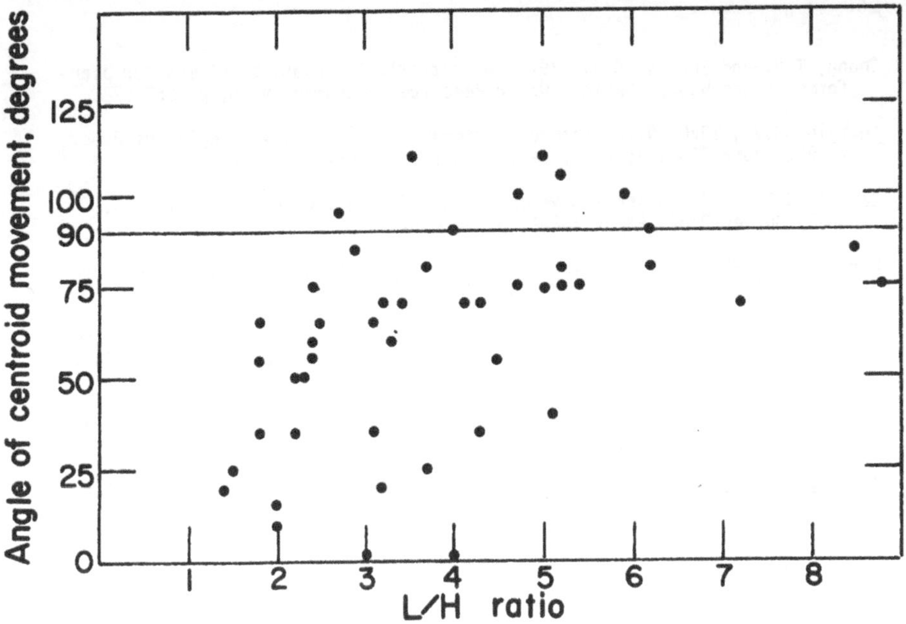

Figure 16
Scatter diagram for L/H ratio versus angular direction of ventroid movement relative to loop axis, Reach E.

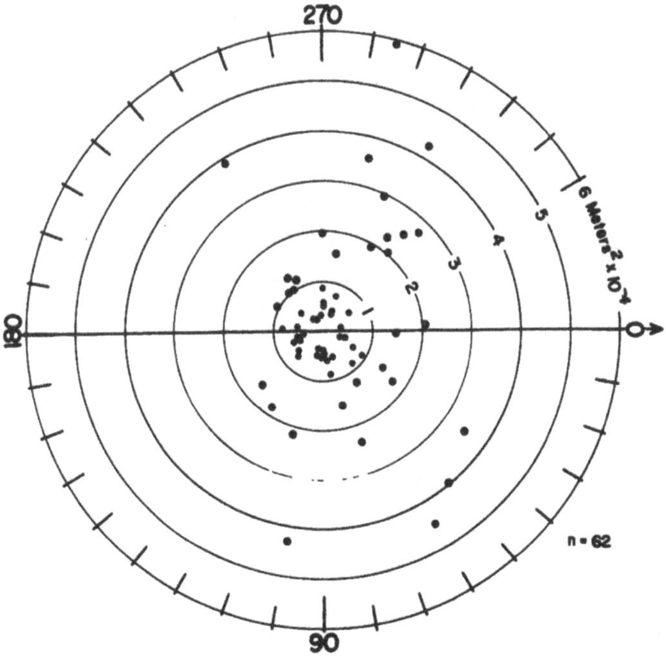

Figure 17
Scatter diagram for azimuth of loop axis in relation to area eroded by migration of loop, Reach E.

Chang, T.P. and Toebes, G.H., 1970, A statistical comparison of meander planforms in the Wabash Basin: Water Resources Research, v. 6, p. 557-578.

Chitale, S.V., 1970, River channel patterns: Am. Soc. Civil Engineers Proc., v. 96, paper 7038, Jour. Hydraulics Div., no. HY1, p. 201-221.

Daniel, J.F., 1971, Channel movement of meandering Indiana streams: U.S. Geol. Survey Prof. Paper 732-A, p. A1-A18.

Dury, G.H., 1961, Bankfull discharge--an example of its statistical relationships: Internat. Assoc. Hydrology Bull., v. 6, no. 3, p. 48-55.

Fisk, H.N., 1944, Geological investigation of the alluvial valley of the lower Mississippi River: Vicksburg, Miss., U.S. Army Corps of Engineers, Miss. River Comm., 78 p.

Friedkin, J.F., 1945, A laboratory study of the meandering of alluvial rivers: Vicksburg, Miss., U.S. Waterways Expt. Sta., 40 p.

Friedman, G.M., 1962, On sorting, sorting coefficients, and the lognormality of the grain-size distribution of sandstones: Jour. Geology, v. 70, p. 737-753.

Inman, D.L., 1952, Measures for describing the size distribution of sediments: Jour. Sed. Petrology, v. 22, p. 125-145.

Krumbein, W.C. and Pettijohn, F.J., 1938, Manual of sedimentary petrography: New York, D. Appleton-Century, 549 p.

Langbein, W.B., and Leopold, L.B., 1966, River meanders--theory of minimum variance: U.S. Geol. Survey Prof. Paper 422-H, p. H1-H15.

Leopold, L.B., Wolman, M.G., and Miller, J.P., 1964, Fluvial processes in geomorphology: San Francisco, W.H. Freeman Co., 522 p.

Leopold, L.B. and Wolman, M.G., 1960, River meanders: Geol. Soc. America Bull., v. 71, p. 769-794.

Rainwater, F.H., 1962, Stream composition of the conterminous United States: U.S. Geol. Survey Hydrol. Atlas 61.

Rohne, Jr., P.B., 1972, Low-flow characteristics of Indiana streams: U.S. Geol. Survey Open-file report, 322 p.

Schumm, S.A., Khan, H.R., Winkley, B.R., and Robbins, L.G., 1972, Variability of river patterns: Nature Physical Sci., v. 237, p. 75-76.

Sooky, A.A., 1964, The flow through a meander flood plain geometry: Ph.D. thesis, Purdue University, Lafayette, Indiana.

Speight, J.G., 1965, Meander spectra of the Angabunga River: Jour. Hydrology, v. 3, p. 1-15.

Thornbury, W.D., 1950, Glacial Sluiceways and lacustrine plains of southern Indiana: Indiana Geol. Survey, Bull. 4, 21 p.

Ulrich, H.P. and others, 1943, Soil survey of Knox County, Indiana: U.S. Dept. Agriculture, Series 1934, no. 23, 95 p.

Water Resources Council, Hydrology Committee, 1968, River mileage measurement: Bull. no. 14.

Wayne, W.J., 1963, Pleistocene formations in Indiana: Indiana Geol. Survey, Bull. 25, 85 p.

Wayne, W.J. and Zumberge, J.H., 1965, Pleistocene geology of Indiana and Michigan <u>in</u> The Quaternary of the United States: Princeton, N.J., Princeton Univ. Press, p. 63-84.

Winkley, B.R., 1972, <u>in</u> Shen, H.W., Ed., River mechanics, v. 1, chapt. 19: Ft. Collins, Colo., Water Resources Publ., 79 p.

9

AGGRADATION OF THE ALEXANDRA-NORTH SASKATCHEWAN RIVER, BANFF PARK, ALBERTA

D. G. Smith

ABSTRACT

Aggradation has been the dominant fluvial condition for the past 2450 years in the two meltwater rivers fed by the Columbia Icefield in Banff National Park. Evidence of past valley filling consists of buried volcanic ash, peat, wood and fluvial sediment found in drill holes. Present aggradation is demonstrated by the presence of buried lower tree trunks and direct measurement of sedimentation with stakes.

The most convincing evidence which supports the existence of aggradation was the buried layer of volcanic ash, which has been dated at 2450 years B.P. Used as a stratigraphic time marker, the profile of ash depth was plotted below the long profile of the present river surface. The difference between the two profiles represents the thickness of alluvium which has been depositing over the past 2450 years. The maximum depth to the ash layer was 25.5 feet.

Aggradation is directly related to one of two types of controls. Upstream control occurs where excessively high sediment loads deposit and form a wedge of sediment over time, thickest near the source and thinning downriver. With upstream control, the dominant sediment is coarse outwash deposited as gravel bars in shifting braided channels. Braids in outwash are typically steep, shallow and, in particular, unstable over time.

In contrast, downstream control, synonymous with base level control, causes a backwater reach, forming a wedge of sediment which thins in the upstream direction from the control point. Aggradation processes caused by downstream control consist of overbank flooding and deposition of fines onto levees, backswamps and floodplain-like surfaces. Distinct from braids in outwash, anastomosed channels which form in backwater reaches are deep, gently sloped channels and shift laterally at very slow rates. Stability of anastomosed channels is attributed to roots of grass and shrub vegetation, which provide a riprap-like protection to silt banks from fluvial erosion. While braided and anastomosed channels have similar patterns, they differ in process.

INTRODUCTION

Much attention in fluvial geomorphology over the last 20 years has been focused on the more common graded or equilibrium type of river. During this time, fluvial up-filling and down-cutting fluvial environments, aggrading and degrading river conditions, have received only modest observation. This neglect of aggradation *per se*, has probably been due to the scarcity of available evidence needed to determine not only the presence of aggradation but also to ascertain patterns and rates of valley-fill. Without such evidence field studies of aggradation are inconclusive. Short term man-induced aggradation can be determined by historical records as was demonstrated by Gilbert (1917) in his study of hydraulic mining debris in California. However, slow, long term aggradation, typical in nature, is very difficult to determine and study without the aid of some sort of buried stratigraphic time marker.

This paper is intended to examine the sediment, rates and processes of aggradation of 10 miles of the Alexandra-North Saskatchewan River system. Two aggrading conditions, caused by upstream and downstream controls, are in agreement with Mackin's (1948) model of aggradation. The relationships between each type of aggradation control and formation of discrete channel patterns, braided and anastomosed, will also be discussed.

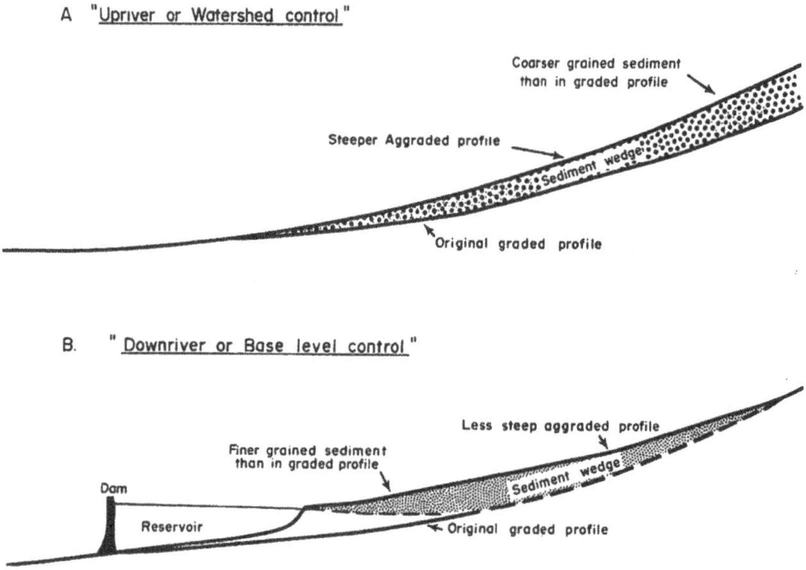

Figure I Idealized sketches of aggradation control types as postulated by Mackin (1948).
A. Upstream control
B. Downstream control

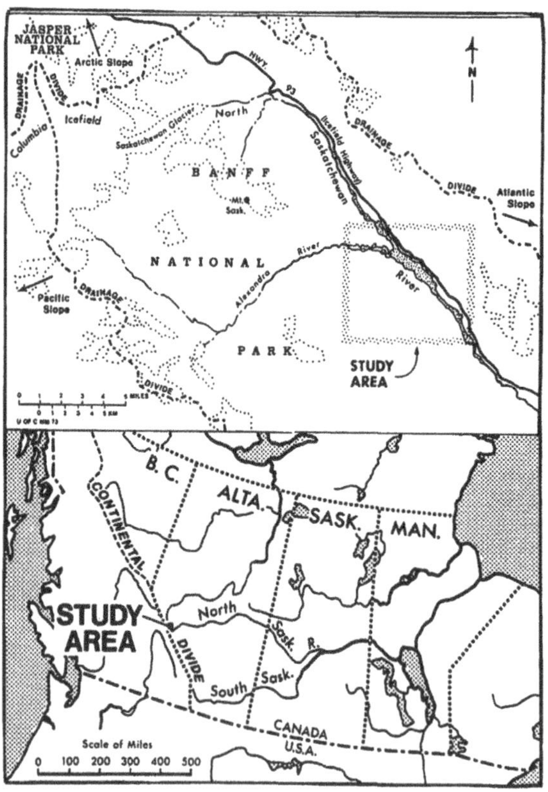

Figure 2 Index map of the study area and adjacent Columbia Icefield in western Canada.

Concept of Aggradation

The concept of aggradation has been best described by Mackin (1948), in which he proposed theoretical controls and criteria. Upstream control occurs where large volumes of coarse sediment are deposited as a wedge thinning downstream from the sediment source area. The classic example of such aggradation control comes from descriptions by Waggoner (1936, in discussion of Stevens, p. 271) of hydraulic mining debris on the western slopes of the Sierra Nevada, California, between 1855 and 1884, cited by Mackin (1948, p. 493).

> ...the deposit...was 20 miles long, had a maximum width of three miles, covering 16,000 acres and containing 600,000,000 yds^3. It was 20 feet deep at the river's mouth, 35 feet deep at the edge of the foothills, and 80 feet deep 5 miles higher up on the Yuba River. The grade of the original bed was 5 feet per mile. After the fill was made, the grade per mile was 2 1/2 feet at the mouth, 10 feet at the middle zone and 20 feet on the upper reaches.

From this description it appears also that grade steepens with continued aggradation.

Downstream control or rise in base level is described by Mackin (1948) to be a second cause of aggradation. This control theoretically should form a wedge of sediment thickest at the control point and thinning upstream (Fig. 1). The most dramatic example of downstream aggradation control was that in which the valley bottom was raised 13 feet near the inlet to Elephant Butte Reservoir on the Rio Grande and about 7 feet just 10 miles upriver, reported by Gottschalk (1964). Under such conditions grade should decrease with aggradation.

In summary, river aggradation occurs when the volume of sediment transported into a reach of river exceeds the amount transported out of the reach. The net result is a continuous rise in elevation and change in slope, either increasing or decreasing, of the river bottomland over a period of time.

Regional Environment

The Alexandra and North Saskatchewan River system is located southeast of the Columbia Icefield in the northwest part of Banff National Park, Alberta (Fig. 2). Both rivers flow within steep-sided glacially deepened valleys through rugged mountains. Along the valley sides high ice-covered mountain peaks rise to 10,900 feet (Mount Saskatchewan), 6,200 feet above the valley floor at 4,700 feet elevation.

All materials on the valley floor consist of fluvial sediments, while adjacent hillslopes are mantled with till and colluvium. Both local bedrock and river sediment are comprised largely of sedimentary rocks, seventy percent limestone, with dolomite, shale and quartzite making up the remainder.

Straddling the continental divide, the Columbia Icefield is the major source area for water in the Alexandra and North Saskatchewan Rivers. As a result river discharge in summer varies with weather conditions over the icefield. The icefield covers 120 mi^2 with about 17 percent of the ice surface extending into the Alexandra and North Saskatchewan watersheds.

The largest flow discharge in two summers (1970-71) at a temporary gauging station several miles downriver from the study area was measured at 5400 cfs. Bankfull discharge was estimated to be 5800 cfs. Of the total discharge measured at various times on both rivers above the confluence, approximately 60 percent of the flow was contributed by the Alexandra with the upper North Saskatchewan making up the balance. In short, the Saskatchewan is a tributary of the Alexandra at the confluence, but transports an estimated 95 percent of the valley gravel fill.

Figure 3 View looking downriver of backswamps, levees and islands of silt. Saskatchewan outwash enters from the left side while the Alexandra backwater is located in the foreground.

RIVER AGGRADATION

Evidence of River Aggradation

Evidence of recent and present deposition in the study reach is provided by: (1) composition in cores from drill holes especially the occurrence of volcanic ash plus datable organic material; (2) buried lower tree trunks; and (3) measurement of present sedimentation rates by means of stakes.

The terms recent and temporary in this discussion will be used to avoid confusion with ancient geologic deposits. Recent deposits are those which have been deposited during recent geologic time. Temporary deposits are the upper several feet of outwash and channel sediments which are temporarily at rest while waiting for the next flow to rework and transport them further downriver.

Drilling Method into Recent Sediments

Drilling was done by means of a hand-powered soil auger with 28 feet of extensions. The drill bit had a screw-shaped tip with a core chamber capacity of 8 inches in length by 1 1/2 inch in diameter. Due to this small capacity the auger had to be withdrawn from the hole every eight inches to remove sediment. Drilling was usually stopped at 24 feet which was the maximum depth at which the author could efficiently operate without breaking either his back or extensions while lifting and withdrawing the auger from a hole.

The reach of river sediment drilled extended 3.5 miles up the Alexandra and 5.5 miles down the North Saskatchewan River from the river confluence. Locations of drill sites were attempted at approximately 1/2-mile intervals downvalley and where possible, 3 or 4 holes were drilled at various sites across the valley floor to provide a check on subsurface sediment type. In a 1.5-mile reach downriver from the confluence, drilling was restricted to 1 hole at 1/2-mile intervals due to the scarcity of silt islands. Several proposed sites were not drilled as deposits of only gravel were present, particularly where the North Saskatchewan River and Rampart Creek tributaries join the main valley. Coarse sand or gravel was encountered at depth in about 30 percent of the attempted holes, which were abandoned if the sediment amounted to more than a lens or if sand sluffing occurred. All successful drilling attempts were in silt islands (Fig. 3) and backswamps of floodplain silt deposits, usually extending to depth greater than 20 feet.

Volcanic Ash, Clastic and Organic Recent Sediments

The most convincing evidence which supports the occurrence of aggradation was a buried layer of volcanic ash. Three layers of recent ash have been found nearby at Saskatchewan Crossing and Sunwapta Pass, buried in loess and bog deposits. Determination of the source area, distribution and dating by carbon-14 of material adjacent to the ash layers has been done by Westgate and Dreimanis (1967) and Westgate, et al (1970). The oldest and thickest ash is termed Mazama ash from Mount Mazama, Crater Lake, Oregon, and dated at about 6600 years B.P. Next is the St. Helens Y ash from Mount St. Helens in southwestern Washington, from an eruption which took place some time between 3000 and 3500 years B.P. The most recent, Bridge River ash, came from an undiscovered vent in the Coastal Range about 100 miles north of Vancouver, near Lillooet, B.C. This ash is dated at about 2450 years B.P. and was found in about 70 percent of all holes drilled in the study reach.

The character and distribution of the Bridge River ash has been discussed by Nasmith, et al (1967). In the study area the ash layer averages about 1/4-inch thick; extreme thicknesses of 1 inch were measured. The ash is white colored, has coarse silt-sized particles, with a few small shards of glass. Besides stratigraphic position, several criteria were used to differentiate Bridge River from St. Helens and Mazama ash. Mazama ash is easily identified by its thickness (3-6 inches) and many, large glass shards. St. Helens Y ash

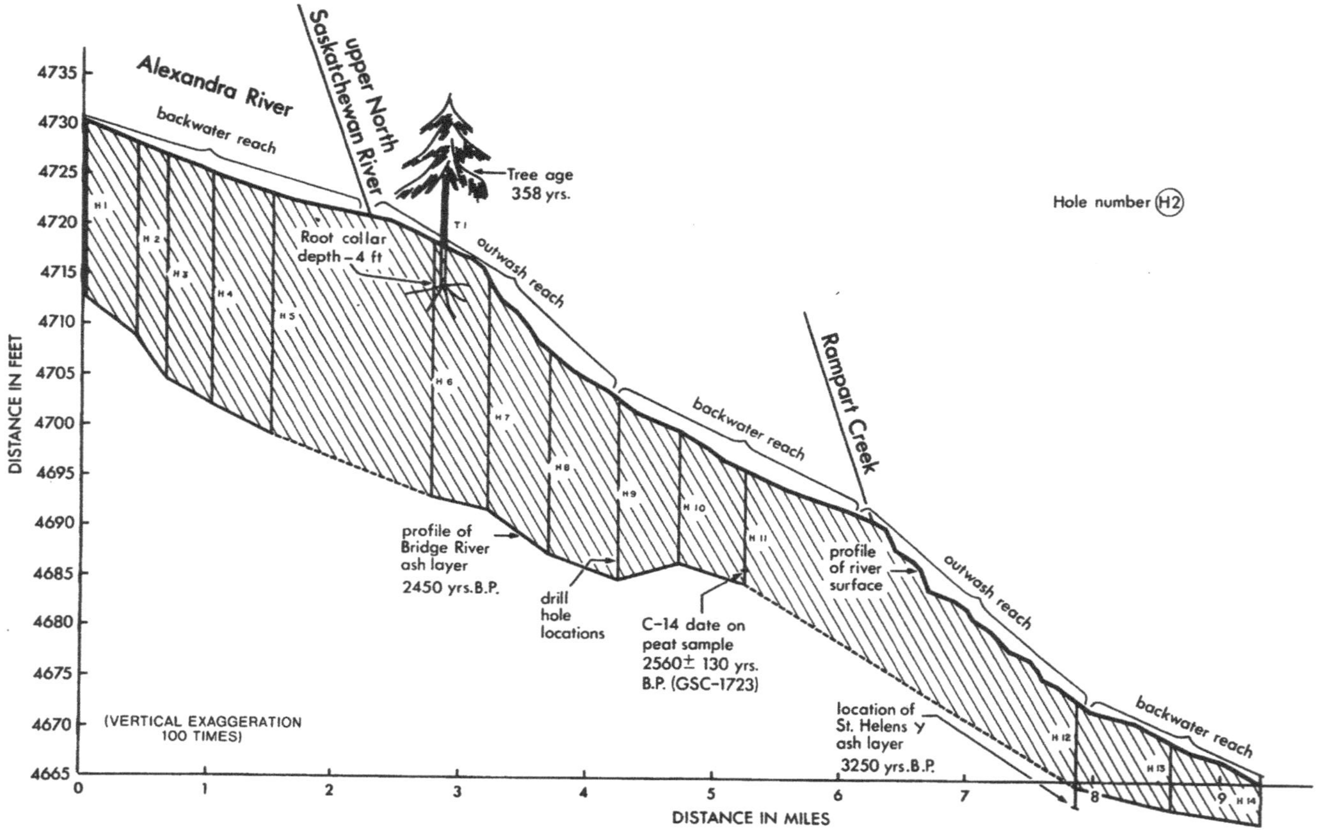

Figure 4   Graph showing the long profile of the Alexandra-North Saskatchewan River surface and buried layer of Bridge River volcanic ash. Note the wedge patterns of aggradation sediment.

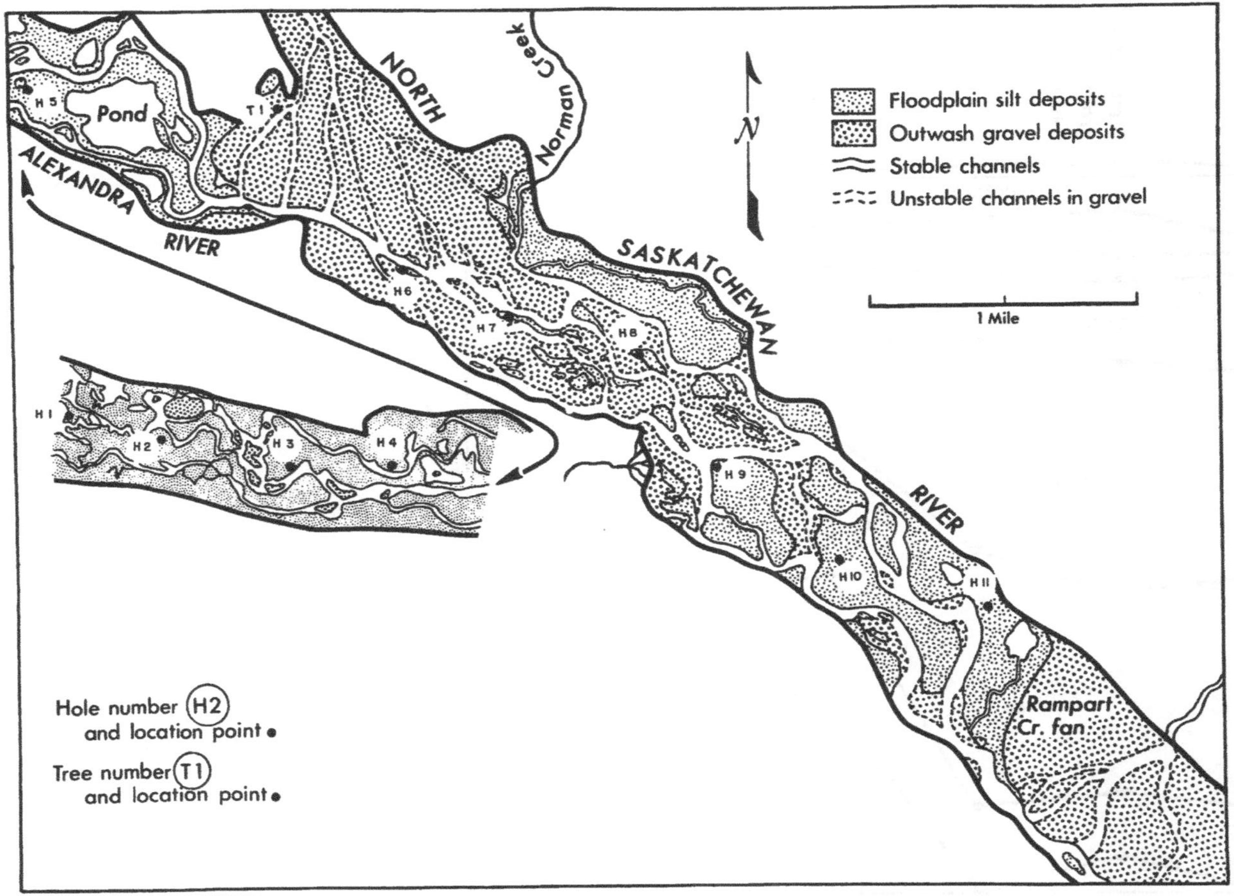

Figure 5 Map showing locations of drill holes, surficial fluvial deposits and location of stable and unstable channels.

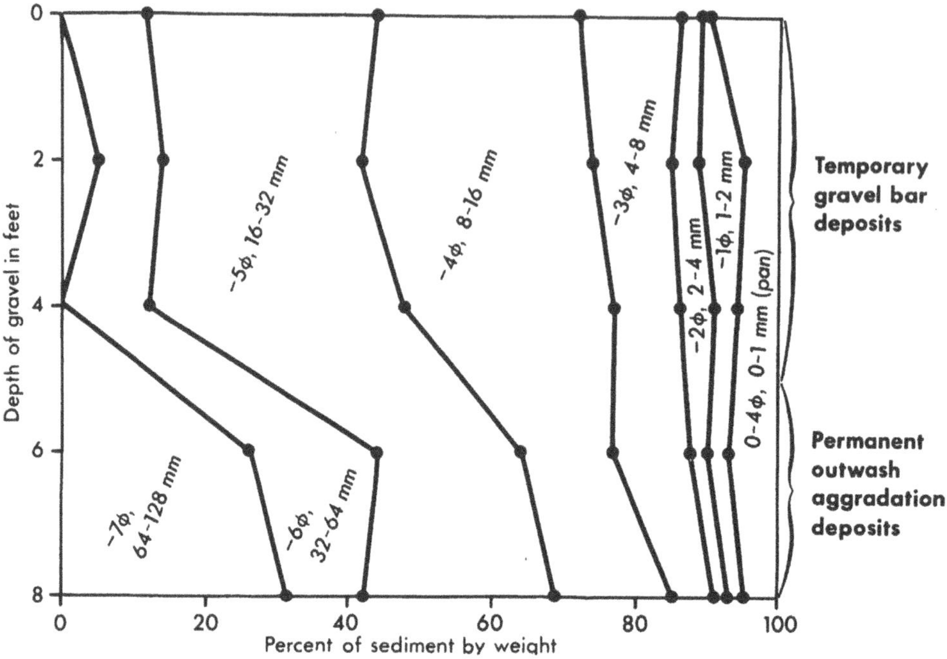

Figure 6 Distribution of gravel size classes from a pit dug 8 feet into outwash gravel.

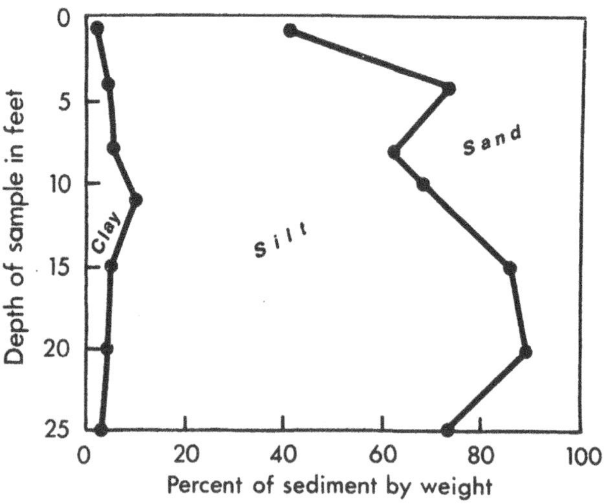

Figure 7 Percentages of clay, silt and sand with depth from one core taken from a silt island 1/2-mile downriver from the Alexandra-Saskatchewan River confluence.

is characterized by its thin layer, between 1/8 and 1/16 inch thick and the very scarce number of distinguishable glass shards. All field samples were compared against index slides of Bridge River ash under a microscope for identification.

Using the layer of Bridge River ash as a stratigraphic time marker, depth to ash was plotted below the long profile of the present river surface (Fig. 4). Locations of drill holes shown in Figure 4 are mapped in Figure 5. The difference between the two profiles represents the thickness of alluvium which has been deposited over the past 2450 years. The maximum depth to the ash layer was measured at 25.5 feet near the confluence of the Alexandra and Saskatchewan Rivers.

Floodplain silts down to depths of 25 feet also indicated the occurrence of aggradation. White it could be argued that the presence of such deposits could be from lake deposition, it would be difficult to explain presence of accompanying peat layers which are frequently present at 2-4 foot intervals. These changes in sediment facies are probably related to changes in paleohydrologic regime, causing periods of vegetation production in backswamps followed by sedimentation from overbank flooding.

One peat sample buried 10 feet beneath the surface and 1.5 feet above the Bridge River ash layer (Fig. 4) was dated at 2560+130 years B.P. the Bridge River ash layer was dated by carbon-14 method (GSC - 1723) by the Radiocarbon Laboratory of the Geological Survey of Canada. Extrapolating from the buried ash gives an estimated date for the peat at 2080 years B.P. Contamination of the carbon-14 sample was possible as it probably has been continuously saturated since its burial. The date is regarded by the author to be in agreement within acceptable limits.

Besides peat, buried wood was frequently penetrated while drilling. The wood was probably drifted onto floodplain backswamps and subsequently buried. This process was observed in the field.

Buried Tree Evidence

The buried lower trunks of standing trees in both outwash gravel and floodplain fines verifies the fact that present aggradation is active in the river reach studied. Many old trees which were dead had their roots well below the water table, death apparently resulting from "root drowning" caused by valley filling and subsequent rise of the water table.

Pits were dug adjacent to trunks of trees down to the root collars (the lowest part of the tree where roots join the trunk) of trees suspected of being partially buried. The distance between the buried collar and the ground surface is believed to represent the depth of sedimentation since tree germination. Pit sites were selected subjectively as distribution and incidence of frost heaving of trees varied throughout the area. Trees growing in clayey-silt soil were rejected as it was found that frost heaving was common in such sediments, which restricted sampling to tree sites in sand and gravel. Only western white spruce (*Picea glauca*) trees were selected; since the only other tree specie growing on the alluvial soils was black spruce (*Picea marinana*), which has a multi-level root collar, unacceptable for determination and measurement of sedimentation.

The range of depths to root collars and ages of trees sampled varied from 0.9 to 4.0 feet and 80 to 358 years respectively. An increment borer was used to determine tree ages. One tree 358 years old had its root collar buried 4 feet beneath the surface. Aggradation rates from buried tree data will be discussed later.

Measurement of Sedimentation with Stakes

Present aggradation was also measured with arrays of stakes on islands of floodplain silt selected at one-mile intervals downvalley from the confluence of the Alexandra and Saskatchewan Rivers. Six steel rods were driven two feet into the ground at each of ten plot sites at the beginning of each field season. Measurements were taken between the tops of each rod

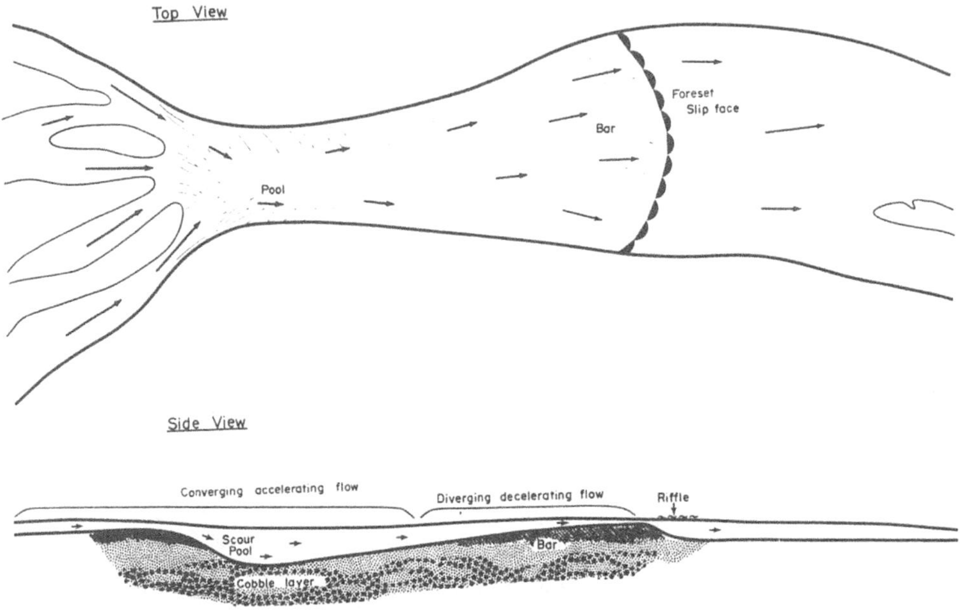

Figure 8 Idealized sketch showing the frequently observed pool and bar sequence in outwash gravel during high discharges.

Figure 9 View looking downriver of the North Saskatchewan outwash during high discharge. The high concentration of suspended load prevents direct observation of the pool and bar process. The broken line marks the forest slope of an active transverse gravel bar.

and ground surface at the beginning and end of two summer field sessions. Deposition occurred on the same three plot sites both summers. Maximum deposition at one rod measured 6.5 inches over a one-year period. However, deposition over the two years in the three plots averaged about 1.5 inches. The plots which showed deposition were the three sites immediately downriver from the Alexandra-Saskatchewan confluence, suggesting that aggradation is presently active in this vicinity. The other seven plots had no deposition.

Rates of Aggradation

Average rates of fluvial aggradation, during the last 2500 years were determined from depths of drill holes in floodplain silts down to the buried Bridge River ash layer (dated at 2450 years B.P.), which was used as a time marker. Determination of specific rates of sedimentation is impossible without data on the frequency and magnitude of fluvial processes, and hence only average rates are considered.

The thickness of sediment above the volcanic ash layer shown in Figure 4 varies considerably along the axis of the valley floor. Thus aggradation rates vary from 1 foot in 98 years to 1 foot in 480 years or when converted to feet per hundred years would change to 1.0 ft./100 years and 0.17 ft./100 years. These two rates represent the most extreme values and were taken from the confluence of the Alexandra-Saskatchewan Rivers and the extreme downriver part of the study area, respectively. The maximum and minimum rates converted to volume would be 280,000 ft.^{3}/mi.^{2}/yr. and 47,500 ft.^{3}/mi.^{2}/yr. When converted to weight (assuming dry sediment weight at 100 lbs./ft.^{3}), maximum and minimum aggradation rates would be 15,000 tons/mi.^{2}/yr. and 2,600 tons/mi.^{2}/yr. Theoretically, rates of deposition should be greatest up the Saskatchewan River from the Alexandra confluence, assuming a thickening wedge of sediment upriver. However, this cannot be confirmed as conditions were unfavorable for the formation and/or preservation of silt islands to enable drilling.

Aggradation rates were also determined from buried trees, as mentioned earlier. The oldest and deepest buried tree found near the Alexandra-Saskatchewan confluence was 358 years old and buried 47 inches down to the root collar. This suggests a rate of burial or deposition of 1.0 ft./93 years. This compares favorably with the most rapid rate determined from buried ash of 1.0 ft./98 years.

AGGRADATION DEPOSITS

Outwash gravel and floodplain silt are the two dominant fluvial sedimentary facies recognized in the study area. These braided river facies will be discussed only in general terms as little new information was gleaned from measurements of textural characteristics.

Outwash Gravel

Pits were dug to assess possible vertical trends of particle size-distribution of outwash and channel gravel. Sites were selected as close as possible to the gravel bar sites at one-mile intervals. Fifty-pound samples were obtained at 2-foot intervals down to 2 feet below the water table, where further excavation was impossible. One pit located one mile up the Saskatchewan River from the Alexandra confluence was 8 feet deep, but most were between 4 and 6 feet deep.

Results from the 8-foot pit in Figure 6 show a fining-upward sequence, which is interpreted as demonstrating two different deposits in outwash aggradation. The upper five feet of gravel is interpreted as being temporary gravel bar deposits, which will be subsequently reworked with future discharge of high flow and shifts of channels. Little variation in size distribution of gravel at the surface, or at two- and four-foot depths suggests that these sediments are from the same gravel bar and result from the same process of deposition.

Below the five-foot level, sediments are generally much coarser with the largest cobbles clustered in horizontal lenses. These clusters of coarse sediment are interpreted as representing cobble pavements which formed on the beds of scour pools during active channel braiding. During active braiding, cobbles are "sorted out" of reworked outwash braid bars and swept into scour pools where they come to rest because the tractive force is only sufficient to transport the finer sediments onto the next bar forming downriver. Thus, the coarser fraction of sediment is permanently deposited in the sequence of Recent aggradation outwash. Discussion of the pool and bar sequence will follow later.

The coarser sediments below the five-foot level are interpreted as representing the uppermost Recent aggradation deposits of a coarsening-upwards sequence, characteristic of upstream aggradation control as postulated by Mackin (1948). Costello and Walker (1972) described a similar sequence in gravel pit exposures of Pleistocene deposits of the Credit River in southern Ontario. Several gravel pits in terraces of Pleistocene outwash near Canmore, Alberta, studied by the author, show the same coarsening-upward trend of size distribution as is inferred in the Saskatchewan outwash. This coarsening trend is considered to be caused by upstream control resulting in a steepening channel slope over time which increases fluvial competency for sediment transport.

Floodplain Sediments

Floodplain facies consist entirely of silt-dominated, fine sediments deposited by overbank flooding processes. These facies occur primarily in backwater reaches. However, 1.5 miles downriver from the Alexandra-Saskatchewan confluence are a series of silt islands which tend to become more numerous and larger downriver, eventually joining the backwater reach. In the two backwater reaches studied an estimated 90 percent of the deposition was observed to have occurred from overbank flows, the remaining 10 percent consisted of channel deposits of gravel.

Silt was the dominant grain size of clastic sediment from drill holes, averaging 65 percent by weight. Figure 7 shows the percentage by weight of clay, silt and sand from a typical hole 25 feet deep, located 1/2 mile downstream from the confluence of the Alexandra and North Saskatchewan Rivers.

In 1948, Mackin postulated that a fining-upward sequence associated with a decreasing channel slope, over time, would occur in reaches caused by downstream aggradation control. An analysis of sediment from bore holes did not confirm this theory, perhaps due to the fact that particle sizes from overbank flooding generally remain constant regardless of channel slope. If a core could have been taken of channel gravels, it is very likely that a fining-upward trend would result since channel slope will certainly determine tractive force and therefore sediment size.

PROCESSES OF AGGRADATION

Pool and Bar Formation

The most frequent process observed in the braided Saskatchewan outwash was the pool and bar process as sketched in Figure 8. The description of the formation and destruction of the pool and bar sequence is based on visual observation including a few measurements of gravel bar development with six-foot steel rod stakes and probing with a 10-foot piece of conduit pipe of scour pools every few hours over a 3 to 5 day period. The observed pools and bars were located between 0 and 3/4 mile up the Saskatchewan River from the Alexandra confluence. Formation and destruction will be described as two phases of development.

In the North Saskatchewan, the pool and bar sequence is most active during peak flow discharges. Before pools begin to form, many abandoned channels must be filled with substantial discharge. When several such channels happen to converge towards a central focus, the stage is set for the process to begin.

The phase of pool scour begins with continued daily increases of discharge which scour the beds of channels near the point of the converging flows. Maximum depth in pools ranges from 4 to 10 feet. Depths of these pools are deceiving and dangerous to the neophyte doing field work. Downstream from the eroded pools, sediment is transported and deposited in the form of gravel bars and sheets.

The bar phase of development in the Saskatchewan outwash occurred most frequently in the form of transverse type bars. These bars are very difficult to locate during their constructional phase, as often the entire outwash is under water in flood (Fig. 9). Direct observation of the bar forming process is very difficult because of the high volume of suspended sediment load of glacial flour, which turns the water white to light brown. However, with some experience one can locate such bars by the riffle pattern on the water surface which occurs just downstream from the foreset slope of a developing bar. A riffle can be seen in the lower center area of Figure 9, marking the location of a developing transverse bar. The depth of water flowing over such active bars will vary from 1 to 2 feet. The foreset slope (tan) of the bar varies from 1 to 2 feet in height.

Some gravel sheets or thin accumulations of gravel 1 to 4 inches thick and similar in shape to transverse bars have been observed following a period of high discharge. Such sheets are believed to form in the downstream direction by foreset slope accretion; however, this was not checked by observation or measurement.

The destructional phase of this process will occur by either a decrease of stream discharge or a shift in position of the channels upstream, cutting off the discharge necessary for continued scour and bar enlargement. When discharge decreases the pool begins to partially fill with cobbles and pebbles while flow over the transverse bar becomes very shallow. Such shallow flow begins to converge into preferred depressions across the bar surface and subsequently erodes several mini-channels into the bar gravels. While the water stage continues falling, the mini-channels eventually dissect the distal portion of the bar into several islands. The complete cycle of pool and bar formation and destruction can occur in several days. Only when discharge and suspended load decrease can the remnants of the pool and bar be observed.

In analyzing the sedimentological structures of braid bars in the Kicking Horse River in nearby Yoho Park, Smith (1973) found no complete gravel bars or "unit bars." The same situation appears to be the case in the Saskatchewan outwash where complete bar forms are scarce. The cobble lenses found in the outwash are believed to have formed in pools near the peak of their formation phase. Thus, cobble lenses, pool filling and parts of dissected bars are suggested to be the origin of Recent aggradation deposits.

Overbank Flooding

Overbank flooding was observed to be the major process of aggradation in the backwater reach, while point bar accretion and channel bed deposition play minor roles. Numerous holes, bored into backwater deposits penetrated thick accumulations (up to 25 feet) of fine sediment, which are interpreted as overbank flood deposits. Active deposition from overbank flows is rather common and can be observed annually on various parts of backwater reaches and islands.

Point bar deposits were surprisingly rare in backwater reaches. Occasionally such deposits were anticipated when drilling near a present active meander. The infrequent occurrence of point bar deposits is attributed to the fact that channels migrate laterally very slowly during aggradation.

BRAIDED AND ANASTOMOSED CHANNEL PATTERNS

Definition and Terminology (of Braided Channels)

The A.G.I. Dictionary of Geological Terms (1962, p. 61) states that "a braided stream is one flowing in several dividing and reuniting channels

Figure 10 Braided and anastomosed channels located in the middle and lower parts of the photo, respectively. View looking upriver between Rampart Creek fan and confluence of Alexandra and North Saskatchewan Rivers.

resembling the strands of a braid, the cause of division being the obstruction by sediment deposition by the stream." In the past, braided channels have been reported as being synonymous with anastomosing channels (A.G.I. Glossary, 1960, p. 10; Fairbridge, 1968; and Leopold and Wolman, 1957). In 1968 Schumm reported that in Australia anastomosing has a somewhat different meaning and refers to low gradient, stable channels which transport flood waters and aggrade slowly from small suspended loads. If channel positions are stable and the reach aggrading, suspended load is presumably deposited by overbank processes. Mollard (1973) further indicated that anastomosing river paterrns have stable, wooded islands surrounded by channel meanders. Both Schumm and Mollard imply that braiding and anastomosing are similar in pattern but differ in process.

In this paper braided channels will be the fluvial terminology used for multi-channel patterns in outwash gravel. Braids in the Saskatchewan outwash are typical of most braided channel patterns reported in the literature (Hjulstrom, 1952; Krigström, 1962; Fahnestock, 1963, 1969; and Church, 1970). Braided channels are commonly unstable, shallow, wide and lined with coarse Sediment, while flow is generally swift, fluctuates daily in summer and transports large volumes of bedload. Numerous temporary gravel bars and abandoned channels are related features which occur most often at low water stages.

In contrast, anastomosed channels, as defined earlier, will be those which are braided in pattern only and are characterized by relatively low gradient, stable channels with stable vegetated islands comprised of silt deposits.

Anastomosing is well-developed in the backwater reach of the Alexandra River, and to a less extent, in the backwater reach of the Saskatchewan caused by deposits of Rampart Creek fan, shown in the lower half of the photo in Figure 10. These two backwater reaches also aggraded primarily by overbank deposition of suspended load.

Causes of Braiding in Outwash

As yet, no single explanation of channel braiding processes and causes have been acceptable to students of braided rivers. The most frequent conditions presented to explain braiding are: variation in discharge and sediment load; steep channel slope; high flow velocity; aggradation; and bank erodibility. In the outwash reach of the North Saskatchewan River, the major causes of channel braiding appear to be the noncohesive quality of channel banks and sufficient annual supply of coarse sediment.

The noncohesive quality of outwash gravel is readily observable during high river discharges in mid-summer as the outside banks of meanders in outwash unravel with ease. Measured rates of lateral erosion on outwash banks averaged one inch per minute over distances from 6 to 21 feet. The banks would slough into the channel as flow undermined the toe of the bank. The eroded gravel was then transported and deposited on the next gravel bar downstream.

A sufficient annual supply of gravel is considered the other major cause of braiding. Without a substantial supply of outwash the smaller pebbles wash downstream leaving a lag of the coarser sediment which would eventually result in a semi-stable channel lined with coarse material. This seems to be the case in many graded rivers in the northern Rocky Mountains.

Several of the parameters (sediment load, steep channel slope, high flow velocity) mentioned earlier as possible causes of braiding are rejected as they are dependent variables. Aggradation and variation in discharge *per se*, both independent variables, must also be rejected. Discharge variation is rejected because of the fact that the same variable discharge of the Saskatchewan River flows through both braided and single meander channels. Aggradation is rejected on the basis of previous work by Leopold and Wolman (1957) and Fahnestock (1963) who reported that braiding can also occur in equilibrium rivers.

Figure 11 Typical anastomosed channels flowing through the densely vegetated silt deposits in the Alexandra backwater. Flow is towards the viewer.

Table I Summary of characteristics of braided and anastomosed channels.

Feature	A. Braided Channels	B. Anastomosed Channels
control	upstream	downstream
sediment texture	outwash gravel	floodplain silt
channel stability	unstable channel	stable channel
sediment trans.	bedload dominant	suspended load dominant
process	pool and bar	overbank deposition
vert. sed. trend	coarsening upwards	fining upwards
aggrading sediment wedge pattern	wedge thinning downriver	wedge thinning upriver
vegetation	none-temporary	dense grass and shrubs

Causes of Anastomosing in Floodplain Deposits

Anastomosed channels in backwater reaches result from aggradation and channel bank stability, contrasting sharply with causes of braids in outwash. Aggradation and bank stability have probably never been considered as major factors since anastomosing in floodplain silts as a distinctly different process has not been described as a unique geomorphic phenomena on this continent except by Schumm (1968).

Aggradation in the anastomosed, backwater reaches is caused by downstream control; the Saskatchewan outwash controls the Alexandra River while the Rampart Creek fan controls the Saskatchewan backwater. The occurrence of aggradation, discussed earlier, has been verified by the presence of a dated buried volcanic ash layer at depth.

If anastomosed channel patterns were not aggrading but were in equilibrium or degrading, a single channel would likely evolve resulting from either lateral channel migration or channel downcutting. However, if anastomosed the channel pattern would remain quasi-stable through time. Such a condition is the case in the anastomosed reaches studied. To demonstrate the quasi-stable channel behavior, numberous holes were drilled adjacent to channels to determine if lateral channel migration had occurred in the past. It was found that some channels had remained remarkably stable, while others had slowly migrated laterally during aggradation. While aggradation is an important cause of anastomosing, channel bank stability is equally significant, if not more so.

Channel bank stability has not been recognized or discussed as a cause of either channel braiding or anastomosing. However, stability and bank cohesiveness of single channels as related to channel geometry (width-depth ratios) have received particular attention from Schumm (1960 and 1969). Schumm devised a percentage of silt-clay parameter (M) from sediment in stream banks, which correlated well with channel measurements of width-depth ratios and sinuosity, suggesting that textural characteristics of channel banks determine channel morphology. Schumm also felt that the regimen of sediment load was also related to channel morphology.

In this study, bank cohesiveness of anastomosed channels was found to be related to the amount of vegetation roots from grass and shrubs; cohesiveness increased with root volume. The lack of bank erosion observed during near flood conditions in the anastomosing reach was at first a casual observation (Fig. 11). During these initial observations it was also noted that a thick mat (approximately two inches thick) or accumulation of roots, flanked the channel banks or sides of islands providing a riprap-like protection from fluvial erosion. These roots were obviously the remains of what were formerly the outer edges of islands from which silts had been washed away, leaving a thickness of roots. In times of high discharge, flow along the sides of silt islands has scoured adjacent gravel to depths of 10 feet adjacent to vertical silt banks with loose vegetation. It is believed that the scouring effect adjacent to such islands prevents gravel from ever washing onto the surface of islands. Following such scour conditions gravel would often refill the scour pool to within 2-3 feet of the islands' surfaces. Erosion of such banks is believed to be very slow due to the reinforcement effect and riprap protection provided by roots. This belief is supported by a comparison of the anastomosed channel patterns on aerial photographs taken in 1950, 1958, and 1967, all of which show no apparent lateral migrations.

SUMMARY

Aggradation in the area studied is confirmed by the presence of a buried layer of volcanic ash, identified as the Bridge River ash, dated at 2450 years B.P. The longitudinal profile and thickness of accumulated sediment overlying the volcanic ash indicates two distinct wedge patterns of aggradation along the valley axis. One wedge of sediment which thins downriver is comprised of coarse sediment and is related to upstream aggradation control. A second wedge pattern, which thins upriver, is dominated by floodplain silts, caused by a backwater from downstream control. Average rates of aggradation downvalley over a 2450-year period computed from depths to the buried ash layer vary from 1 ft./98 yrs. to 1 ft./480 yrs. or, when converted to weight per area, 15,000 tons/mi.^{2}/yr. and 2,600 tons/mi.^{2}/yr.

Channel braids in outwash gravel and anastomosed channels in floodplain silts are related to upstream and downstream control, respectively. Braided channels result from a pool and bar process in non-cohesive outwash. Anastomosed channels have remarkably stable banks comprised largely of silt and root of vegetation. Erosion of such banks is believed to occur very slowly due to the reinforcement effect and protection provided by the roots.

Finally, Table 1 summarizes the features common to each type of channel pattern.

ACKNOWLEDGEMENTS

The author is very grateful to Professor M. Gordon Wolman, who provided stimulating discussion and many valuable suggestions. Assistance with finance, equipment and housing was provided by the Department of Geography and Environmental Engineering of the Johns Hopkins University, the Geological Society of America (Grant 1498-71), the Glaciology Division of the Inland Waters Branch, Department of Energy, Mines and Resources of Canada, and the Canadian Youth Hostel Association. Permission from the National and Historic Parks Branch, Department of Indian Affairs and Northern Development of Canada to do field work in Banff Park is appreciated.

REFERENCES

Church, M., 1970, Baffin Island Sandar: A Study of Arctic Fluvial Environments: unpublished Ph.D. thesis, Department of Geography, University of British Columbia, 536 p.

Costello, W.R. and Walker, R.G., 1972, Pleistocene Sedimentology, Credit River, Southern Ontario--A New Component of the Braided River Model: Jour. of Sed. Pet., v. 42, p. 389-400.

Fahnestock, R.K., 1963, Morphology and Hydrology of a Glacial Stream--White River, Mountain Rainier, Washington: U.S. Geol. Survey Prof. Paper 422-A, A70 p.

Fahnestock, R.K., 1969, Morphology of the Slims River: Icefield Ranges Research Project Scientific Results: Amer. Geog. Society and Arctic Inst. of N. ., v. 1, p. 161-172.

Fairbridge, R.W., 1968, Encyclopedia of Geomorphology: Reinhold Book Corporation, p. 90-93.

Gilbert, G.K., 1917, Hydraulic Mining Debris in the Sierra Nevada: U.S. Geol. Survey Prof. Paper 105, 154 p.

Gottschalk, L.C., 1964, Reservoir Sedimentation in Handbook of Applied Hydrology: ed. by V.T. Chow, McGraw-Hill Co., p. 17-1 - 17-67.

Hjulstrom, F., 1952, The Geomorphology of the Alluvial Outwash Plains (Sandurs) of Iceland, and the Mechanics of Braided Rivers: International Geographical Union, 17th Congress, Washington, 1952, Proceedings, p. 337-342.

Krigstrom, A., 1962, Geomorphological Studies of Sandur Plains and Their Braided Rivers in Iceland: Geografiska Annaler, v. 44, p. 328-346.

Leopold, L.B., and Wolman, M.G., 1957, River Channel Patterns: Braided, Meandering and Straight: U.S. Geol. Survey Prof. Paper 282-B, 85 p.

Mackin, J.H., 1948, Concept of the Graded River: Geol. Soc. America Bull., v. 59, p. 463-511.

Mollard, J.D., 1973, Airphoto interpretation of Fluvial Features: 9th Canadian Hydrology Symposium (In Press).

Nasmith, H., Mathews, W.H. and Rouse, G.E., 1967, Bridge River Ash and Some Other Recent Ash Beds in B.C.: Can. J. Earth Sci., v. 4, p. 163-170.

Schumm, S.A., 1960, The Shape of Alluvial Channels in Relation to Sediment Type: U.S. Geol. Survey Prof. Paper 352-B, p. 17-30.

Schumm, S.A., 1968, Speculations Concerning Paleohydrologic Controls of Terrestrial Sedimentation: Geol. Soc. America Bull., v. 79, p. 1573-1588.

Schumm, S.A., 1969, River Metamorphosis: Am. Soc. Civil Engineers Proc., Jour. Hydraulics Div., v. 95, No. HY-1, p. 255-273.

Smith, N.D., 1973, Sedimentology and Bar Formation in the Upper Kicking Horse River, A Briaded Meltwater Stream: (In preparation).

Westgate, J.A. and Dreimanis, A., 1967, Volcanic Ash Layers of Recent Age at Banff National Park, Alberta, Canada: Can. J. Earth Sci., v. 4, p. 155-161.

Westgate, J.A., Smith, D.G.W. and Tomlinson, M., 1970, Late Quaternary Tephra Layers in Southwestern Canada, From Early Man and Environments in Northwest North America: The Students' Press, Calgary, p. 13-34.

10

KNIK AND MATANUSKA RIVERS, ALASKA: A CONTRAST IN BRAIDING

R. K. Fahnestock and W. C. Bradley

ABSTRACT

These glacial rivers, the main tributaries of Knik Arm of Cook Inlet, about 40 miles from Anchorage, Alaska, have distinctly different flows, sediments, patterns and morphology.

The Knik has a valley train well over a mile in width, except where it is "confined" to about 2000 feet by an alluvial fan of a small Matanuska distributary. The Lake Fork of the Knik for more than 40 years until 1966 was subject to the annual breakout floods from Lake George, which is dammed by the Knik Glacier.

During the 20 years of record, the breakout peaks ranged from 165,000 cfs to 360,000 cfs and lasted about one week before reverting to normal summer meltwater flows of about 20,000 cfs

There are no important sources of coarse material in the first 16 miles below the Knik Glacier which dams the lake. The size reduction of about 90% in the coarse fraction over this distance is produced primarily by selective sorting. The sediments moved during the flood flows appear to have been largely sand and fine gravel. The coarse upstream bars appear to have changed little. The stream pattern and many cracks in the river bars formed by the Alaska Earthquake were only slightly modified by the three subsequent breakouts.

The pattern changes surprisingly little during the floods and over a period of years the main channels lie in the same places with only minor surface changes on the bars. The bars are as large as 5 or more miles in length extending diagonally across the valley. They are overtopped by less than 5 feet of water at peak flood. Most of the fall of the river is in the first mile below the glacier with a gradient for the lower reaches of about 0.001 (ft/ft).

In contrast the Matanuska River has a slightly smaller summer peak flow of about 15,000 cfs that modifies the appearance of the valley train rather radically each year. The pattern is much more actively changing. There are more channels and the areas of active change shift from one side of the valley to the other and back to the middle in a matter of days. Only in rare floods is the entire valley train covered by flow. There is obvious movement of all sizes of gravel present on the valley train.

The river appears to be transporting coarser sediments. The gradient is about 0.005, much steeper than the Knik.

The apparent response to the repeated catastrophic floods of the Knik is stability of pattern and channel form on a

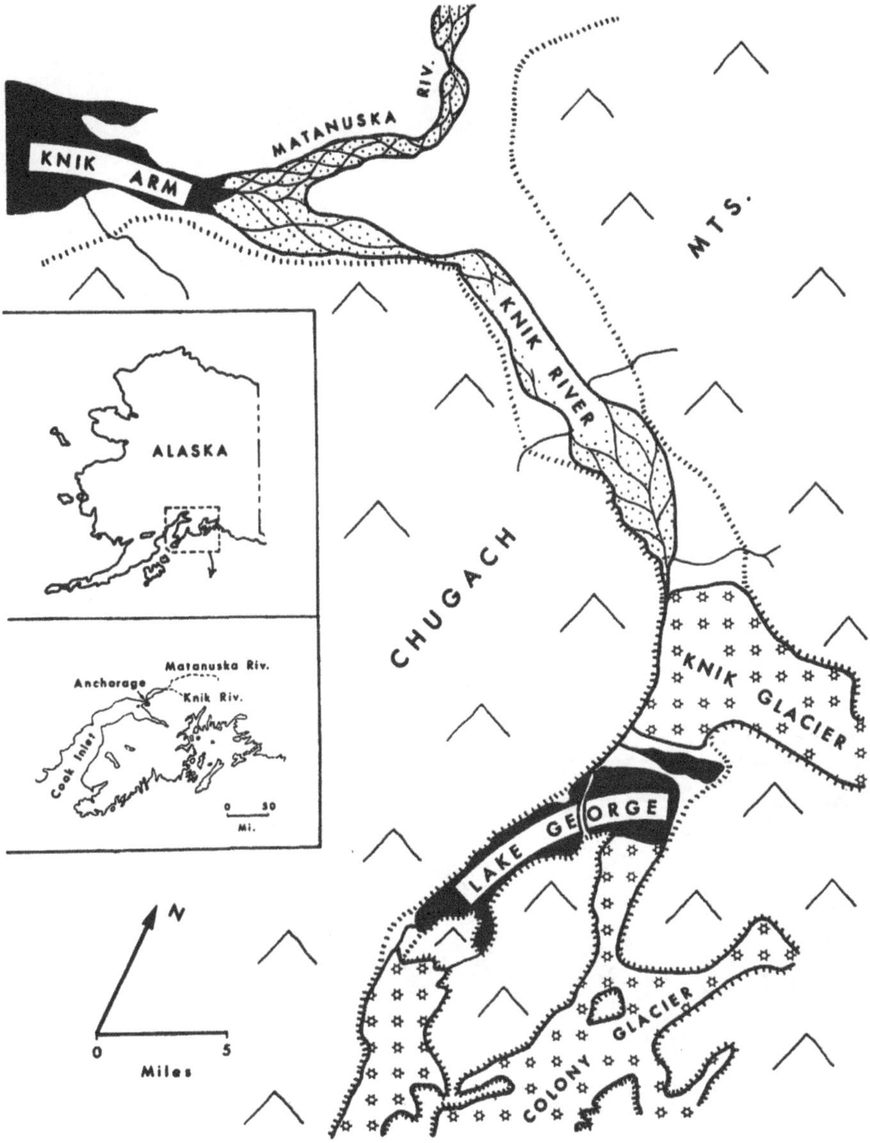

Figure 1 Sketch map of the Lake George-Knik River-Mantanuska River area, southern Alaska. At low water an end moraine of Colony Glacier divides Lake George into three bodies. At high water there is a single large lake. (From Bradley and others, 1972)

Figure 2 Looking up Knik River toward the glacier during the summer high flows of 1967. Not all of the many occupied channels are active. The main flows follow the left bank from the glacier, cut across the valley to Friday Creek fan on the right bank (left of photo), and diagonal in a series of subparallel channels to the left bank in the foreground. Figure 6 shows that this has been the pattern for the 20 years an air photo record has been available. During the breakout floods the entire unvegetated valley floor was flooded.

relatively low gradient. A much wider range of flows is capable of modifying the Matanuska. This appears to be a response to a large amount of coarse bed load.

INTRODUCTION

This paper deals primarily with the channel patterns, bar forms, and sedimentary structures of the Knik and Matanuska Rivers of Alaska. An earlier paper (Bradley and others, 1972) also based on research reported here, was primarily concerned with the nature and mechanisms of the size and shape changes of the different lithologies present in the Knik alluvium.

The Knik River seemed a natural site to study the effect of frequent large floods on stream equilibrium, sediment transport, and bar and channel development because it had received the annual breakout flood of Lake George for the past 40 years. The adjacent Matanuska River with a more conventional hydrology was included to provide a basis for a comparison.

The funding by the National Science Foundation of a 2-year project starting in 1967 may have been the kiss of death for these flows because the Lake George floods ceased that year. Although it proved to be impossible to study the direct effects of the large flows and the changes they produced, we were able to study the Knik valley train as a record of past floods.

A comparison of the Knik and Matanuska Rivers permits additional testing of the proportionality[1] $S \sim \frac{1}{Q_w}, Q_s, D_s, \frac{w}{d}$, a description of the interaction of the dependent and independent variables of a water system.

BACKGROUND

Previous studies of the Lake George breakout floods have been those of Stone (1963a, 1963b), and members of the U. S. Geological Survey (U. S. Geological Survey, 1969; and unpublished reports and charts in the files of the Water Resources Division, Anchorage). Some additional information may be found in Knudsen (1951), Alseth, (1952), and Meier (1960). All of these studies have concentrated on either Lake George and the Gorge or on the floods as they passed the U. S. Geological Survey gaging station on the lower Knik River. Little or no information has been compiled on the valley train.

In our earlier paper on the Knik, it was concluded that the valley train was aggrading and that this, with the flow variations of the Knik, provided ideal conditions for size and shape sorting and documented the 330 to 44 mm decrease in the mean diameter of the coarser fraction of the alluvium that takes place in a distance of 16 miles. (Bradley and others, 1972).

The Knik and Matanuska rivers are building a compound delta into the head of Knik Arm of Cook Inlet (Fig. 1). The Matanuska enters from the northeast with June-August flows ranging from approximately 6000-21,000 cfs from a drainage basin that includes the northern flank of the Chugach Mountains and the southern flank of the Talkeetna Mountains. The Matanuska drainage basin includes some large glaciers and is about 2100 mi^2 in area, slightly less than twice as large as the Knik basin.

---

1 S = slope, Q_w = water discharge, Q_s sediment discharge, D_s sediment size, $\frac{w}{d}$ = width/depth ratio. (based on Gilbert (1877), Rubey (1933), Mackin (1948), Lane (1955).

Table 1

LAKE GEORGE BREAKOUT AND FLOODS 1958-1973

		Lake George				Flood crest of Knik River at Palmer Highway Bridge		
		Maximum lake level						
Year of Record	Date of Breakout	Date Measured	Ft. Above Sea Level	Fall in lake Level (ft)	Water Released (acre-feet)	Date	Height (ft)	Rate of flow (cfs)
1958	---	7/13	345.5	160	1,800,000	7/18	25.3	359,000
1959	6/26	6/26	300.2	115	900,000	7/1	20.8	223,000
1960	7/12	7/14	319.6	135	1,200,000	7/17	24.4	328,000
1961	7/20	7/23	326.7	142	1,400,000	7/26	24.3	355,000
1962	6/26	6/27	281.1	96	600,000	6/29	18.5	165,000
1963	No breakout--lake did not form--Gorge remained open.							
1964	6/26	6/28	283.0	98	700,000	7/1	20.0	216.000
1965	7/8	7/9	290.0	105	900,000	7/11	21.4	236.000
1966	6/22	6/22	286.5	---	560,000	6/24	17.9	144,000
1967	No breakout--lake did not form--Gorge remained open.							
1968	No breakout--lake did not form--Gorge remained open.							
1969	No breakout--lake did not form--Gorge remained open.							
1970	No breakout--lake did not form--Gorge remained open.							
1971	No breakout--lake did not form--Gorge remained open.							
1972	No breakout--lake did not form--Gorge remained open.							
1973	No breakout--lake did not form--Gorge remained open.							

From U. S. Geological Survey, 1969, p. 14-15.

Table 2 Discharge and Suspended Sediment Data, Knik River Breakout Floods 1962 and 1964. (From U. S. Geological Survey, 1965, 1970)

Date	Mean discharge cfs	Suspended Sediment: Mean concentration-ppm	Suspended Sediment: Tons per day
1962 June			
26	9,200	2800	70,000
27	19,600	6290	334,000
28	74,800	2640	518,000
29	144,000	1260	490,000
30	99,400	992	266,000
July 1	41,500	1150	129,000
2	28,700	1080	83,700
1964 June			
27	7,100	2200	36,000
28	9,970	4100	100,000
29	46,500	4000	500,000
30	153,000	2300	950,000
July 1	159,000	1700	730,000
2	51,800	1100	150,000
3	30,400	960	79,000

Figure 3 Knik Glacier (foreground) and Lower, Inner, and Upper Lake George (in sequence, looking south). Colony Glacier calves bergs into the Inner Lake which is more than 150 ft deep at low water. Most bergs are contained by the moraine even when the lake reaches maximum stage just prior to breakout. Breakout occurs when the river undermines the glacier along its contact with Mt. Palmer to the right. Lower and Upper Lake George are quite shallow. (1968 photo by Lewis Lively).

Figure 4 Knik Glacier (left) and the Lake George outlet Gorge (center) against Mt. Palmer. Note the ice marginal lake in the foreground formed by the stagnation and collapse of this part of the glacier margin. The remnant of outwash fan between this lake and the Lake Fork in the front center has a higher slope than Lake Fork and was probably formed by normal flows prior to the beginning of the breakout floods. (1968 photo by Lewis Lively)

Entering from the southeast is Knik River with June-August flows ranging from approximately 8000-31,000 cfs and a drainage basin lying entirely in the Chugach Mountains. The average discharge for the Knik River is approximately 7000 cfs and about half of that for the Matanuska River. The Knik basin has the greater discharge because it is located closer to the Gulf of Alaska, the primary source of moisture. Knik River above the old highway bridge drains an area of approximately 1200 mi^2, about 40 percent of which is ice covered. The Knik River area of most intense investigation is shown in Figure 2.

Knik River, until recent years, has borne a brief, large summer flood from the abrupt emptying of ice-dammed Lake George, 30 mi upstream from Knik Arm (Fig. 3). Knik Glacier, some 200 mi^2 in area, flows 25 mi from the Chugach crest to block a partially deglaciated tributary valley to the south, creating Lake George. The drainage basin tributary to Lake George, some 350 mi^2 in area, is 65 percent covered by glacier ice. In summer the lake swelled with meltwater until it breached its glacier dam, then quickly emptied through a five-mile-long Gorge between bedrock and glacier. Flood flows so eroded the front of Knik Glacier that ensuing drainage easily kept the outlet channel open for the remainder of the melt season. Before each melt season, for more than forty years prior to 1967, Knik Glacier advanced across the outlet channel to reseal the basin, thus completing the cycle.

In 1963, the Gorge was not sealed off by the glacier, no lake filled, no breakout flood occurred. In 1964, the Gorge was observed open as late as April 4, following the Alaska earthquake (Hembree, written communication). It closed before the beginning of the melt season, Lake George filled, and the breakout flood occurred in late June. The glacier again sealed the Gorge before the melt seasons in 1965 and 1966 and breakout floods occurred during those summers. It is possible that the 1964 earthquake had some affect on the Knik Glacier resulting in the closure of the Gorge in 1964, 1965, and for the last time in 1966. Data on the breakout floods is provided in Tables 1 and 2.

The formation of the ice dam for Lake George is dependent on the flow rate and thickness of the Knik Glacier terminus. Several lines of evidence suggest that Knik Glacier is in failing health. The most obvious is that since 1966 the glacier has failed to seal the Gorge sufficiently to store any significant quantities of water in Lake George. In both 1967 and 1968 there was ice on both sides of the Gorge when visited in July indicating that the glacier had pushed forward against Mt. Palmer. The failure to seal off the flow which subsequently undermined portions of the terminus, had left ice stranded on the bedrock wall of the Gorge.

A mile-long lake was present within the glacier margin along the right portion of the terminus (Figure 4). There was noticeable enlargement of this lake from 1967 to 1968.

It is thought that the glacier closes the Gorge for a shorter distance along Mt. Palmer than it did in the days of the annual breakout floods prior to 1967. The shift from breakout floods at 10 to 15 year intervals to annual floods beginning about 1914 may reflect an earlier thinning of the glacier.

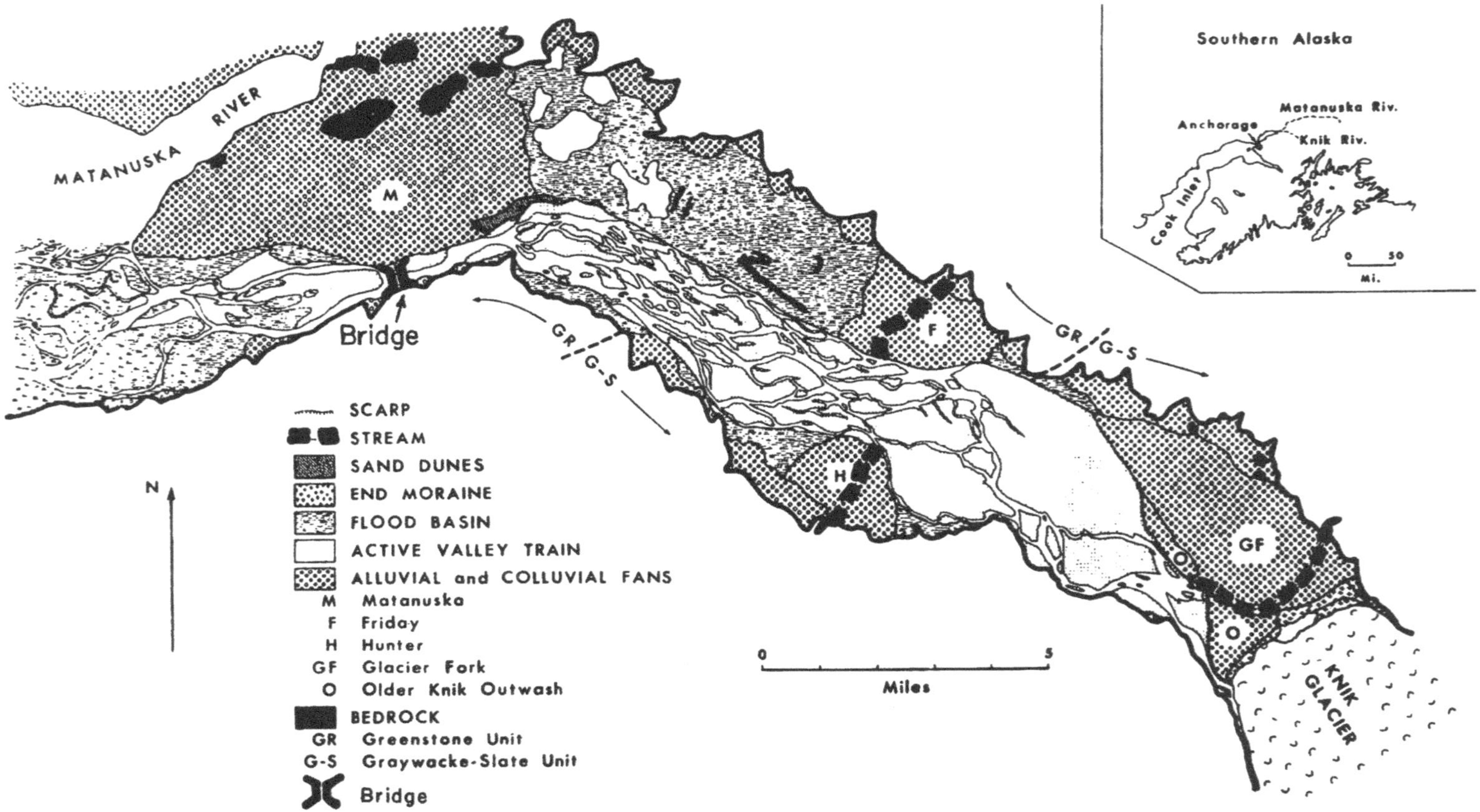

Figure 5 Surficial and bedrock geology of the Knik valley bottom (Bradley and others, 1972). Tributary streams are shown only for the major fans. Hunter Creek presently discharges about a mile upstream from where it is drawn. Note that several of the channels downstream of the Matanuska Fan flow from the Mantanuska to the Knik or from the Knik to the Matanuska depending on which river is higher.

METHODS OF INVESTIGATION

Knik River bars were studied over the entire valley train of the Knik and the portions of the Matanuska shown in Figure 8. An Everglades-type air boat allowed full river navigation so that the entire valley train could be examined. In addition to flying over the bars, a trail bike was used to examine the larger Knik bars for details. Standard surveying techniques were used to measure water surface slopes and valley train cross sections and to map and compare bar forms. A depth sounder used by the U. S. Geological Survey during the 1965 and 1966 flood measurements was used in 1967 and 1968 to record bedforms and channel configuration. The methods used to measure grain size and shape changes are discussed in Bradley and others, 1972.

VALLEY FLOOR FEATURES

An understanding of river deposits and processes requires some knowledge of the other surficial deposits of the valley floor. Figure 5 is a map of surficial and bedrock geology in the Knik valley bottom (from Bradley and others, 1972). All surficial deposits shown are Neoglacial in age, and many are truly modern. All probably were deposited within the last few hundred years. The area lies about 50 mi behind the late Wisconsin (Naptowne) terminal moraine near Anchorage (Miller and Dobrovolny, 1959; Karlstrom, 1964).

Bedrock in the Matanuska valley floor in the reach considered in this paper is exposed only at the narrows (Fig. 8) approximately 8 mi upstream of junction with the Knik River and is the reason for the existence of the narrows. Elsewhere the Matanuska is cutting laterally into unconsolidated glacial deposits which have not been mapped in detail. There are no tributary fans of significance. The construction of the old Glenn Highway and the protection for the Alaska Railroad have cut off distributaries of the Matanuska fan which constricts the Knik. The distributary interrupted by construction of the old Glenn highway was reoccupied by the 1971 Matanuska flood which overtopped and washed out the highway embankment and caused considerable damage to property on the fan (Russ Dow, personal communication).

End moraine.

The end moraine lies only a few yards beyond the active ice of the Knik Glacier. On the basis of form and vegetation and proximity to ice, it was correlated with a moraine of the nearby Colony Glacier (Fig. 1) which is younger than 200 radiocarbon years (Levin and others, 1965, p. 391). Erosion of the moraine has provided a source for a small part of the materials of the Knik River valley train. The coarsest materials on the valley train are lag concentrations of large boulders which have not been transported far from their original positions in the moraine.

Alluvial and Colluvial Fans.

Small fans with locally derived coarse detritus are common along the flanks of the Knik valley. Most are not in contact with the valley train being separated from it by the flood basins. There are five large fans, however, which do affect Knik River and its alluvium. The largest fan occurs downvalley (M in Fig. 5) and was deposited by the Matanuska River (Stump and others, 1959; Trainer, 1960) as evidenced by granites and other rock types foreign to the Knik drainage, a steep, fan-shaped southward slope, and a grain size four times larger than Knik alluvium (Bradley and others, 1972). The fan has multilevels and is generally inactive, although some parts receive water (but little sediment) from high Matanuska flows as in 1971.

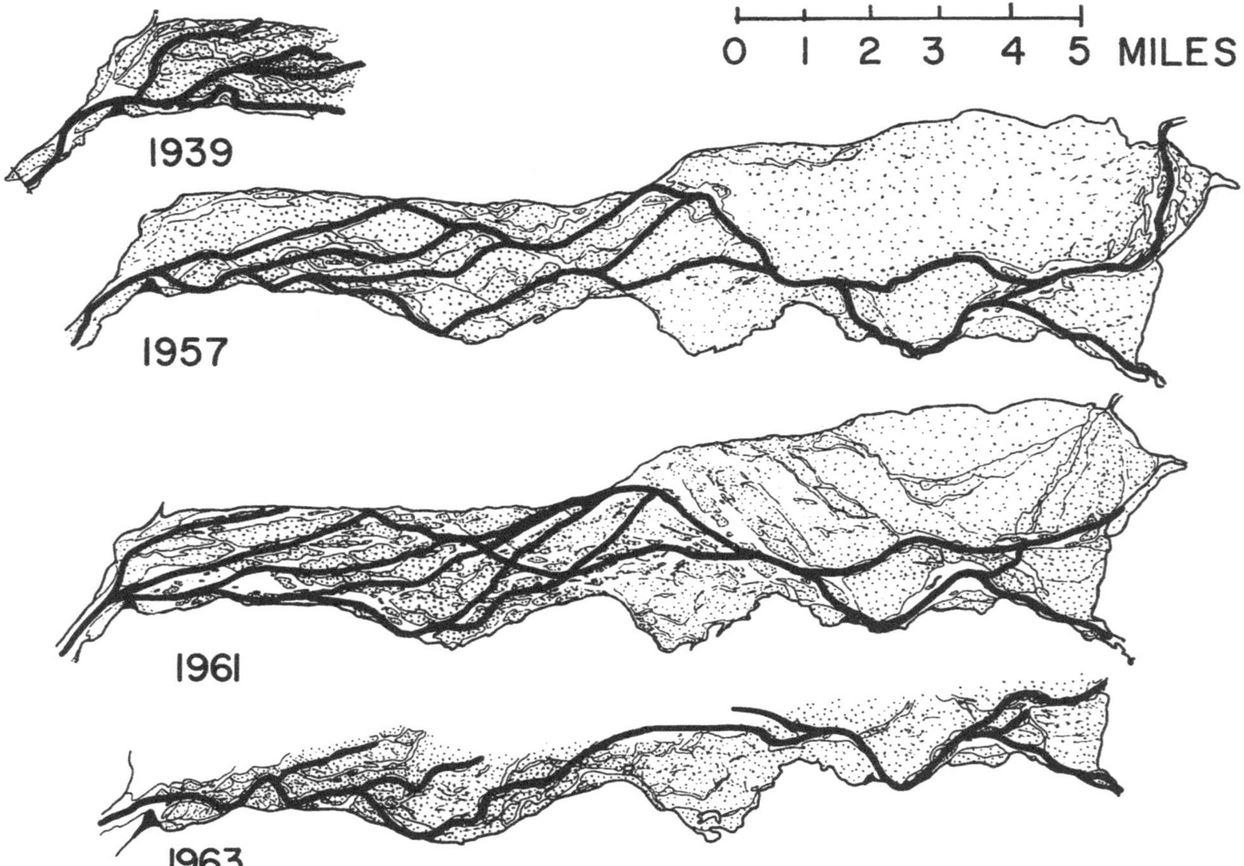

Figure 6 Patterns of the Knik River 1939-1963. Lake Fork of the Knik, which is subject to the Lake George breakout floods and Glacier Fork join to form the Knik River. Note the location of the main channels has not changed.

This fan creates the lower Knik narrows (Fig. 5) by forcing the river to flow in a single, 2000 ft wide channel along the bedrock of its left bank. Here on the old Glenn highway bridge the U. S. Geological Survey has its gaging station. Figure 5 shows this reach is about 1/4 as wide as the average width of the valley train which explains why gaging data other than discharge, cannot be extrapolated to the wider reaches. Persistence of the narrows, although the fan is inactive, indicates that even Knik flood flows have difficulty in reworking the fan alluvium.

The second largest fan is that of Glacier Fork at the head of the valley. Although parts are inactive, it is largely a modern feature. This fan and river together provide one of the largest sources of sediment for the Knik and keep the main river along the left side of the valley in its upper course.

Friday and Hunter Creeks are building fans which severely constrict the valley train, but at present are not significant sources of new sediment to the river (Bradley and others, 1972).

The prominence of these fans which have been only slightly modified by the Knik River supports the conclusion reached in the study of the Slims River that the ability of a braided stream to cut laterally into fan materials is quite limited (Fahnestock, 1969). The fans play a major role in determining valley train width. Widening by burial rather than lateral erosion is more likely in the situation of an aggrading stream like the Knik.

The fifth fan is composed of older Knik outwash which was deposited by meltwater streams flowing directly from Knik Glacier when the latter was constructing its end moraine (Fig. 4). It is now inactive and separated from the valley train by an erosional scarp which is more than 30 ft high at mile 0, the mouth of the Gorge, but decreases in height downvalley until at about mile 4 it disappears. Correlation of older outwash and end moraine and the steeper slope of this fan all indicate probable deposition before the Lake George floods became an annual affair.

Sand Dunes.

Strong downvalley winds have blown sand from the valley train into a dune fringe where Knik River turns to enter its lower narrows. Active dunes form a valley rim along the flood basin (Fig. 5) and help to confine floods to the central valley floor. They act with natural levees along tributaries-distributaries to confine the distribution of the lower breakout floods to the main channel. Active wind blown dunes are the dominant surface features on the large right bank bar just above the narrows. The truncated bedding now stabilized by moisture may have been deposited during a time of a lower water table.

The active dunes receive their sand from the surface of the large bars of the lower river above the narrows especially those along the right bank. There is a 10 ft high dune on the downstream end of the last bar where slip faces continually are undermined by the stream and most of the sand is returned to the river. The right bank channel of this portion of the valley train carries nothing but sand. It is especially treacherous to navigate as much of it is less than 6 inches deep and in places has an extremely soft bed.

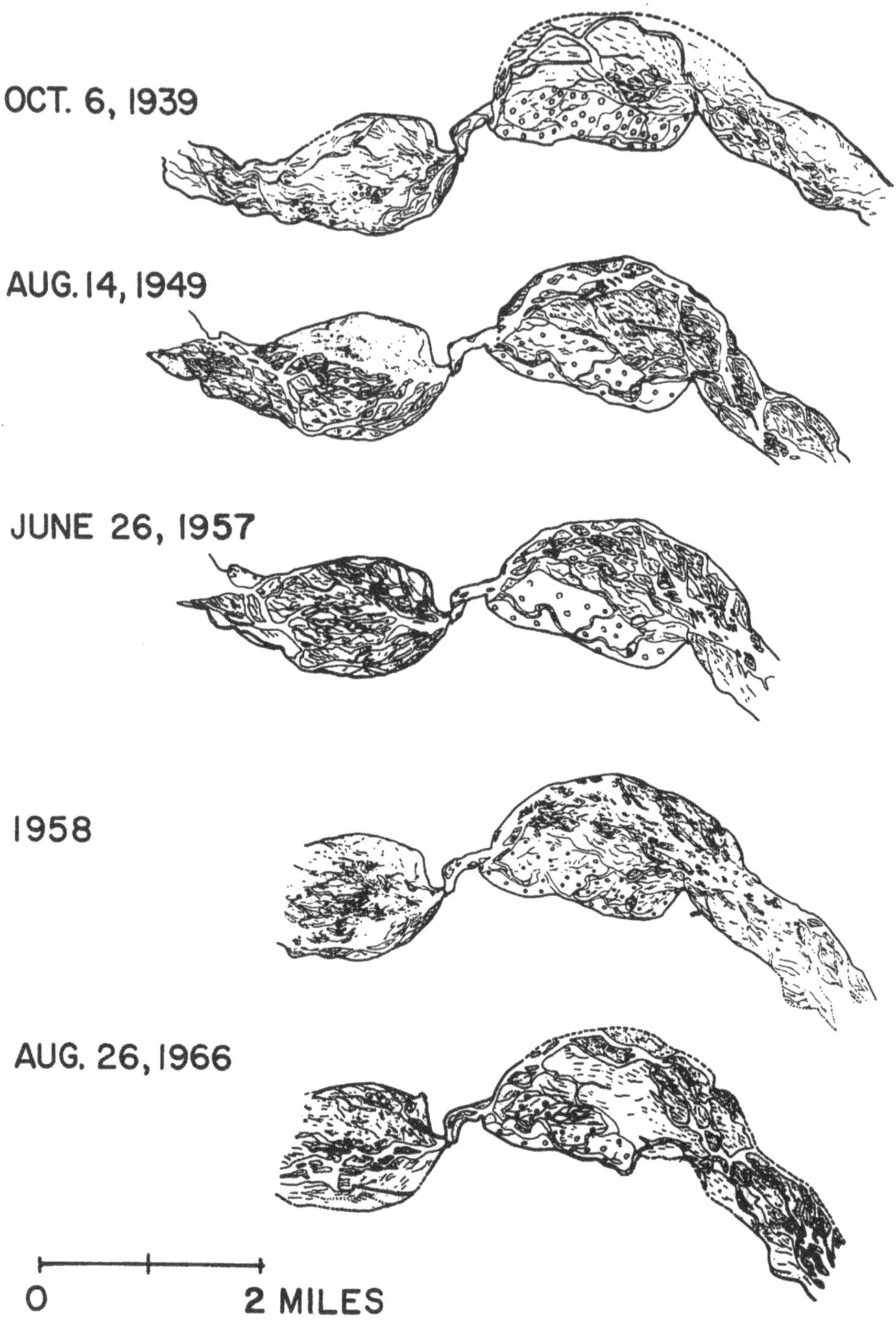

Figure 7 Patterns of the Matanuska River at the narrows 1939-1966, taken from uncorrected air photos. The differences in the number of channels are largely due to differences in discharge. Changes in the position of the main channels is due to erosion and deposition which may take place within a few days or weeks.

Stable dunes within the right bank flood basin (Fig. 5) are evidence of a rise of the water table probably as a result of alluviation of the valley train. One has a parabolic dune form.

The only group of dunes associated with the Matanuska River is located at the top of a cliff 2 mi upstream from the Palmer narrows bridge. According to Trainer (1961, p. 7, 10) the Matanuska River dunes are remnants of earlier dunes now undergoing erosion and do not have the present river bars as a source.

Flood Basins.

Flood basins are lateral basins which smooth out the irregularities caused by fans and bedrock spurs and give the valley train its even margins. The large right-bank basin between Matanuska Fan and Friday Fan illustrates the variety of sedimentary facies that may be present; lake and back-swamp deposits, levees and channel deposits of local streams (not shown in Fig. 5), and eolian sands and silts; all facies are locally interbedded with fan deposits.

These basins receive water and suspended sediment during peak breakout flows, but at other times are separated from the valley train by natural levees and eolian sand dunes. As much as 10 percent of the peak breakout flood discharge enters these basins and is released slowly once the flood has passed. (Unpublished data, U. S. Geological Survey).

These floods are distributed over the flood basins by a temporary reversal of flow in the streams which normally drain the flood basin lakes. The short period of reversed flow, however, is dominant in the formation of these streams. The influx of flood water and sediment has built levee systems which decrease in elevation from 5 or more ft at the Knik to inches above the low water surface at the flood basin lakes. Deltas have been built into these lakes during the floods but when the Knik is not in flood, the flow is from lake to Knik and the distributaries serve as tributaries. Flood basins occur also as interdistributary basins on the Knik River delta, 20 miles downstream from the glacier.

Flood basins are not known to exist along the Matanuska reach. Some tree covered areas along the active valley train are periodically invaded by active channels but because of the gradient, no large lakes are formed and no significant quantities of water are stored.

VALLEY TRAIN FEATURES AND PROCESSES

Introduction.

Both the Knik and Matanuska Rivers are actively depositing valley trains although, until recent years, under radically different flow conditions. The Matanuska provides a model for understanding what rivers, like the upper Plattes of Colorado and Wyoming, must have been like when they had alpine glaciers at their heads. The Pleistocene equivalent of the Knik is more difficult to picture--more water, less sediment, repeated floods. Presented here is a picture of these two rivers.

Valley Train Zones.

Between the Gorge and Matanuska fan, the Knik valley train appears to have the same three zones reported from other simple valley trains (Krigstrom, 1962; Fahnestock, 1969). An upstream zone (mile 0 to mile 6 from the Gorge) contains a main, somewhat sinuous, locally incised channel, and large,

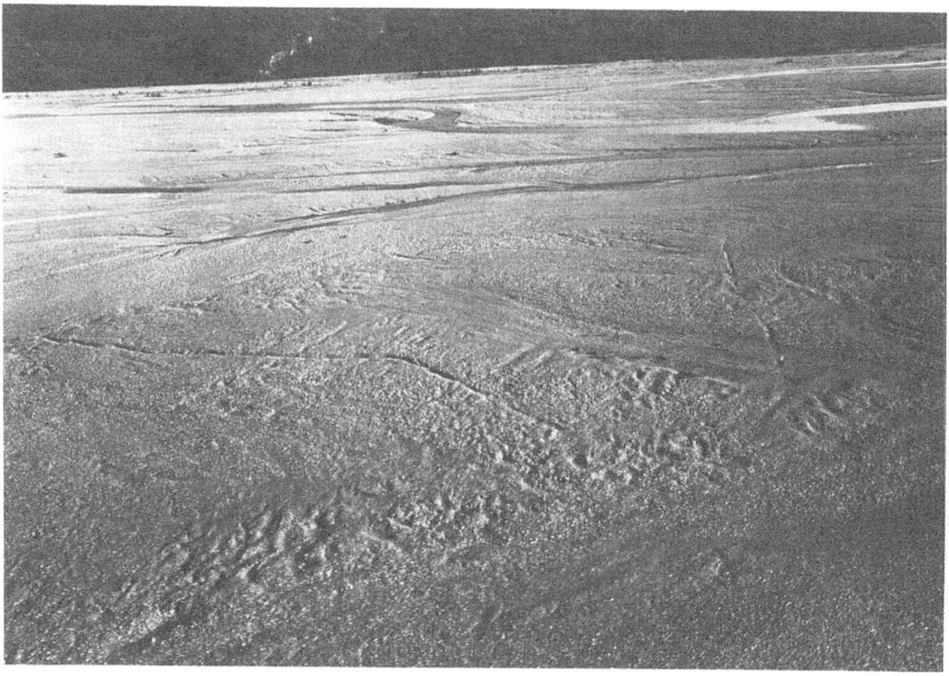

Figure 8 Bars of the intermediate zone of the Knik. Flow is from left to right. Note the lack of logs or other floating debris, a possible exception to this is the holes which appear to have been left by the melting of stranded ice bergs. The lack of vegetation indicates that the bar surface was modified during the high flows of recent years. The modification of the 1964 earthquake cracks indicates that this bar, although overtopped has not been involved in active sediment transport over its entire surface. (Lewis Lively photo, 1968)

high bars. An intermediate zone (mile 6 to mile 14) contains many channels and smaller, somewhat lower bars. A downstream zone (mile 14 to start of the narrows) has broad, shallow sheets of water and low bars. All bars are composed of sand and gravel, but gravel decreases downvalley in both size and abundance.

The entire Matanuska reach studied is characteristic of the intermediate zone. The Matanuska is a much longer river with many large tributary contributions of water and sediment and is much harder to classify but in the reach above its junction with the Knik it is characterized by many active channels.

Valley Train Slope.

Slope of the Knik valley train is slight, ranging from 5 ft/mi at its lower end to 10 ft/mi toward the glacier (Bradley and others, 1972). However the slope in the reach constricted by the Matanuska fan is about 3 ft/mi, about half that of the wider valley train upstream. This decrease in slope is thought to result from the greater sediment transporting efficiency of the narrower and deeper channel and in spite of the increase of the size of the materials on the bed.

This raises the question of the role of size and amount of load in determining the slope of the channel. As there are no active channels on the fan and the widening of the Knik cross sections has been relatively slight in recent years, the volume of sediment added at the fan is negligible. The coarser sediments of the fan may provide materials to armor and stabilize the bed but apparently are not present in sufficient volume to force the river in its narrow section into a steeper slope to provide transport.

The activity of the Matanuska takes place on a slope 4 to 5 times as high as that of the Knik and is accompanied by the movement of large and audible quantities of coarse bed load. This activity is not restricted to a few days of high flow but takes place on many of the days of summer. The few samples of Matanuska gravels in the reach considered have a median diameter of about 170 mm to 200 mm and show no systematic change in size in the downstream direction. There is no apparent change in slope in this reach of the Matanuska but detailed water surface profiles were not run.

Pattern Development.

Flood flows from Lake George completely submerge the Knik valley train. The breadth of the valley train is so great that the large increases in discharge can be accommodated by increases in water level of only 3 to 10 ft, barely overtopping the highest bars. During the rest of the summer Knik River has a braided pattern similar to the ones shown in Figure 5 (adapted from 1957 air photos) and Figure 6.

Knik River braiding is on a scale commensurate with the large flows. This point emphasizes a striking fact; the major bar and channel development has taken place during only a few days each year when the breakout was on. (For 40 years uniformitarianism for Knik River has consisted of a succession of catastrophies). Normal summer flows cause only minor changes in the pattern. For that matter, pattern evolution occurs slowly even with the large flows. Nine sets of air photos taken over a 30 year period show that the main elements have stayed approximately the same during that time, although details have changed.

Figure 9 Several of the large Knik bars formed by the movement of bed load from the adjacent main channels toward a central downstream channel. Flow is from left to right and toward the viewer. This type of movement from the channel over the upstream bar margins toward the center is thought to be the most common. (1968 photo, Lewis Lively)

Figure 10 Knik bar formed by the spreading of the flow left to right from the central channel toward the downstream bar margins. The pockmarked surface results from the scour around, or partial burial of, stranded icebergs. (1968 photo by Lewis Lively)

Figure 6 shows the stability of the Knik pattern. In the long-term pattern, flow follows a main channel that hooks to a left bank position opposite Glacier Fork, then heads obliquely across to Friday fan where it breaks up into several major diagonal channels joined by smaller crossover channels, finally converging into a very broad main channel near the left bank prior to entering the lower narrows. The pattern of parallel diagonal channels connected by smaller cross over channels has also been observed in Icelandic sandur streams.

Pattern development by the large flows has now stopped. In time Knik River will rework its valley train to establish a pattern more appropriate to its normal summer size. The limited extent to which this has been done in the active channels indicates the new pattern, developed for the most part within the existing Knik channels, will be more intricate (Figure 10). The new pattern will also have less relief allowing further changes in pattern to proceed more rapidly perhaps in a manner similar to the Matanuska River. Glacier Fork of the Knik has always had a rapidly changing, intricately braided pattern. The rate of development of such a pattern will depend largely on the sediment supply which is still unknown.

The Matanuska at normal summer high flows is actively changing the details of its complex pattern, vigorously attacking and undercutting a tree laden bank or island, depositing bars and levees across channel mouths, blocking flow from active channels or raising the local bed elevation to cause the reactivation of abandoned channels (Fig. 7). Even at the gaging station at the bedrock narrows, the rattle and hiss of gravel bed load movement can be heard and discharge measurements on successive days show the changes in channel cross section. Sometimes the deepest section and the fastest current are against the gage on the left bank, at other times relatively quiet water flows around the gage with the main current in midchannel or near the far shore. The angle of the current and the number of channels also change.

Bar Characteristics.

Knik river bars in the upstream and intermediate zones are large smooth forms with a total relief between the highest points on the bars and the nearby water surface of 5 ft or more (Figure 8, 9, and 10). Most Knik bars formed during breakout floods are thousands of feet to miles in length. Matanuska bars are only a few hundred yards long with a relief less than half those of the Knik and are more intricate (Figures 11 and 12). In post breakout channels of the Knik, bars of similar size and intricacy form but they have a lower relief and finer sediment than those of the Matanuska.

Bedforms and Sediment Transport During Normal and Breakout Flows.

As part of a study of flood scour around bridge piers by the U. S. Geological Survey for the Alaska Highway Department in 1965, a series of cross sections and longitudinal profiles were taken using depth sounding equipment. Water and sediment discharge measurements were made in a reach extending 2000 ft up and downstream from the old Highway bridge (Leveen, 1967). Only in this reach and in a similar reach 7 miles downstream at the new bridge, have data been taken during a breakout flood. With the aid of the U. S. Geological Survey, additional data was taken during the "normal" flows of the summers of 1967 and 1968 at both of these reaches. The net scour in these reaches was slight although the water surface drop around the bridge piers was 5 ft at the flood peak.

Figure 11 Dry channels of the Matanuska looking upstream on the left side of the valley train. The picture is taken about 2 miles above the first channel connecting the two rivers. The river is at normal summer meltwater stages but is active along the right side of the valley train. The inactive channel at the right of the picture carries only minor flow. The logs are indications of active bank erosion upstream. (Lewis Lively photo, 1968)

Figure 12 The Matanuska River looking downstream about 3 mi. upstream of Figure 11, both taken on the same flight. Here, at this time, the river covers one half of the valley train. The water surface is indicative of shallow, rather fast flows of depths ranging from inches to a maximum of 3 to 5 feet. It is obvious why boatmen, including those with air boats, avoid the everchanging Matanuska. Logs are again indications of active bank erosion upstream. (1968 Lewis Lively photo)

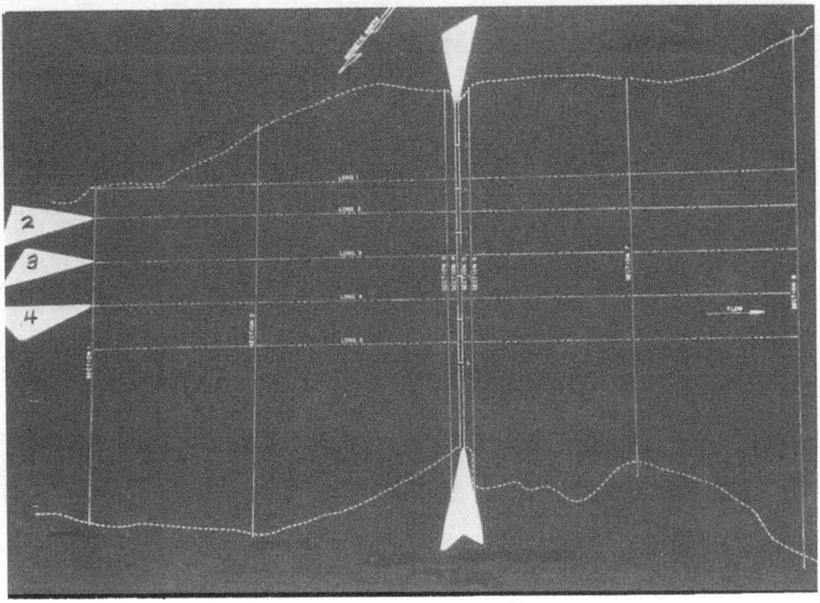

Figure 13 Location map of depth sounder cross sections and profiles at the old Knik River highway bridge. The flow is from left to right. The cross section lines are 1000 ft. apart. A portion of profile 4 is reproduced as Figure 14.

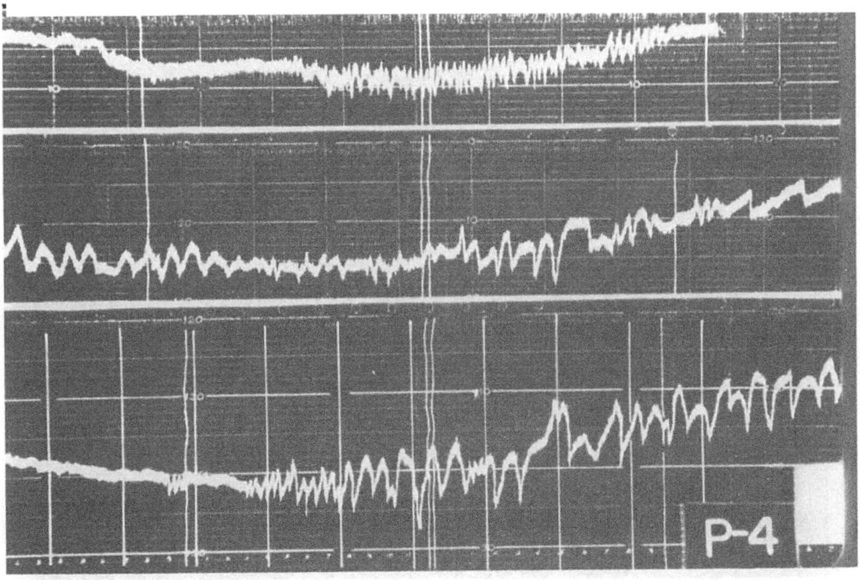

Figure 14 A reproduction of depth sounder profile number 4 from the Knik River at the old Highway bridge. The bar is 10 ft of depth. Each vertical line 20 seconds of recording time, or 250 to 350 ft, depending on the velocity of the boat. The darker vertical lines are about 1000 ft apart. Bedforms of a variety of heights and wave lengths characterize this reach during a breakout flood. The profiles were taken near the center of the channel. The lower profile is taken at the 1965 flood peak, the middle profile one day later, and the upper one three years later. Note the adjustment of the height and length of the bedforms to the change in flow.

Figure 15 Sedimentary structures

A Planar and cut and fill bedding of a Knik River bar. The cut and fill bedding may have resulted from channel migration and the development of a bar like that shown in Figure 10.

B. Detail of the stratification of the right side of Figure 15 A.

The local scour hole around a bridge pier increased 1½ ft in depth from 4 to 5½ ft during the passage of the flood peak, in spite of a difference in elevation between the pile up of water upstream of the pier and draw down around it of about 5 ft. This is considerably less scour and fill than that involved in the passage of a sediment wave in the unobstructed channel. Of course it is possible that the effects of much higher discharges might be more severe.

The median diameter of the bed material was measured at 1.3 mm two days before the flood peak, 2.5 mm on the day of the flood peak, and 0.9 mm the day after the peak. The D_{90}, the diameter larger than 90 percent of the bed material, changed from 11 mm to 55 mm to 9 mm as the flow varied from 8000 to 236,000, to about 70,000 cfs.

The changes in bed material size during the passage of a flood emphasize the extreme difficulty in obtaining size data in gravel deposits that can be related to the processes of transportation and deposition. For example, perhaps samples should be taken only from avalanche faces to obtain an average sample unbiased by possible lag concentrations on the eroding upper surface of the sediment wave.

The depth sounder was used to measure the deepest portions of the large channels upstream during 1968. The deeper portions of these channels were in the few places where the current was directed against the outside of a cutbank in more resistant materials such as a coarse gravel bar or the margin of one of the tributary fans. In the middle of one broad expanse of flowing water a depth of more than 15 ft was encountered. Such wide channels are usually less than 5 ft deep. Closer examination revealed the convergence of three channels. The channels were separated by slightly submerged bars and all could be waded. The convergence and pinching of the current created the second deepest spot on the valley train. The deepest, 25 ft, was against the left bank bedrock at the upstream end of the narrows.

The occurence of such a mid-valley-train deep without the presence of resistant materials suggests the possibility of a greater depth of scour and reworking of valley train materials than had been expected by the investigators.

At the new highway bridge, about 7 mi downstream, depth soundings during the 1966 breakout flood showed dunes with an amplitude of 4 ft and a wave length of about 40 ft in water about 12 ft deeph. Boils were quite evident throughout the section during the peak discharge of about 72,000 cfs. There are several other channels also sharing the flood waters. Surface velocities ranged from 5 to 8 fps with the mean velocities in a vertical from 4.5 to 6.5 fps. The intermediate diameters of the bed material in this reach were smaller than at the old bridge upstream, 0.6 mm increasing to 1.8 mm during the passage of the flood. The D_{90} of the coarsest sample was about 18 mm.

Thus we can see that in response to changing discharge and sediment load there is a significant change in the form of the bed and in the scale of the internal structures of the deposits. Evaluation of the relative importance of changes in sediment size and mode of transport, depth of flow, discharge of water and of sediment and width/depth ratio of the channel will require the collection of much more data. As this is both time consuming and expensive, it may require the utilization of studies where such immediate returns as a knowledge of bridge scour can pay the bill.

Figure 16 Cross strata thought to have resulted from the movement of dunes similar to those shown on the depth sounder profile in Figure 14. The surface of the bar shows more subdued "washed out" dunes thought to represent the transition between dunes and plane bed sediment transport conditions (Simons and others, 1965; Harms and Fahnestock, 1965).

Longitudinal depth profiles run at the flood peak and one day later showed a variety of bedforms. These profiles were taken by running the boat on the straight courses through the center of the openings between bridge piers as shown in Figure 13. They are not run exactly parallel to the current. There is a probability that they may differ in lateral position from the mapped line by some tens of feet at the ends of each profile. Figure 14 is a reproduction of the actual sounder chart.

The smallest sediment wave that this equipment was capable of resolving is 10 or more ft long and 6 in or more high. Thus the ripple type of bedform cannot be recorded and it was not possible to use the equipment effectively in the shallow depths and swift, tortuous channels of the Matanuska.

The 2000 ft wide reach in which data have been taken during the breakout flood is in marked contrast to the 16 mi of valley train upstream. Bar forms and flow conditions are different in the 1½ mi wide, braided, upper reaches of the Knik than they are in the single channel downstream. It is unfortunate that depth sounder data is not available at flood flows for the entire valley train. The primary reason for the sudden narrowing of the Knik River is the alluvial fan of gravels from the Matanuska River which has forced the Knik River against the bedrock of the south valley wall.

The single channel is characterized by mile long gravel bars attached to alternate banks. The surface of parts of these bars is marked by the washed out remains of some of the same types of bedforms recorded by the depth sounder during the passage of the 1965 breakout flood (Fig. 16). These forms are chiefly deposits of sand and fine gravel with little coarser materials.

The water surface slope in this narrow reach is about 0.0006 or about 3 ft/mi. This is significantly less than the 0.001 slope of the much broader reach upstream.

The bedforms generated at peak flow were up to 11 ft in height with 7 to 9 ft being more common in the larger forms. The average length was on the order of 85 ft. It is not known if the downstream face of these forms was a true avalanche face, but the possibility of extensive development of large scale cross beds is apparent. As each crest is followed by a trough, only in rare instances would even a major portion of the full height of the crossbedding be preserved. The average velocities in each vertical of the discharge measurement ranged from 7 fps near the left bank to 11 fps in midstream to 5 fps near the right bank at the flood peak. On the falling stage the depth was about 14 ft, the height of the larger forms 4 to 6 ft and the length about 70 ft.

A profile made in 1968 at normal flows of about 15,000 cfs showed depths of about 6 ft, lengths of about 30 ft and heights of about 2 ft for the larger forms. From this small sample the height of the forms appears to be about one half the depth, and the length of the form is about 5 times the depth showing an adjustment to the flows.

No bedforms with repeated crests two or more ft in height were found on the surface of the Knik River bars. In some places crossbeds more than 1 ft in height were observed but these are thought to have been formed by other bedforms.

Figure 17 Gravel levee deposits.

A These deposits result from the movement of gravel from a main channel across the surface of a bar.

B Where there is quiet water over the center of the bar to nearly the same elevation as the water in the source channel, a slightly sinuous avalanche face results.

C When there is a significant difference in water surface elevations, for example on falling stages, concentration of flow and scour result in the development of small deltas which protrude beyond the former slipface.

Data collected by the U. S. Geological Survey on the change in suspended load during a breakout flood indicates the suspended sediment concentration of the Knik peaked before the maximum flood discharge (Table 2). There apparently was a diluting effect of the full flow from Lake George. The maximum flow for 1963 when there was no breakout was 39,100 cfs on Aug 26, the maximum concentration (3700 ppm) and total suspended sediment transport of 391,000 tons/day took place on this date. These figures are within the range of those encountered during breakout floods.

Sedimentary Structures and Related Bar Surface Features.

The sedimentary structures produced by the movement of bed forms, sediment waves and bars are difficult to expose in coarse noncohesive materials because of the frequent collapse of banks and the difficulty of hand excavations. Occasionally the Knik River produced a cutbank giving access to cross sections of the gravel deposits.

In keeping with the large size of the bars, most of the sedimentary structures encountered were quite extensive, usually occupying the entire exposure (Fig. 15). This is in striking contrast to the rapid horizontal variation encountered by Williams and Rust (1969) and the probable, though unknown, distribution of facies in the Matanuska bars.

Some exposures show cut and fill structures thought to be related to the shifting of channels and/or the passage of dunes and their associated troughs (Figs. 15 and 16).

Other exposures show thick tabular sets of avalanche cross beds which must be related to the advance of the face of a levee or gravel wave into quiet and deep water in the center of a bar (Fig. 17). These levees are prominant features of many of the broad bars of the Knik. Figure 18 and 19 show the size and form of such a gravel levee. In some ways they resemble the ridge and runnel systems of the near shore zone described by Hayes (1972). Sediment, in this case gravel, moves from the main channel in plane bed transport across the bar surface to avalanche into deeper quiet water toward the center of the bar. The importance of the relatively quiet water of the interior of the bar to the formation of the avalanche faces and resulting cross bedding is apparent when one considers the destruction of the face by concentrated flows on the high, local slopes created during falling stages shown in Figure 17 C.

The planar structures of some of the bar surfaces (Figs. 15 and 19) are thought to represent rather delicate adjustments between depth, water surface slope and sediment supply. The depth of water over the bar must be shallow enough or the transport rate great enough to prevent the formation of dune bedforms and deep enough to maintain flow and transport. The water surface slope depends on the level of water in the channel supplying the sediment and the level of the central drainage channel. Both of these vary with the total flow of the river as well as the movement of water and sediment in the channels that supply them.

Other exposures show only planar structures with little or no crossbedding and are thought to have an origin similar to the planar surfaces of some of the bars. In this case there may have been no deeper water into which a slip face could form and advance.

Figure 18 The tabular cross stratification produced by the advance of a slipface of such a feature as the gravel levee. The planar stratification at the top is thought to result as sheets of gravel are transported to a nearby slipface.

Other structures which are thought to be most often associated with breakout floods are the scour pockets shown in Figure 19. These were formed by scour around and, in some cases, by the partial burial of stranded ice blocks that calved from the glacier. Much larger features of this type have been observed on the near glacier portions of an Icelandic sandur (Skeiðarársandur) which is subject to the glacier bursts from Grimsvötn. Similar structures might also be formed by scour around river ice during spring floods.

The Knik bars present a series of bar, bedforms and bedding types that are thought to occur more frequently in a valley train environment subject to repeated flood flows.

CONCLUSIONS

The primary differences between these rivers are the slopes on which deposition is taking place, the intricacy of the Matanuska deposits as opposed to the continuity of the Knik deposits, and the relative stability of the Knik as opposed to the rapid changes of the Matanuska. Perhaps the most unexpected conclusion was the lack of pattern change of the Knik in spite of the catastrophic nature of the annual flood. This emphasized the role of coarse bed load in changing the pattern of a braided stream. The little-altered earthquake cracks and the pattern stability indicate that the amount of bed load movement during recent Knik floods is relatively small and local.

The adjustments of the gradients and patterns of the two streams to the sediment and water provided supports the use of the proportionality $S \sim \frac{1}{Q_w}, Q_s, D_s, \frac{w}{d}$ in predicting the effects of a change in the conditions of stream equilibrium. While the adjustments to flow and sediment loads and changes in w/d ratio are in the directions predicted, much further work is necessary to quantify these relationships.

The difference in these streams as well as the radical difference of an anastomosing stream described by Darrel Smith in this volume emphasize the need for the classification of streams by their controlling factors in order to understand more subtle relationships between stream processes and form.

Streams with similar sediment loads and flow regimes as well as pattern and channel characteristics should be investigated very carefully before comparison is made between radically different streams, to reveal all that is to be learned by such comparisons. Comparisons between streams of unknown characteristics must thus be of little help in understanding river processes.

ACKNOWLEDGMENTS

Funds for the project came from National Science Foundation grants GA-696, Ga-1442. Members of the Water Resources Division of the U. S. Geological Survey, Anchorage, were particularly helpful in the loan of vital field equipment and unpublished data; we offer thanks especially to Harry Hulsing, L. S. Leveen, and M. J. Slaughter. Field assistance was provided by W. L. Haushild, Jr., and T. C. Silver. Important help, particularly with logistics and river navigation, came from a number of Alaskan residents especially Russ Dow, Mike Vasili, Eugene Webb, and Charles Akers.

Figure 19 Scour holes formed around ice bergs stranded during a Lake George breakout flood. View looking upstream.

REFERENCES CITED

Alseth, I. B., 1952, Self-emptying lakes: Nat. History, v. 61, p. 8-13.

Bradley, W. C., Fahnestock, R. K., and Rowekamp, E. T., 1972, Coarse sediment transport by flood flows on Knik River, Alaska: Geol. Soc. America Bull., v. 83, p. 1261-1284.

Fahnestock, R. K., 1969, Morphology of the Slims River: Icefield Ranges Research Proj. Sci. Results, v. 1, p. 161-172.

Gilbert, G. K., 1877, Report on the geology of the Henry Mountains: U. S. Geol. Survey Rocky Mtn. Region (Powell), 160 p.

Harms, J. C., and Fahnestock, R. K., 1965, Stratification, bed forms, and flow phenomena (with an example from the Rio Grande): in Middleton, G. V. (ed.), Primary sedimentary structures and their hydrodynamic interpretation: Soc. Econ. Paleontologists and Mineralogists Spec. Pub 12, p. 84-115.

Hayes, M. O., Owen, E. H., Hubbard, D. K., and Abele, R. W., 1973, The investigation of form and processes in the coastal zone, in Coates, D. R. (ed.), Coastal geomorphology: Publications in Geomorphology, State Univ. of N. Y., Binghamton, p. 11-41.

Karlstrom, T. N. V., 1964, Quaternary geology of the Kenai lowland and glacial history of the Cook Inlet region, Alaska: U. S. Geol. Survey Prof. Paper 443, 69 p.

Knudsen, D. C., 1951, Alaska's automatic lake drains itself: Natl. Geog. Mag., v. 99, p. 835-844.

Krigström, Arne, 1962, Geomorphological studies of sandur plains and their braided rivers in Iceland: Geog. Annaler, v. 44, p. 328-346.

Lane, E. W., 1955, The importance of fluvial morphology in hydraulic engineering: Am. Soc. Civil Engineers Proc., v. 81, paper no. 745, 17 p.

Leveen, L. S., 1967, Investigation of scour at selected bridge crossings in Alaska: U. S. Geol. Survey open-file report.

Levin, Betsy, Ives, P. C., Oman, C. L., and Rubin, Meyer, 1965, U. S. Geological Survey radio-carbon dates VIII: Radiocarbon, v. 7, p. 372-398.

Mackin, J. H., 1948, Concept of a graded river: Geol. Soc. America Bull., v. 59, no. 5, p. 463-511.

Meier, M. F., 1960, The outbreak of a glacier-dammed lake (abs): Jour. Geophys. Research, v. 65, p. 1315.

Miller, R. K., and Dobrovolny, Ernest, 1959, Surficial geology of Anchorage and vicinity, Alaska: U. S. Geol. Survey Bull. 1093, 128 p.

Rubey, W. W., 1933, Equilibrium conditions in debris-laden streams: Am. Geophys. Union Trans. 14th Ann. Mtg., pp. 497-505.

Simons, D. B., Richardson, E. V., and Nordin, C. F., 1965, Sedimentary structures generated by flow in alluvial channels; in Middleton, G. V. (ed.), Primary sedimentary structures and their hydrodynamic interpretation: Soc. Econ. Paleontologists and Mineralogists Spec. Pub. 12, p. 34-52.

Smith, D. G., 1974, Aggradation of the Alexandra-North Saskatchewan River, Banff Park, Alberta, in Morisawa, Marie, (ed.), Fluvial geomorphology: Publications in Geomorphology, State Univ. of N.Y., Binghamton.

Stone , K. H., 1963a, The annual emptying of Lake George, Alaska: Arctic, v. 16, p. 26-39.

______, 1936b, Alaskan ice-dammed lakes: Assoc. Am Geographers Annals, v. 53, p. 332-349.

Stump, R. W., Handy, R. L., Davidson, D. T., Roy, C. J., and Thomas, L. A., 1959, Silt deposits in the Matanuska Valley, in Davidson, D. T., Roy, C. J., and associates, The geology and engineering characteristics of some Alaskan soils: Iowa State Univ. Eng. and Experiment Sta. Bull. 186, p. 3-32.

Trainer, F. W., 1960, Geology and groundwater resources of the Matanuska Valley agricultural area, Alaska: U. S. Geol. Survey Water-Supply Paper 1494, 116 p.

______, F. W., 1961, Eolian deposits of the Matanuska Valley Agricultural Area, Alaska: U. S. Geol. Survey Bull. 1121-C, 35 p.

U. S. Geological Survey, 1965, Quality of surface waters of Alaska 1961-1963: U. S. Geol. Survey Water-Supply Paper 1953, 95 p.

U. S. Geological Survey, 1969, The breakout of Lake George: U. S. Geol. Survey non-technical leaflet, 15 p.

U. S. Geological Survey, 1970, Quality of surface waters of the United States, 1964-Parts 12-15, Pacific slope basins in Washington and upper Columbia River basin to Alaska: U. S. Geol. Survey Water-Supply Paper 1959, 428 p.

Williams, P. F., and Rust, B. R., 1969, The sedimentology of a braided river: Jour. Sed. Petrology, v. 39, p. 649-679.

PART IV

The papers in Part IV represent three different approaches to the problem of river equilibrium since each is basically concerned with the concept of balance and adjustment in an equilibrated river system.

A river not only adjusts its hydraulic variables of width, depth and velocity to accomodate available discharge, but it also grades its slope. The longitudinal profile may seem smooth but, looked at in detail, is represented by a series of pools and riffles. Pools are deep stretches with very low gradients where velocity is slow, and riffles are shallows with steeper slopes and faster velocity. The water slopes and energy gradients differ also, being greater over the riffles. Pools and riffles are spaced so that they are approximately 5-7 channel widths apart. This regular spacing between pools and riffles seems to be a further manifestation of the dynamic equilibrium of a river system.

Keller and Melhorn liken the pool-riffle, point bar sequence in gravel stream beds to the ripple, dune and antidune forms on a sand river bottom, because the forms remain the same even though individual particles may move on downstream. They believe these forms are a response to the amount of sediment in transport and the discharge, and the spacing of the forms represents a constraint on the meandering pattern of a river. One wonders, however, whether this represents a constraint on meandering or whether both represent an expression of the mutual response of channel bottom and river pattern in adjustment of the river. It has been shown, for example, that the Mississippi River assumed the same braided bar (pool-riffle) spacing after cut-offs as it had in the meandering pattern before the cut-off. Keller and Melhorn discuss the origin of pools and riffles and feel that the ultimate reason for their existence is maintenance of the equilibrium of energy to work.

Most of the energy of a river is dissipated in friction. However, some small fraction is available for use in the work of erosion and transport of debris. A prevalent concept in modern thinking about fluvial geomorphology is that a river tends toward an even distribution of energy expenditure and a minimum rate of work. In achieving this state the morphometric variables characterizing the channel change regularly in a downstream direction, but each changes as little as possible.

Yang and Stall discuss the dynamic equilibrium of a river as a balance of the rate of work done to the rate of load imposed on the stream, or the balance of total sediment concentration with unit stream power. Unit stream power is defined as time rate of potential energy expenditure to the unit weight of water. Transforming data from Schumm and Gilbert to their terms, they show that the balance of sediment transport with unit stream power is maintained even as bedforms change from ripples to dunes to antidunes, and as the channel pattern is altered from straight to meandering to braided.

The tendency of a stream in equilibrium is, thus, to balance the energy expended while doing a minimum of work. In this state there is a reaction against change towards stability. General response of an equilibrated river is toward negative feedback since the adjustment made is towards negation of the change. Moreover, since there are many variables involved, the exact reaction during and after adjustment may vary from river to river. So we may expect different rivers to react differently to the same stimulus.

This is the theme of Schumm's paper as he discusses thresholds of response in watersheds, and the complexity of that response. He explains that there may be extrinsic (outside) and intrinsic (within the system) influences that cause changes in the system. Rivers respond to each of these and it may or may not be possible to discern the threshold of response in each case. The adjustment may be difficult to predict since in a large river basin the different parts of the system may reach thresholds at different times. Also we must differentiate between immediate feedback in the system and final response. Since hydrologic events are discontinuous river adjustment may be sudden and rapid and high magnitude events may not have the same results everywhere in the system. These considerations make the work of a fluvial geomorphologist difficult but, hopefully, not impossible.

11

BEDFORMS AND FLUVIAL PROCESSES IN ALLUVIAL STREAM CHANNELS: SELECTED OBSERVATIONS

E. A. Keller and W. N. Melhorn

ABSTRACT

Better understanding of morphology, morphometry and spatial relations of characteristic bed forms in alluvial stream channels may substantially aid deductions about the processes that have produced the forms. However, it is essential to recognize that bed forms such as pools, riffles, and point bars are apparently produced at relatively high channel-forming flows and are modified only at low flow. Therefore, characteristic cross-sections surveyed at low flow may depict remnant forms.

Investigation of channel morphology and morphometry in gravel-bed stream channels containing regularly spaced pools and riffles suggests that the tendency of fluvial processes to produce these forms is a significant factor in both the development of straight stream reaches and general development of the entire fluvial system.

Consideration of relations between fluvial processes and channel form suggests that the inherent ability of many gravel-bed streams to build pools and riffles may be physically explained by considering: 1) convergence and divergence of flow; 2) kinematic wave relations between the concentration and rate of transport of bed-load material; and 3) dispersive stress. However, as a strictly philosophical consideration, the origin and significance of pools and riffles may be related to the distribution of energy in the fluvial system. A balance of the time-rate of potential energy expenditure per unit mass of water through the pool-riffle sequence may exist, and this must facilitate the maintenance of processes and forms which tend to minimize energy expenditure. Furthermore, it appears that it is neither processes in alluvial stream channels which entirely control channel form, nor form which entirely controls process. Rather, form and process evolve together in harmony as feedback mechanisms inherent to open systems approaching dynamic equilibrium.

INTRODUCTION

If alluvial stream channels are considered as natural channels in which bed and bank material consists of sediment that has been deposited by the stream sufficiently recently to be still unconsolidated, then alluvial channels must be the product of processes produced by the interaction between flowing water and moving sediment. Furthermore, it is assumed that characteristic channel forms reflect the processes that produced them. Therefore, we believe that an understanding of morphology, morphometry, and spatial relations of the characteristic forms in alluvial channels will facilitate deductions about the processes which interact to produce the forms.

It is conceded that no two streams are exactly alike. The possible combinations and spatial arrangement of controlling variables such as climate, bed and bank material, and slope make each stream unique. Therefore, generalizations concerning fluvial processes are nebulous and the more general the statement the more exceptions may be found. However, when a certain type or group of streams such as gravel-bed, alluvial streams are considered, the number of exceptions to generalized statements may decrease. For purposes of this paper, let us consider gravel-bed, alluvial channels and define them as these in which a conspicuous part of the bed material is larger than 2 mm in diameter.

This paper is intended, in part, as a philosophical review covering various aspects of knowledge about gravel-bed alluvial streams. Although some of the discussion also applies to other types of alluvial streams, most of the relations between form and process considered herein are best observed and studied in gravel-floored channels.

The scope of this paper is to 1) introduce basic concepts and definitions necessary to understand relations between form and process in gravel-bed alluvial streams; 2) discuss pool-riffle spacing as it relates to straight, sinuous, and meandering channels; 3) discuss the development of straight stream reaches; and 4) discuss, in a basically hypothetical framework, the origin and significance of pools and riffles.

SOURCE AREAS

The source areas for much of the data in this paper are four gravel-bed alluvial streams:

Dry Creek near Winters, California is an entrenched meandering stream. Stream flow is intermittent with an average of 29 days of flow a year and mean annual flood of 785 cfs (Keller, 1969, p. 27). The 1.30 mile studied reach is located upstream from California highway 128 bridge, SW corner Sec. 21, T8N, R1W. (The intermittent flow and classical pool-riffle sequences in Dry Creek combine to make this stream an ideal field laboratory).

Durkee Run, Lafayette, Indiana is a small, ungaged intermittent stream with fairly well-developed pool-riffle sequences. The 0.38 mile studied reach is located upstream from where the stream crosses 9th Street in Lafayette, E 1/2 Sec. 33, T23N, R4W. The channel bed and bank material is primarily alluvium with minor flood plain development.

South Fork Wildcat Creek near Dayton, Indiana is a perennial stream with fairly well-developed pool-riffle sequences. The 2.80 mile reach studied is located downstream from Indiana highway 38 bridge, SE corner Sec. 4, T22N, R3W. The channel bed and bank material is alluvium with a well-developed flood plain. Average discharge over a 26 year period is 227 cfs (Water Resources Data for Indiana, 1969, p. 66).

Wea Creek near Lafayette, Indiana is an ungaged perennial stream with well-developed pool-riffle sequences. The 1.05 mile reach studied is located upstream and downstream of Indiana highway 25 bridge, SW 1/4 Sec. 1, T22N, R5W. The channel bed and bank material is primarily alluvium with a fairly well-developed flood plain.

A summary of general channel characteristics for the source areas is shown in Table 1.

The advantage of studying natural alluvial channels cannot be emphasized too strongly. It has been recognized in simulation that much of what has been considered as functional relations among hydraulic variables of flow in alluvial channels may simply be due to constraints caused by the flume (Maddock, 1969). The problem is that flumes are not natural channels, and the variability of natural alluvial channels cannot always be studied in flumes. This results, in part, because the rigid sides of flumes will not allow flow to contract and expand to produce the river-like fluvial hydraulics of convergent and divergent flow. This statement should not be construed as an appeal to abandon flume studies, but rather is an appeal to balance them with more and varied field observations and analyses.

BASIC CONCEPTS AND DEFINITIONS

Channel Pattern

Channel pattern refers to the configuration of a river as it appears in plan view. There appear to be two main patterns: 1) braided channels characterized by the existence of numerous islands and bars which divide and reunite the channel; and 2) channels that are not braided. Because long, straight reaches are relatively rare, channels that do not braid may be designated as sinuous. However, sinuous channels commonly contain relatively short, straight reaches. These straight reaches may be produced in two principal ways: 1) meander cutoffs which reduce sinuosity and thus locally increase channel slope; and 2) lateral migration of adjacent meander bends, which tends to increase sinuosity and thus locally decrease channel slope. Furthermore, although a straight channel has straight banks it implies neither a uniform stream bed nor a straight thalweg (Leopold and Wolman, 1957). Meandering channels may be considered as a special type of sinuous channel characterized by a series of relatively symmetrical curves with sinuosity greater than 1.5. However, the definition of symmetry necessary to designate a channel as meandering has not been established. Therefore, it seems advantageous to use the term "meandering" rather loosely, and apply the term to any sinuous channel with sinuosity greater than 1.5, as suggested by Leopold *et al* (1964, pp. 295-296).

Bed Forms

Excluding the fluid phase, the most obvious forms in an alluvial channel are bed forms. A bed form is any irregularity produced on the bed of an alluvial channel by the interaction between flowing water and moving sediment (Simons and Richardson, 1966, p. V). There appears to be two main types of groups of bed forms: 1) pools, riffles, and point bars, which tend to give some gravel-bed channels their basic morphology; and 2) ripples, dunes, and antidunes, which are controlled by the fluid phase and are not generally a significant part of the basic channel morphology in gravel-bed channels. However, if sufficient sand

Table 1 Summary of Channel Characteristics for the Studied Streams

Stream	Type of Stream	Bank Material	Sinuosity	Channel Length (mi)	Channel Width (ft)	N*
Dry Creek	Intermittent	Alluvial	2.40	1.30	33	38
Durkee Run	Intermittent	Alluvium	1.13	.27	14	18
Wea Creek	Perennial	Alluvium	1.38	1.05	67	16
Wildcat Creek	Perennial	Alluvium	1.42	2.80	82	30

*N is the number of pool-riffle sequences.

is available, slowly migrating ripples and dunes at low flow may be superposed on and partly mask the more stable pools, riffles, and point bars.

Pools, riffles, and point bars may be defined on the basis of their morphology (Keller, 1971b).

Pool: A topographically low area produced by scour and which generally contains relatively fine-grained bed material. Pools are usually associated with a point bar.

Point bar: An accumulation of relatively coarse-grained bed material on the concave side of the thalweg and adjacent to a pool. The point bar and pool together produce an asymmetrical cross-profile.

Riffle: A topographic high area produced by the accumulation of relatively coarse-grained bed material. Ideally, the inflection point of the thalweg is located on the riffles between successive pools. At the inflection the cross-profile is generally symmetrical.

Pools and riffles have also been defined on the basis of descriptive hydraulic characteristics (Leopold, et al, 1964, p. 206). At low flow, pools are recognized by relatively deep, slow-flowing water with low water-surface gradient, whereas riffles are recognized by relatively shallow, fast-flowing water with a steeper water-surface gradient. As the water level rises the difference in gradient tends to disappear until, at about bankfull stage, the pool-riffle sequence is "drowned out".

Quantitative definitions for pools and riffles have been suggested by Dolling (1968) and Yang (1971b). Dolling used the velocity to depth ratio as a criterion, whereas Yang used energy gradients to distinguish pools from riffles. Quantitative definitions valid for both high and low flow will be valuable for a mathematical treatment of pools and riffles. However, present quantitative definitions based on hydrology or energy gradient appear to be valid only at low flow.

Fluvial Hydraulics

In many gravel-bed alluvial channels, pools, riffles and point bars appear to be produced at relatively high, channel-forming flows and are modified only at low flow. Under these conditions, conventional hydraulics apply at low flow, when the channel is essentially a rigid container for the liquid phase. However, at high flow when appreciable sediment is being transported by the stream, conventional hydraulics are no longer applicable because many of the variables are not unique (Maddock, 1969, p. 67). Leliavsky (1966, p. 98) also distinguishes fluvial hydraulics from hydraulics in general as a necessity to understanding natural streams.

Three important principles of fluvial hydraulics emphasized by Leliavsky are 1) in no part of alluvial channels are contiguous stream lines parallel to one another or to the bank of the river, 2) the greater the curvature of the horizontal projections of the stream trajectories, the deeper the channel scours below them, and 3)(perhaps the most significant principle in fluvial hydraulics), de Leliavsky's (1894) convergence-divergence criterion:

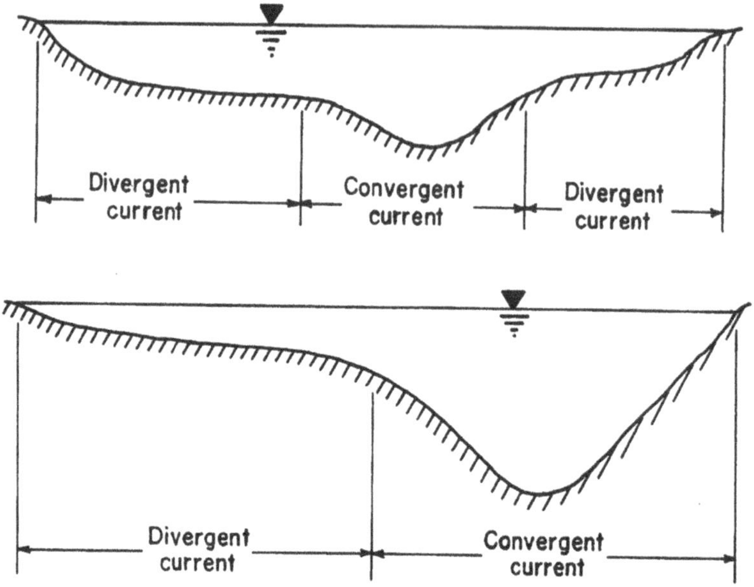

Figure 1 Illustration showing de Leliavsky's convergence-divergence criterion for common river cross sections (▼ represents the water surface).

After Leliavsky, 1966

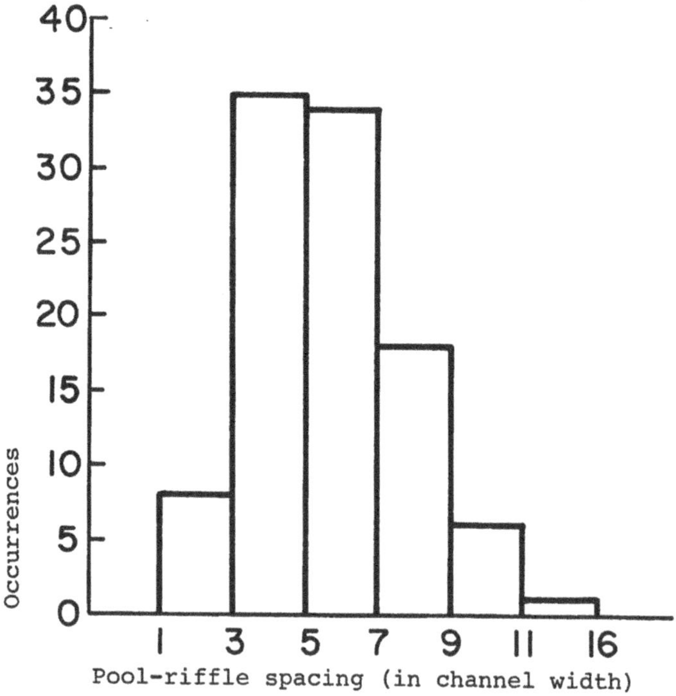

Figure 2 Frequency of pool-riffle spacing for the 102 pool-riffle sequences in Dry Creek, Durkee Run, Wea Creek and Wildcat Creek.

"... It was an inference, i.e., a general law, drawn from a very large number of individual facts, that is observations, derived from nature and from engineering practice. This aspect of de Leliavsky's criterion is best described by the term "ex flumine nato", i.e., born from the river, for it was the natural structure of the erodible river channel and the distribution of the velocities therein, which suggested its basic idea; namely, that erosion was always correlated with, and due to, convergent flow, while on shoals the flow was always divergent."(Leliavsky,1966,p.162)

Typical river sections showing de Leliavsky's criterion are shown in Figure 1. In general, at high flows, convergence of flow is typical of pools, whereas divergence of flow is characteristic of shallows (riffles and point bars). This criterion is a powerful tool that facilitates analysis of the relationship between morphology and process, and allows inferences about hydrology and sediment transport.

POOL-RIFFLE SPACING

In this paper we consider only gravel-bed, alluvial channels with regularly spaced pools and riffles. In these channels, pools and riffles remain in approximately the same locations over a series of flows. Therefore, these channels may be considered as morphologically stable. However, there exists, at the other end of the continuum, alluvial channels lacking pools and and riffles and with little morphological stability over a series of flows. Furthermore, intermediate zones exist for channels that have poorly developed pools and riffles. These intermediate channels have some characteristics of both morphologically stable and morphologically unstable channels. This argument is consistent with Lokhtine (1909), who attempted to define morphological stability in terms of a coefficient of fixation, defined as the ratio of the average diameter of the bed material to the slope of the free water surface. High values (166.0 cm for the Dniester River) reflect morphologically stable channels whereas low values (3.1 cm for the Vistula River) reflect morphologically unstable channels. The coefficient of fixation for all other streams studies by Lokhtine fell between these two extremes.

Analysis of Pool-Riffle Spacing Data

Pool-riffle spacing is measured along the channel from center point to center point of consecutive pools. The deepest point in the pool is considered the center point. This helps remove some of the subjectivity of deciding where the pool or riffle starts or stops, and the periodicity is not affected as long as a consistent reference point is used. The average pool-riffle spacing for the four gravel-bed, alluvial stream channels studied is: Dry Creek near Winters, California, 5.9 times channel width; Durkee Run, at Lafayette, Indiana, 6.0 times channel width; Wea Creek, near Lafayette, Indiana, 5.3 times channel width; and South Fork Wildcat Creek near Dayton, Indiana, 5.0 times channel width. The pool-riffle spacing in Dry Creek was also considered in terms of channel pattern. This showed that average spacing for pools was 5.9 times the channel width for the straight and sinuous reaches inclusive, 5.7 for the straight reaches, and 6.1 for the sinuous reaches. This suggests that spacing of pools and riffles is independent of channel pattern. The average spacing for all four streams (102 pool-riffle sequences) is 5.6 times the channel width (Fig. 2). These data strongly confirm the conclusion of Leopold et al (1964, p. 203), that pools and riffles are spaced at 5 to 7 times the channel width.

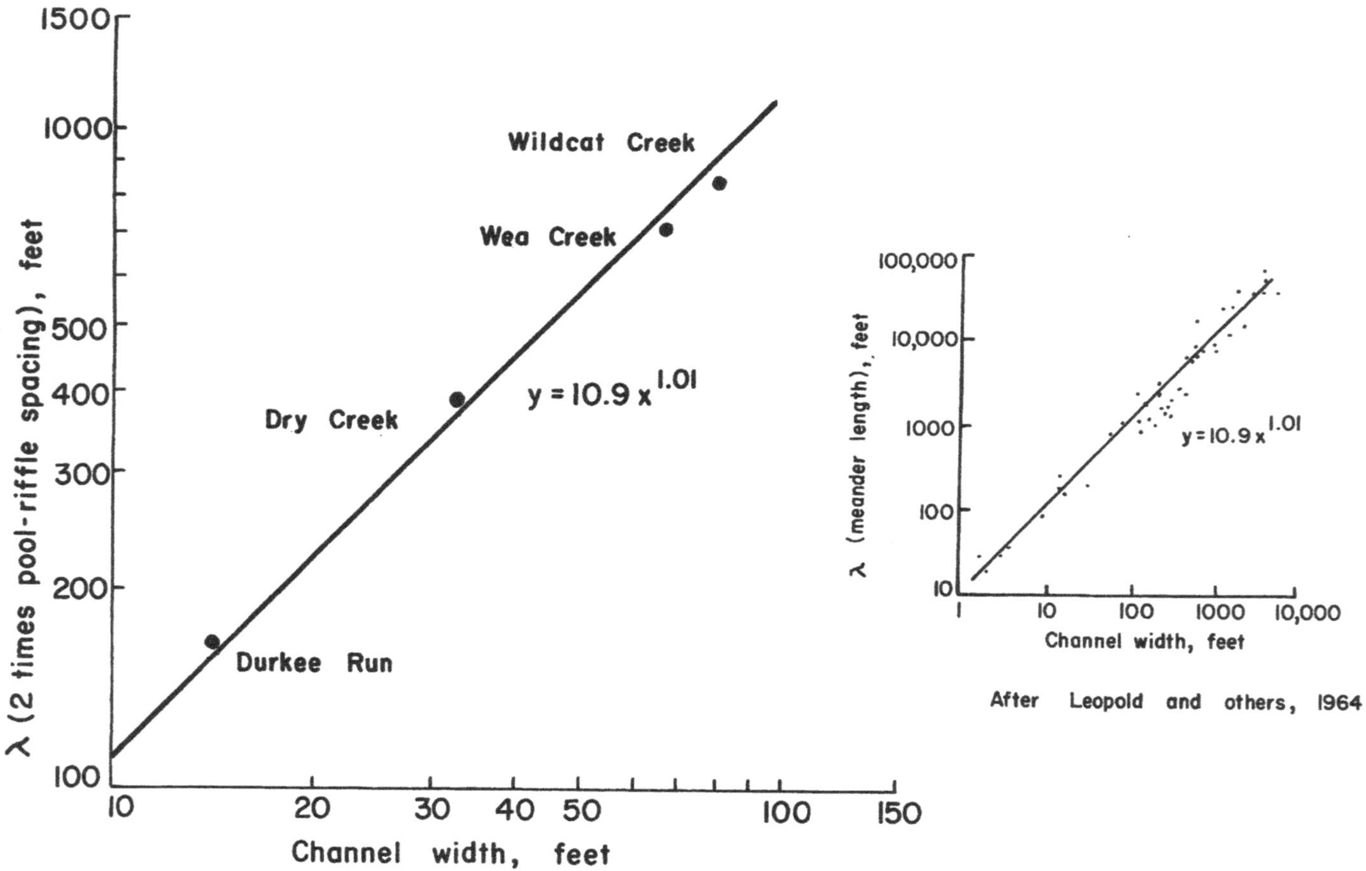

Figure 3 Comparison of pool-riffle spacing to spacing of meander length.

The conclusion, from a pool-riffle spacing consideration, is that in many gravel-bed alluvial streams, one of the primary fluvial processes is that of building pools and riffles. Two lines of evidence suggest this may be valid: 1) pool-riffle spacing appears to be independent of channel pattern, and therefore, given dynamic equilibrium conditions, as straight and sinuous reaches merge there is no change in the spacing of the pools and riffles, and 2) the relationship between meander length and channel width (Leopold et al, 1964, p. 296) is very similar, if not the same, as the relationship between twice the pool-riffle spacing and channel width (Fig. 3). We believe that spacing of pools at 5 to 7 times the channel width may impose a constraint on the meandering such that one-half the wave length of symmetrical meanders rarely exceeds this value significantly. Therefore, similarity in spacing of pools and riffles in both straight and meandering channels suggests that processes which create the tendency for formation of pools and riffles are present in many alluvial streams, and the mechanism, which is associated with some types of wave phenomena, may produce a meandering channel. This last statement is consistent with the concept that the tendency for meandering is present in straight channels (Leopold et al, 1964, p. 203).

An important unanswered question remains: Why should pools and riffles be spaced at about 6 times the channel width? We speculate that this question will remain unanswered until we know more about the concept of the unit cell which controls the development of channel networks. Studies concerning hexagonal hierarchies by Woldenberg (1969) and development of the unit hexagonal cell by Coffman (1972) may help in this endeavor.

DEVELOPMENT OF STRAIGHT STREAM CHANNELS

Lateral Migration of Meander Bends: An Example From Dry Creek

Two straight reaches in Dry Creek (A-A' and B-B' in Fig. 4) appear to have developed by lateral migration and downward erosion of meander bends. Evidence consists of the well-developed slip-off slope on the inner bank of the large bend between the two straight reaches. Data on rates of lateral migration of bends in Dry Creek are scanty, but erosion of one bend downstream from the straight reaches can be estimated. A local rancher stated that the fence that spanned part of the meander bend in 1969 (Fig. 5) was on solid ground in 1967. Allowing for some original curvature of the fence, the rate of lateral erosion of this bend is about 2 to 3 feet per year. However, the water years of 1966-1967 and 1968-1969 produced the largest amount of run-off since 1960 when stream flow data were first recorded for Dry Creek. Therefore, the average yearly rate of migration of meander bends over a longer period may be considerably less.

In Dry Creek it was necessary to distinguish two types of pools, which we have designated primary and secondary. Primary pools are these that exhibit deep scour (commonly to older, partly-consolidated alluvium), are generally found on bends, and are always associated with a point bar. Where a primary pool is on a bend, the point bar is on the inside of the bend and slightly upstream of the pool. Secondary pools are scoured to a lesser depth than primary pools, and may not always be associated with a well-developed point bar. Locations

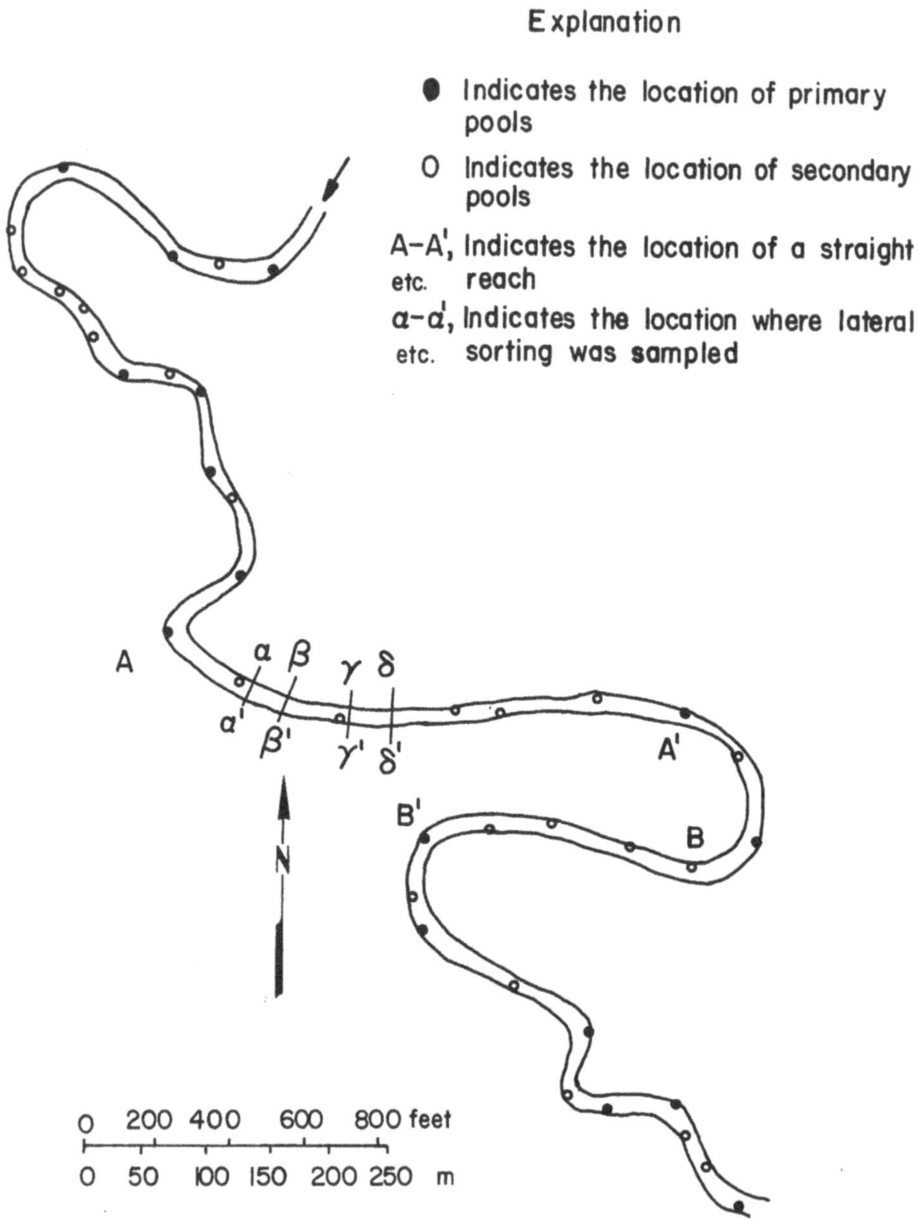

Figure 4 Location of primary and secondary pools for the study reach of Dry Creek. Channel width is exaggerated for illustrative purposes.

Figure 5 Lateral migration of a meander bend in Dry Creek. (Photograph taken in 1969.)

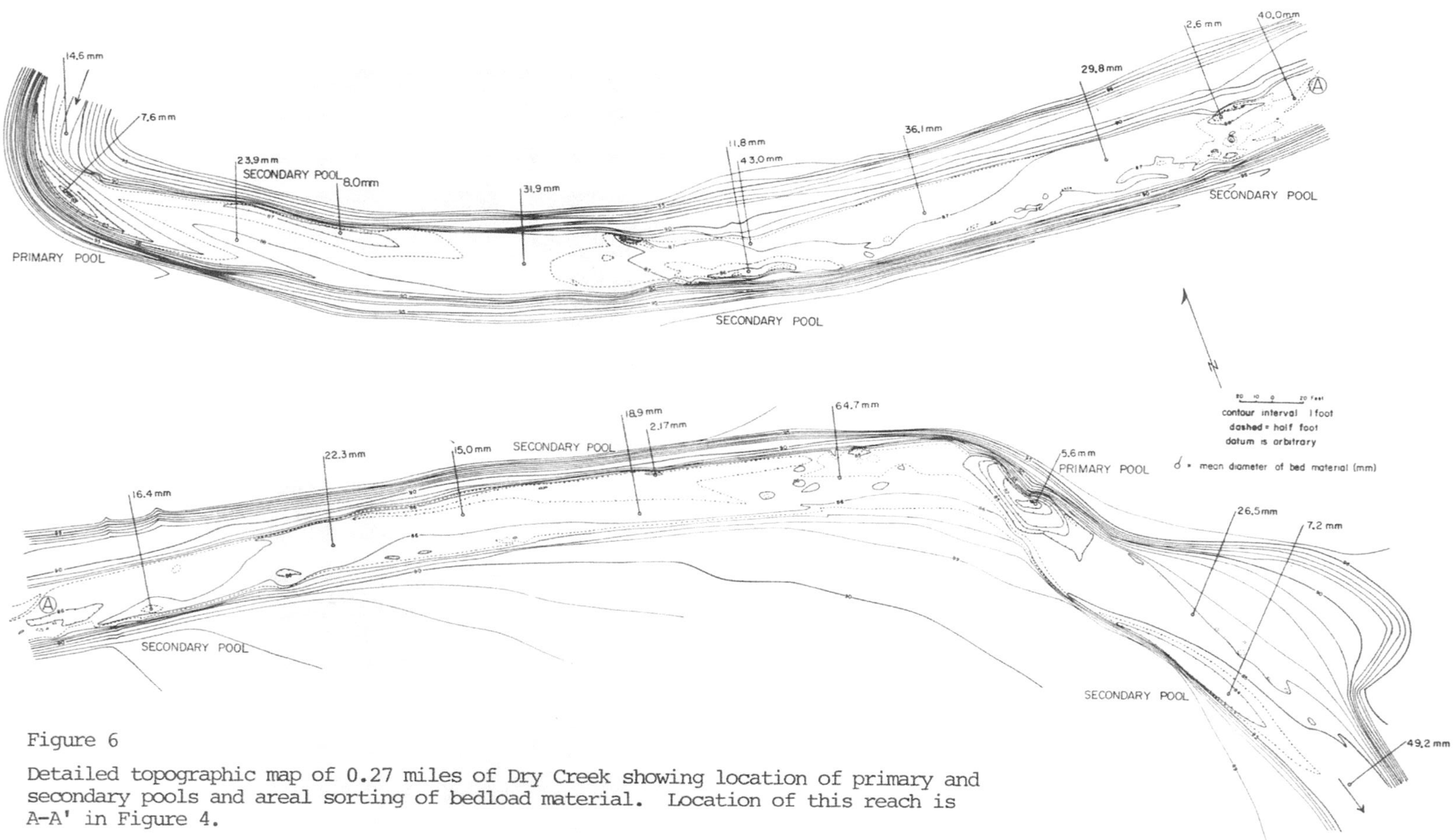

Figure 6

Detailed topographic map of 0.27 miles of Dry Creek showing location of primary and secondary pools and areal sorting of bedload material. Location of this reach is A-A' in Figure 4.

of primary and secondary pools, as mapped in 1969 for 1.30 miles of channel along Dry Creek, are shown in Figure 4. A detailed topographic map (Fig. 6) of .27 miles of Dry Creek channel (A-A' on Fig. 4) shows the morphology of a straight reach between two primary pools. The secondary pools between the bends alternate from one bank to the other. The mean pool-riffle spacing of this reach is about 7 times the channel width. The bed profile (thalweg) shown on Figure 7 is for the same series of pools mapped on Figure 6.

The distribution of pool-riffle spacing for Dry Creek (Fig. 8) shows that although mean spacing is 5 to 7 channel widths, the most frequent spacing is 3 to 5 times the channel width. The abundance of pools spaced at 3 to 5 channel widths, and relative absence of pools spaced at greater than 9 times the channel width may suggest that, for the development of this type of straight reach, pools are not initiated at much less than 3 channel widths and are unstable when spaced at more than 9 channel widths (Keller, 1972a). Therefore, as channel length between primary pools increases owing to lateral migration of meander bends, new secondary pools may be formed to maintain a pool-riffle spacing of approximately 6 times the channel width. A small scour area about 70 feet upstream from the downstream primary pool shown in Figures 6 and 7 is spaced at about 2 channel widths from the adjacent pools, and may become a new secondary pool as the meander bend continues to move laterally. It must be remembered that addition of new pools will not occur until a considerable increase in sinuosity has taken place. For example, assuming pools are stable if spaced at 6 channel widths and with sinuosity of 1.25 to 1.50, then sinuosity would need to increase to about 2.00 before pools will be spaced greater than 9 channel widths apart and new pools are added.

The number of secondary pools in Dry Creek is a function of the mean spacing and channel length between primary pools (Fig. 9). The correlation coefficient for this relation is 0.94 and a Student "t" test of significance for $n=15$ at $n-1$ degrees of freedom yields a value of $t=4.43$, setting the confidence level at 99 percent. Relations shown in Figure 9 also suggest that if there are no secondary pools between consecutive primary pools, the spacing is the same. This may support the argument that secondary pools were produced at 3 to 5 times the channel width as channel length between consecutive primary pools increased beyond a threshold of about 9 channel widths.

In 1969, one of the perplexing problems associated with the straight reach in Dry Creek (A-A' in Fig. 4 and Fig. 6) was the apparent stability of the banks. With the exception of the bends, the stream did not appear to be eroding its banks as actively as might be expected. An investigation of the local geology suggested that there was no apparent structural control, and that the bed and bank material were similar to that along the rest of the stream.

Hypothetically, the stability of these straight reaches, which have developed owing to lateral migration of meander bends, may be related to the fact that, as the straight reach develops, the sinuosity locally increases but slope decreases. This is consistent with the argument presented by Schumm and Khan (1972) that threshold values of slope may exist that partly control channel pattern. A generalization of one of their conclusions is that stable, sinuous channels may have a greater slope than stable, straight channels, but that the change in pattern will not take place until threshold values of

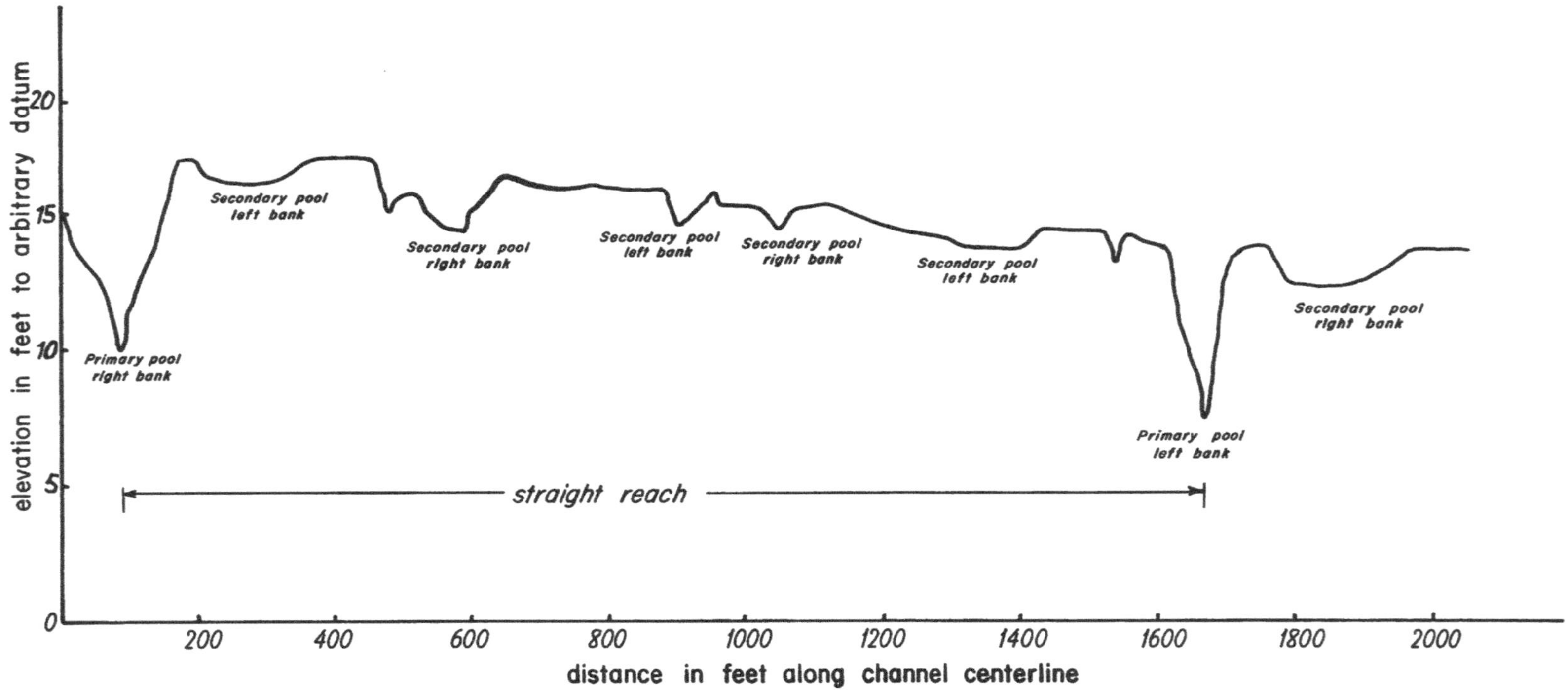

Figure 7 Bed profile (thalweg) of the mapped channel in Figure 6, showing primary and secondary pools.

slope are exceeded. Therefore, in a local sense, the reduction of slope, the result of development of some straight reaches, may produce a more stable channel than one that is more sinuous. However, the channel slope of the straight reach (A-A' in Fig. 4) is 0.0025, which is the same as the slope for the entire study reach. Nevertheless, as the straight reach developed channel slope must have decreased, and this may explain why the reach appears relatively stable. However, the bends are still eroding, indicating that the stream is still actively developing.

Meander Cutoff: An Example from Wildcat Creek

As an alluvial channel migrates back and forth across the floodplain, new straight, meandering, and braided reaches are constantly being created, maintained, or destroyed. New channel may be created quickly when a meander bend is cut off by overbank flow; this results in a local reduction in sinuosity and channel length, and creates a locally steeper channel slope.

The South Fork of Wildcat Creek produced a meander cutoff in the Spring of 1971. Local residents stated that the cutoff, which produced 1,320 feet of new channel, happened during a single high flow. The path of the new channel initially transected a cornfield and was partly controlled by one rill of a previously existing rill system which drained the flood plain. Figure 10 shows the channel morphology of the abandoned channel compared with the new channel as mapped in the summer of 1972. Table 2 summarizes channel characteristics before and after the cutoff.

Table 2 Channel Characteristics Before and After a Meander Cutoff in the South Fork of Wildcat Creek, Near Dayton, Indiana.

Channel Characteristics	Before Spring 1971	After Spring 1971
Length	2,497 feet	1,320 feet
Width	82 feet	130 feet
Slope	0.0013	0.0024
Pool-riffle spacing*	5.8	2.5**

* pool-riffle spacing is measured in channel widths
** incipient pools

The cause of the cutoff is not entirely known, but is probably related to partial damming of the channel by a large log jam near the downstream end of the meander bend (Fig. 10). This may have backed up water sufficiently to permit overbank flow and cutting of a new channel where the bank was least stable. Aerial photography taken in 1968 clearly shows the log jam, and a lack of trees along the flood plain where overflow first breached the channel bank. It is not known if the log jam was produced by the inability of the braided channel to move the logs, or if the braided channel is the result of the log jam.

Channel morphology in the now abandoned meander bend is quite different from that in the new channel. Pools in the old channel are spaced at about 6 times the channel width, whereas incipient pools in the new channel, as mapped in 1972, were spaced at about 3 channel widths. Observed channel changes since 1971 suggest that as of April, 1973 the channel development

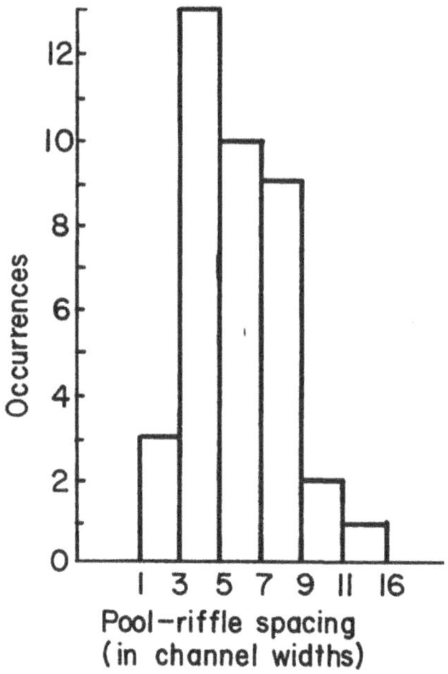

Figure 8 Frequency of pool-riffle spacing for Dry Creek.

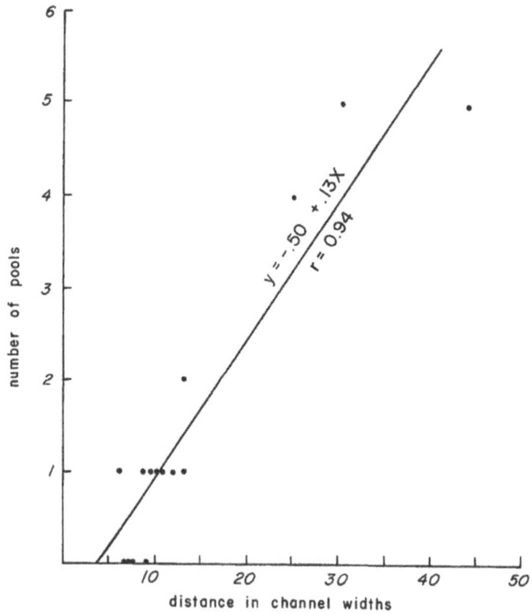

Figure 9 Number of secondary pools between consecutive primary pools as a function of distance between consecutive primary pools (in channel widths).

is proceeding as anticipated by Keller (1972a): the asymmetric shoals which alternate from bank to bank (Fig. 10) are being converted to point bars as pools and riffles develop. Three of the original four incipient pools are still developing and it is expected that the reach will eventually have three well developed pools, each associated with and slightly downstream from a point bar.

Other important differences between the two channels are slope and width. The new channel is nearly twice as steep and about 60 percent wider than the old channel. The increase in channel width is at least partly a result of the lack of trees along the new channel. The banks can be eroded easily, and therefore a wider channel is produced. However, the increased channel width may also be related to the steeper slope, as suggested by Schumm and Khan (1972). Therefore, the cutoff in Wildcat Creek, which resulted in a locally much steeper slope and therefore greater tractive force, produced a new channel that was not stable with a channel width of less than about 130 feet. It is expected that as slope locally decreases with time, the trees encroach upon and begin to stabilize the banks, the channel width will slowly be reduced to about 82 feet, which is the average width of channel upstream and downstream from the cutoff. Furthermore, even with no further increase in sinuosity, the pools, following the reduction of channel width, will be spaced at about 5.3 channel widths which is nearly the average spacing for the rest of the channel.

ORIGIN AND SIGNIFICANCE OF POOLS AND RIFFLES

A complete explanation for the origin and significance of pools and riffles must consider not only fluvial processes which produce the forms but also related problems such as: 1) how the generally observed areal sorting of bed material, i.e., relatively fine-grained bed material in pools and relatively coarse-grained bed material in riffles, relates to the processes which produce pools and riffles, and 2) why, in streams that occasionally scour to older alluvium in pools, there may be no bed material observed at low flow, whereas the adjacent point bar and riffles contain abundant bed material. On the other hand, the explanation for the origin and significance of pools and riffles must also consider why, from a theoretical view, these forms should be produced at all. In other words, we must not only explain how pools and riffles are physically developed but why, in terms of what streams are doing, they are produced.

Processes suspected as being important in the origin of pools and riffles are: 1) convergent and divergent flow; 2) kinematic wave relations between concentration and rate of transport of bed-load material; and 3) dispersive stress.

Convergence and Divergence of Flow

Channel topography for Dry Creek (Fig. 6) and Seneca Creek, Maryland (Leopold, et al., 1964 (pp. 204-205) suggests that asymmetric cross-profiles for pools are associated with convergent flow, and the more symmetrical profile between the pools (riffles) are associated with divergent flow as proposed by de Leliavsky (Fig. 1). This indicates that pools are produced by convergent flow which causes increased tractive force and scour at high stages, whereas riffles are produced by divergent flow and deposition. However, the entire process is considerably more complex. It is emphasized that scour takes place only over a relatively restricted area in the channel. Furthermore, these convergent-divergent relationships may not hold for low flow conditions. This results because secondary scour channels may develop on riffles at high flow, even though the flow is primarily divergent. These are not significant at bankfull

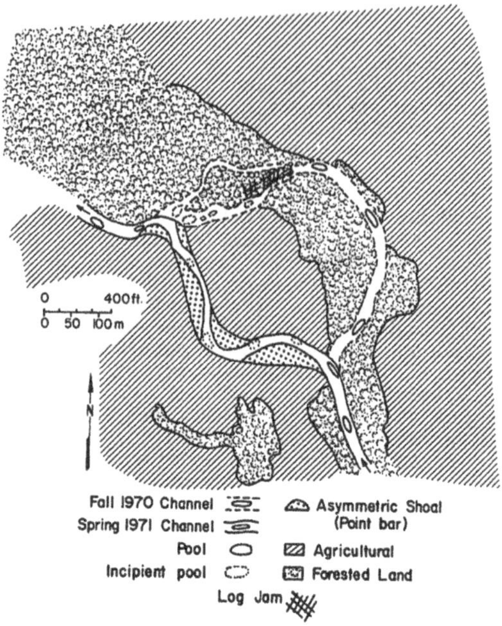

Figure 10 Channel morphology before and after a meander cutoff in Wildcat Creek.

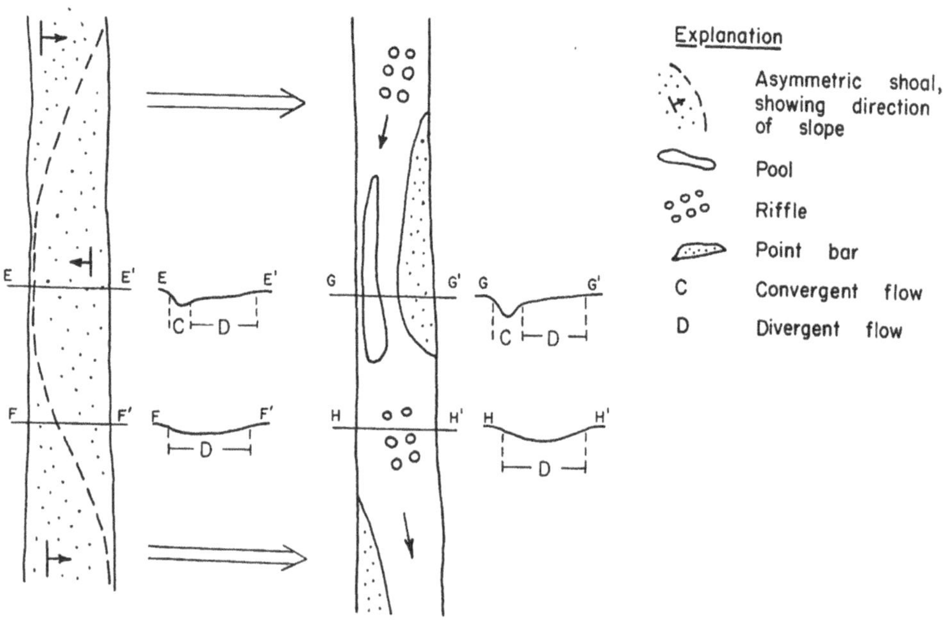

Figure 11 Idealized transformation from asymmetric shoals to pools, riffles and point bars.

stage, but at low flow with a much reduced channel width these channels may actually converge flow. If one of these channels heads into a large scour pool formed at high flow, then water at low flow will actually diverge over the pool. This is just the opposite of what occurs at high flow. Therefore, bankfull morphology measured at low flow depicts conditions at high flow, but the water at low water flow may be at the mercy of constraints which formed at bankfull stage.

Assuming that pools and riffles are produced by convergence and divergence of flow, fluvial processes which may produce a typical pool-riffle sequence can be hypothetically described by a two stage-transformation. Stage one is characterized by transformation from the original condition, whatever that may be, to a series of asymmetric shoals which alternately slope first toward one bank, then toward the other. Asymmetric shoals or "skew shoals", as defined by Quraishy (1944), form early in the development of alluvial channels (Keller, 1972a). They are distinguished from point bars by stage of development; otherwise, the two terms are analogous. Stage 2 is characterized by transformation from asymmetric shoals to pools, riffles, and point bars. Although the development of asymmetric shoals is not completely understood, they may be related to Yang's (1971a) law of least time rate of energy expenditure. Therefore, if streams must decrease channel slope to minimize energy expenditure as Yang proposes, the channel morphology must be compatible with processes which tend to reduce slope. Asymmetric shoals thus may develop because they are the most probable bed form to initiate processes which facilitate slope reduction by increase in channel length. Furthermore, for asymmetric shoals, it is neither form which produces process nor process which produces form. Rather, development of the shoals may be considered as a metamorphosis resembling a feedback mechanism in which process and form evolve in harmony. An idealized diagram showing process for the second stage is shown in Figure 11. The thalweg is the dashed line, which denotes the boundary between asymmetric shoals. The morphology of the shoals produces an asymmetric profile at the widest part of the shoal, and this tends to converge flow in a restricted area where the water is the deepest. The water, after leaving the area of convergent flow, starts across the channel toward the opposite bank. Here flow is divergent and, therefore, the cross profile is symmetrical. This may produce incipient pools and riffles: pools where scour is greatest and riffles where deposition is greatest. As this process continues, pools and riffles grow at the expense of asymmetric shoals. The end result is a well developed pool-riffle sequence within a meandering thalweg system. What is left of the asymmetric shoal is referred to as a point bar. This does not imply that point bar formation is primarily an erosional process. As pools scour part of the original shoal, and riffles cover up other parts, flow is still divergent on the upper part of the shoal, and deposition of material continues there. If this were not the case, the stream channel would become wider but there would be little increase in channel length. This argument is further supported by the fact that meander bends migrate laterally, meanwhile maintaining a relatively constant width, a condition which would be impossible if there was not a balance between scour on the outside and deposition on the inside of the bend.

It is important to know if pools and riffles retain their basic morphology at channel-forming flows (bankfull). Flume studies, because they do not normally use gravel channels do not produce well developed pools and riffles. However, the

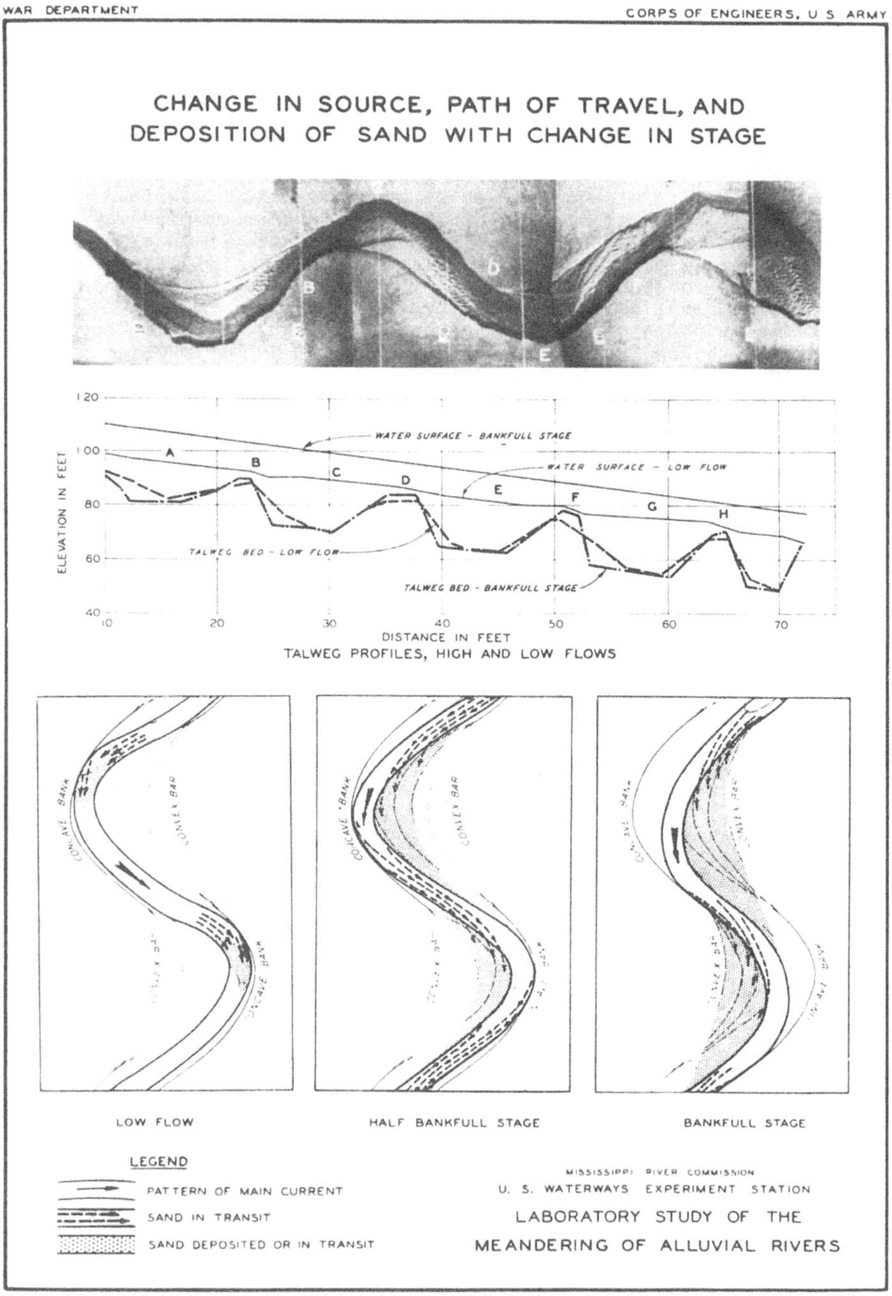

Figure 12 Change in morphology and thalweg profile with change in stage (after Friedkin, 1945).

characteristic spacing of shallows and deeps has been observed (Friedkin, 1945), and this is significant because it may reflect periodic scour and deposition characteristic of pools and riffles. Figure 12 summarizes Friedkin's experiment. The thalweg profile indicates that the basic form of the channel bottom, in a vertical scale, does not change with increasing discharge up to bankfull stage, where the forms are considered drowned-out. Furthermore, the tendency of the main current to migrate to the inside of the bends with increasing discharge explains why pools are nearly always located slightly downstream from the point of greatest curvature of the point bars. A conclusion which may be proposed, on the basis of Friedkin's experiment, is that pools and riffles do not apparently change their basic morphology at channel-forming discharge. This conclusion is necessary if convergence and divergence of flow is to maintain the pools and riffles.

Dispersion and Kinematics

Two important processes that must be considered as significant in gravel-bed alluvial streams with pools and riffles are the effect of dispersive stress on bed material (Bagnold, 1968) and the influence of kinematic waves on channel morphology (Langbein and Leopold, 1968).

In riffles, the largest bed material commonly forms a single layer on the surface (Keller, 1971a). Bagnold (1968) explained the tendency for large material to be deposited on the surface of riffles by a dispersive stress. This stress is proportional to the square of the particle size and, therefore, the relatively large particles will migrate toward the region of least shear at the upper surface of the bed. As large particles reach the surface of the bed, hydraulic and kinematic action may concentrate them on the riffles. Yang (1971b) has used the dispersive stress to explain how pools and riffles form. However, even though dispersive stress may facilitate the formation of pools and riffles it cannot produce all of them. This results because dispersive stress cannot cause the observed complete scour to older alluvium in some pools.

Langbein and Leopold (1968) concluded that riffles in gravel-bed streams could be considered as kinematic waves produced by particle interaction during fluvial transport. They further concluded that stability of riffles is related to the interaction between two opposing factors: increasing water velocity, which tends to increase the wave length of the kinematic wave; and decreasing amplitude owing to erosion, which tends to decrease the wave length. The balance between these factors results in riffles which do not appreciably migrate downstream. This explains Lokhtine's (1909) observation for channels with a succession of pools and riffles, that although the bed material is continuously moving through the channel, the location of the pools and riffles remains fixed. This may be considered as a particle by particle replacement process in which the form remains unaltered. Although kinematic theory explains why pools and riffles are relatively stable, it does not completely explain the origin of the forms. We might, for example, argue that the concentration of coarse-grained bed material is a function of fluvial hydraulics rather than conversely.

Areal Sorting of Bed-Load Material

Areal sorting of bed-load material, as discussed by Leopold et al., (1964, p. 203) is characterized by the tendency for bed material on riffles to be coarser-grained than in pools. Areal

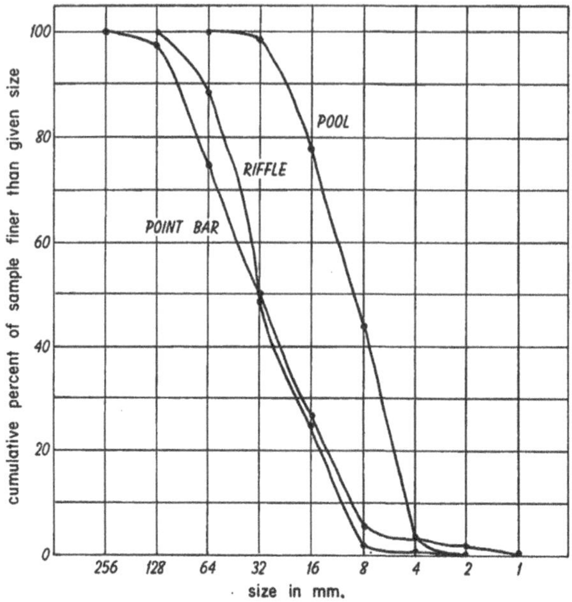

Figure 13 Frequency distribution of bed-load material over a pool, adjacent point bar, and downstream riffle. The location of the sampled area is in the vicinity of γ-γ'in Figure 4.

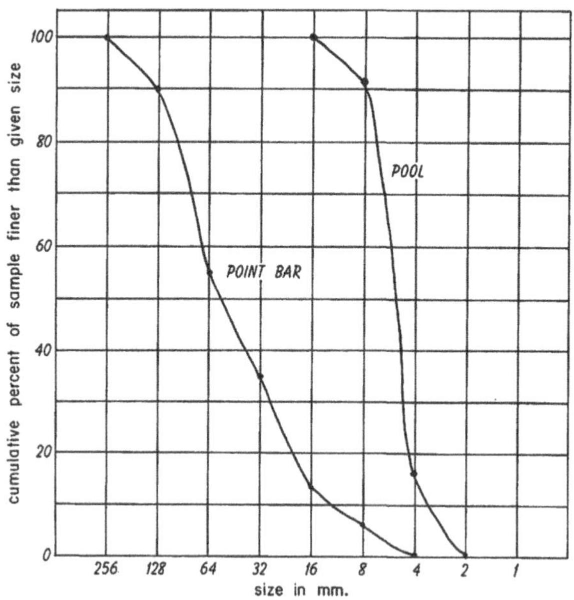

Figure 14 Frequency distribution of bed-load material over a pool and adjacent point bar. Location of the sampled area is in the vicinity of A' in Figure 4.

sorting in Dry Creek is especially well developed. The detailed topographic map (Fig. 6) shows mean diameters of bed material for pools, riffles, and point bars as determined by Wolman's (1954) method of sampling coarse-grained bed material. Frequency distributions (Fig. 13 and Fig. 14) further demonstrate the tendency for relatively coarse particles to accumulate on riffles and finer particles to concentrate in pools. Lateral sorting across the channel is also characteristic of Dry Creek. This is demonstrated by measuring the ten largest size bed-load particles at three foot intervals along section lines (α-α', etc., Fig. 4). The average size of the ten particles in each interval is shown in Figures 15 and 16. These graphs show that the largest material on a point bar gradually decreases in size across the channel to the bottom of the pool. The largest bed material on riffles is located near the center of the channel. The distribution of bed material, together with the topographic map, suggests that there is good correspondence between areal sorting and channel form.

The origin of areal sorting of bed-load material has been a subject of debate (Keller, 1971a, 1972b, and Teleki, 1972). Any solution must explain not only the areal distribution of bed material but also the lack of bed material in some pools. Figures 17 and 18 are a downstream view of a well-developed pool and point bar in Wea Creek, Indiana. The light area in the center of the pool is scoured into an underlying basal glacial till. Although the point bar and adjacent riffles contain abundant bed material, the center part of the pool is completely void of bed material. A similar condition is shown for Dry Creek near Winters, California (Fig. 19 and 20). It is suggested that the very coarse-grained material downstream from the pool can be explained by some process that produces greater tractive force in the pool than in the adjacent riffle. The pool is scoured to older, partly consolidated alluvium and contains no coarse-grained bed material. It is assumed that the fan-shaped deposit of coarse-grained bed material was deposited at high flow when convergent flow in the pool diverged on the adjacent downstream riffle.

It seems obvious that in many alluvial channels the velocity distribution at low flow, with little or no sediment transport, is such that bottom or bed velocity in pools is significantly less than in adjacent riffles. At this low stage normal hydraulics apply and de Leliavsky's criterion may not be applicable. The riffles at this stage act as small dams to the slower-moving, deeper water in the pools. There may not be sufficient traction to transport the bed-load, and the only moving sediment may be relatively fine-grained material which is winnowed off the riffles and settles in the pools. Therefore, at low flow the steeper slope of the riffle is more significant in producing tractive force than any convergence of flow in the pools. However, it also seems obvious that pools have to scour at some flow, and if not at low flow, then certainly at high flow. Along this line of thinking, Carey and Keller (1957) stated, "It may develop that in effect the bends are scoured out at high stages, and that much of the scoured material is simply moved to the first crossing." They also concluded that slope (in a local sense) is a result rather than a cause, and suggested that bed-load movement in bends (pools) may be completely different from bed-load movement in crossings (riffles). If it is assumed that pools are scoured during high flows, perhaps bankfull stage, then the hydraulics are much different than at low flow. Thus, if characteristic cross-sections, surveyed at low flow, depict relict forms produced at high flow, then pools converge flow and riffles diverge flow. Therefore, the principal conclusion remains that at some stage the tractive force in pools is greater than for riffles, and this in conjunction with dispersive stress and kinematic wave relations between the

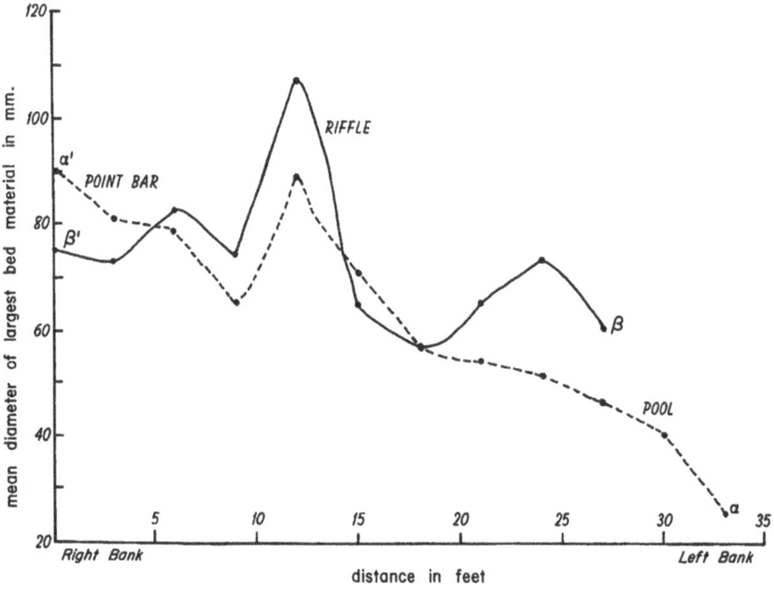

Figure 15 Lateral sorting of largest bed-load material across a point bar-pool and adjacent riffle. Location of the sampled areas (α-α' and β-β') are shown on Figure 4.

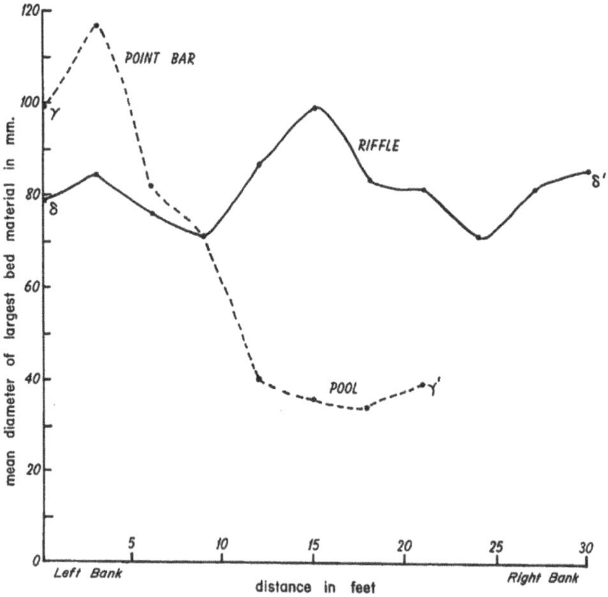

Figure 16 Lateral sorting of largest bed-load material through a point bar-pool and adjacent riffle. Location of the sampled areas (γ-γ' and δ-δ') are shown on Figure 4.

concentration and rate of transport of bed-load material must produce the areal sorting.

Why Pools and Riffles Form: Some Hypotheses

The formation of pools and riffles in alluvial streams appears to be a self-adjustment mechanism that helps minimize the time rate of potential energy expenditure per unit mass of water (Yang, 1971b). This is consistent with Dolling's (1968) conclusion that the pool-riffle sequence somehow maintains a balance in energy expended by the stream, and therefore is an important element in maintaining a state of quasi-equilibrium. However, neither of these authors was able to completely define conditions that allow the proposed energy relationships to be established in the pool-riffle sequence.

If it is assumed that stream length must increase to minimize the time rate of energy expenditure, then the morphology of the channel must facilitate such an energy distribution. Furthermore, if the only way a stream can increase its channel length, with the exception of headward erosion or progradation, is by lateral erosion which increases sinuosity, then pools, riffles, and point bars may have the most probable morphology to facilitate lateral migration while the time rate of energy expenditure is minimized. The only way channel length between two points can increase is for a stream to concentrate energy expenditure on one bank or the other. If this is done by converging flow, according to de Leliavsky's criterion, then the cross-section must be asymmetrical. This appears to violate Yang's (1971a) law of least time rate of energy expenditure. However, if the entire pool-riffle sequence is considered, the increase in energy expenditure may be compensated for by divergent flow on the riffles and point bars. Therefore, there may be a balance of the time rate of energy expenditure through the pool-riffle sequence which allows channel length to increase until a minimum value of time rate of energy expenditure is reached. The minimum value for any particular stream reach is determined by external constraints imposed by geology, discharge, valley slope, etc. This argument is valid only for relatively high flow conditions when the forms are being produced. It appears likely that constraints produced by the pool-riffle sequence may best satisfy two premises of the entropy concept: uniform distribution of energy, and minimum energy expenditure.

It is concluded from a consideration of the cause of pools and riffles that there is no general tendency for alluvial streams to meander. Meandering is merely one of the more probable ways by which a stream can minimize energy expenditure. The significant variables which tend to control channel pattern are mode of sediment transport, resistance of bed and bank material to erosion, and channel slope. Streams may minimize the time rate of potential energy loss per unit volume of water in three ways: 1) by decreasing slope (meandering); 2) by increasing channel width; and 3) various combinations of decreasing slope and increasing width (Yang, 1971a). Streams with easily erodable bank material (i.e., predominantly sand) will tend to minimize energy loss by increasing channel width, and therefore such streams will be characterized by low sinuosity and large width to depth ratio. Streams with more resistant bank material (i.e., with little sand) tend to minimize energy loss by decreasing slope (meandering), and therefore such streams will be characterized by high sinuosity and relatively small width to depth ratio. The assumption made

Figure 17 Downstream view of a riffle-pool sequence in Wea Creek near Lafayette, Indiana.

Figure 18 Close up view of the pool in Figure 17. The light area is the central part of the pool. There is no bed material in this pool, whereas the adjacent downstream riffle contains abundant course-grained bed material.

Figure 19 Upstream view of a pool in Dry Creek near Winters, California. This pool is scoured to older alluvium and contains essentially no bed material, whereas the adjacent downstream riffle contains abundant coarse-grained bed material.

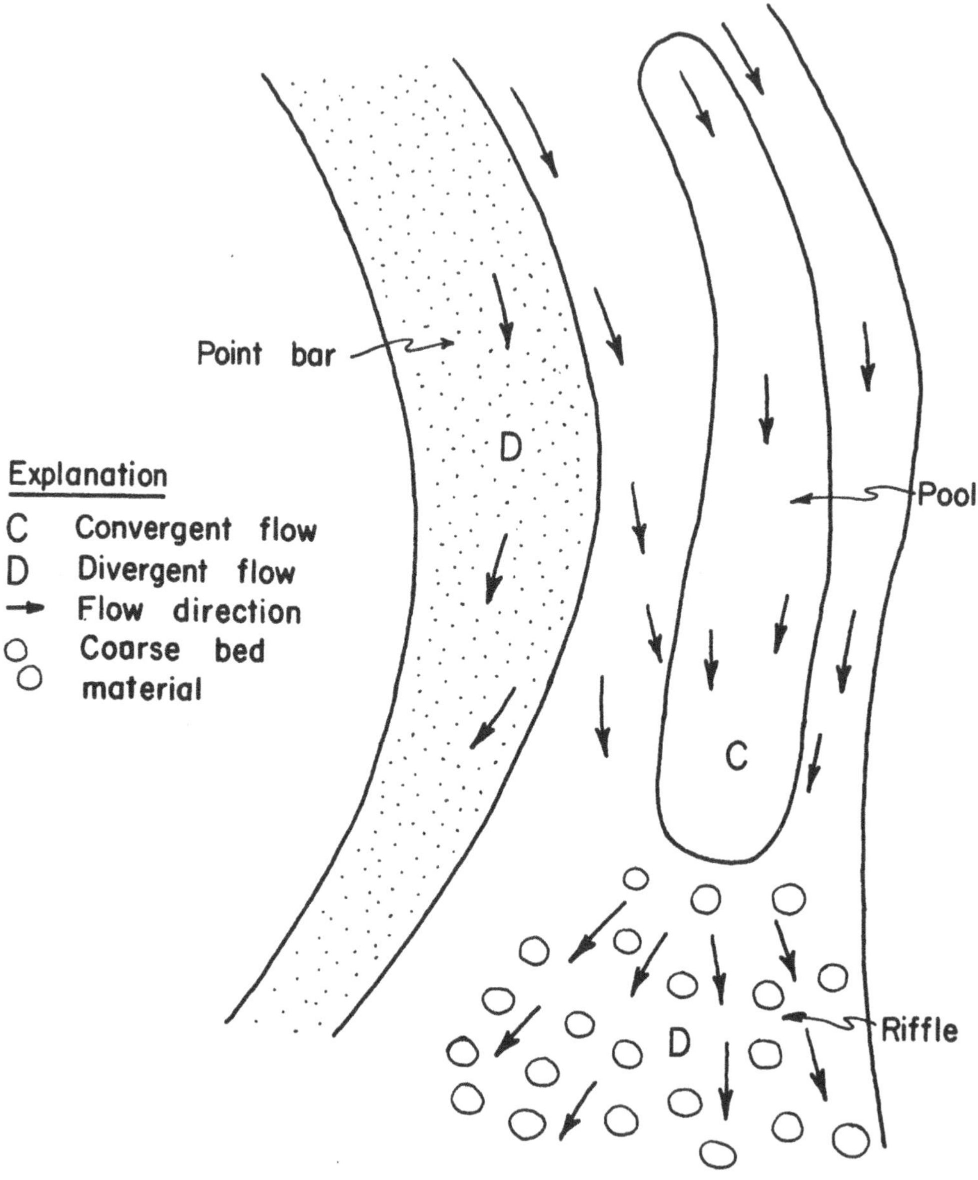

Figure 20 Idealized illustration of convergent and divergent flow which may have facilitated the development and maintenance of the pool and riffle in Figure 19.

is that channels with bed and bank material finer-grained or coarser-grained than sand will be morphologically more similar than channels with predominantly sandy banks. This results because sand apparently is easier to erode than either finer or coarser material. This argument, for channels with less than 20% gravel and coarser-grained material, is consistent with Schumm's (1968, pp. 37-41) classification of alluvial river channels.

CONCLUSION

The evidence necessary to understand alluvial streams can be found in the field; a pool with ripples on the stream bed, dunes superposed on a point bar at low flow, or observed relations between pool-riffle spacing and channel pattern all suggest that there is order in the genesis and spatial relations of the forms. Furthermore, although the study of morphology and morphometry can be utilized to increase our understanding of fluvial processes, we must remember that it is not necessarily one process which entirely produces the forms. For example, kinematic wave processes alone probably cannot produce pools and riffles, but kinematic processes combined with dispersive stress and convergent and divergent flow may indeed produce the observed forms. Conversely, although form can to some extent control process, as for example the tendency for scour and convergent flow in pools to be partly controlled by the asymmetric cross-profile, morphology alone cannot produce all observed processes. Therefore, from a philosophical view, we must accept that form and process evolve in harmony as feedback mechanisms inherent to open systems approaching dynamic equilibrium.

ACKNOWLEDGMENTS

We gratefully acknowledge the suggestions and critical reading of this manuscript by William B. Bull and Frank A. Packard. Stimulating discussions and helpful suggestions from James Brice, Charles G. Higgins and Marie Morisawa are also appreciated. We are indebted to John Sullivan, Gerald C. Kukal and Richard A. McBane for assistance in the field. Financial support was provided in part by the University of California, Deans Grant for Research, and the Water Resources Research Center, Purdue University, OWRR Project A-021-Indiana.

REFERENCES CITED

Bagnold, R.A., 1968, Deposition in the process of hydraulic transport; Sedimentology, v. 10, pp. 45-56.

Carey, W.C. and Keller, M.D., 1957, Systematic changes in the beds of alluvial rivers; Proc. Amer. Soc. Civil Engrs. 83 No. HY4, pp. 1331-1 to 24.

Coffman, D.M., 1972, Advances in the theory of classification and modeling of stream channel networks; unpublished Ph.D. Thesis, Purdue University, 214 pp.

de Leliavsky, Nicholas, 1894, Currents in streams and the formation of stream beds, Sixth International Congress on Internal Navigation, The Hague, The Netherlands; in Leliavsky, Serge, 1966, An Introduction to Fluvial Hydraulics; Dover Publications, Inc., New York, New York, 257 pp.

Dolling, Ray, 1968, Occurrence of pools and riffles: an element in the quasi-equilibrium state of river channels; Ontario Geography, No. 2, pp. 3-11.

Friedkin, J.F., 1945, A laboratory study of the meandering of alluvial rivers; Waterways Exp. Sta., U.S. Army Corps of Engrs., Vicksburg, Mississippi, 40 pp.

Keller, E.A., 1969, Form and fluvial processes of Dry Creek, near Winters, California; unpublished M.S. thesis, California University, Davis, California, 73 pp.

Keller, E.A., 1971a, Areal sorting of bed-load material: the hypothesis of velocity reversal; Geol. Soc. America Bull., v. 82, pp. 753-756.

Keller, E.A., 1971b, Pools, riffles, and meanders: discussion; Geol. Soc. America Bull., v. 82, pp. 279-280.

Keller, E.A., 1972a, Development of alluvial stream channels: a five stage model; Geol. Soc. America Bull., v. 83, pp. 1531-1536.

Keller, E.A., 1972b, Areal sorting of bed-load material: the hypothesis of velocity reversal: reply; Geol. Soc. America, v. 83, pp. 915-918.

Langbein, W.B., and Leopold, L.B., 1968, River channel bars and dunes - theory of kinematic waves; U.S. Geol. Survey Prof. Paper 422-L, 20 pp.

Leliavsky, S.B., 1966, An Introduction to Fluvial Hydraulics; Dover Publications, Inc., New York, New York, 257 pp.

Leopold, L.B., and Wolman, M.G., 1957, River channel patterns; braided, meandering, and straight; U.S. Geol. Survey Prof. Paper 282-B, pp. 53-59.

Leopold, L.B., Wolman, M.G., and Miller, J.P., 1964, Fluvial Processes in Geomorphology; W.H. Freeman and Company, San Francisco, 522 pp.

Lokhtine, 1909, Sur le Mechanisme de Lit Fluvia, St. Petersburg; Flamant, Hydraulique, Paris, p. 287, in Leliavsky, Serge, 1966, An Introduction to Fluvial Hydraulics; Dover Publications, Inc., New York, New York, pp. 8-10.

Maddock, Thomas, Jr., 1969, The behavior of straight open channels with movable beds; U.S. Geol. Survey Prof. Paper 622-A, 70 pp.

Quraishy, M.S., 1944, The origins of curves in rivers: Current Sci., v. 13, pp. 36-39.

Schumm, S.A., and Khan, H.R., 1972, Experimental study of channel patterns; Geol. Soc. America Bull., v. 83, pp. 1755-1770.

Schumm, S.A., 1968, River adjustment to altered hydraulic regimen - Murrumbidgee River and Paleochannels, Australia; U.S. Geol. Survey Prof. Paper 598, 65 pp.

Simons, D.B., and Richardson, E.V., 1966, Resistance to flow in alluvial channels; U.S. Geol. Survey Prof. Paper 422-J, 61 pp.

Teleki, P.G., 1972, Areal sorting of bed-load material: the hypothesis of velocity reversal: discussion; Geol. Soc. America, v. 83, pp. 911-914.

U.S. Geological Survey, 1969, Water Resources Data for Indiana, (prepared in cooperation with the U.S. Corps of Engrs. and the State of Indiana), p. 66.

Woldenberg, M.J., 1969, Spatial order in fluvial systems: Horton's Laws derived from mixed hexagonal hierarchies of drainage basin areas; Geol. Soc. America Bull., v. 80, pp. 97-112.

Wolman, M.G., 1954, A method of sampling coarse river-bed material; Am. Geophys. Union Trans., v. 35, No. 6, pp. 951-956.

Yang, T.C., 1971a, On river meanders; Journal of Hydrology, v. 13, pp. 231-253.

Yang, T.C., 1971b, Formation of riffles and pools; Water Resources Research, V. 7, pp. 1567-1574.

12

UNIT STREAM POWER IN DYNAMIC STREAM SYSTEMS

Chih Ted Yang and John B. Stall

ABSTRACT

As water flows downgradient, it spends its energy in transporting sediment and carving its own channel into a particular geometry and profile. During this process of landscape evolution, a stream system constantly adjusts its geometry and profile to keep a dynamic equilibrium between the rate of work done and the rate of load imposed. This dynamic equilibrium is maintained during the process of sediment transport with different stages of bed form development. Dynamic equilibrium can also be found during different stages of channel pattern development from straight to meandering to braided.

A stream system can be considered an open system under dynamic equilibrium. This model enables us to use the concept of entropy for the explanation of some morphological phenomena observed in stream systems. Observed data indicate that the average stream fall ratio between different order streams in a river basin is unity. The cumulative result of all the adjustments made by a stream is the minimization of its unit stream power in the downstream direction.

INTRODUCTION

As water flows downgradient, it releases its potential energy to transport sediment and carve its own channel. As this process continues, the channel geometry and longitudinal stream bed profile are constantly readjusted. The rate of sediment transported by running water also varies with respect to time and space. It is difficult for us to predict precisely the behavior of a stream at a particular time and location. However, if enough observations are made, we can safely say that even though a stream never stops adjusting its geometry and profile, its general pattern is predictable in a statistical sense.

The present paper attempts to explore the concept of dynamic equilibrium. This dynamic equilibrium is shown to exist in the consistent pattern of hydraulic geometry as well as during the process of sediment transport. The process of sediment transport and the general pattern of a stream system will be explained by the concept of unit stream power.

DYNAMIC EQUILIBRIUM

There are two different schools of thought in the study of the geomorphic processes of a stream system. The first school founded on Davis' (1889) idea of geographic cycle considers a stream system as a closed system. Hack (1960) made a good evaluation of Davis' theory. The second school considers a stream system as an open system and uses the concept of dynamic equilibrium to explain landscape evaluation. This paper emphasizes the physical meaning of dynamic equilibrium and its implications.

Figure 1 Channel pattern of the Skuna River, Mississippi, before and after channel improvement (courtesy of Brien R. Winkley).

A system is called an open system if the energy and material can be transported into and out of the system. A closed system is a special case of an open system in which the inflow and outflow of material and energy are zero (Chorley, 1962). A system is in a static equilibrium condition when there is an exact balance between opposing forces. This balance should be independent of time. A stream system cannot be maintained at a static equilibrium condition because a stream system never stops adjusting its geometry and profile. During the process of reaching its final static equilibrium (which may never occur), a stream adjusts itself in such a manner that an approximate balance is maintained between the rate of work done and the rate of load imposed. This situation is called dynamic equilibrium. A dynamic equilibrium stream system may make some minor adjustment of its geometry and profile from time to time. However, the statistical mean patterns remain almost unchanged with respect to time. If there is any outside interference which upsets this balance, a stream system will readjust itself to restore the balance.

Figure 1 shows an example of a natural stream's ability to return to its original dynamic equilibrium condition after a man-made change is made which interrupts the balance. This aerophoto shows the old meandering Skuna River above Grenada Reservoir, Mississippi, as well as the man-made straight channel which replaced the natural meandering river. Shortly after straightening the channel, sand bars developed along both channel banks. These sand bars created a meandering thalweg within the straight channel as shown by the sinuous dashed line. The alternate bars in the artificial channel are linked by dashed lines with corresponding bends of the original Skuna River. It is very interesting to note that the distances between these sand bars have the same values as the wave length of correlated meanders of the original river. This is an indication that the man-made straight channel is in the process of returning to its original natural condition after its dynamic equilibrium condition was upset by man.

A stream system can be considered as an open system in the sense that energy and material can be transported into and out of the system. Potential energy is released as water follows down-gradient through a stream system. This potential energy is used in carving channels and transporting sediment. For a unit mass of water flowing through a stream system, the difference between energy inflow and outflow is negligible when it is compared with the energy used in a stream system. Thus a stream system also has the basic property of a closed system in the sense that there is no net change of energy in the system. This is the consequence of the dynamic equilibrium of a stream system. There is no perfect model which can encompass the whole complex of a stream system. If our interest is to study the dynamic behavior of a stream system, the best approach is to consider it as an open system under dynamic equilibrium. This model enables us to assume that the total relief of a system is constant and the concept of entropy can be applied to the study of stream morphology. No detailed theoretical derivation shall be made in the remaining sections. The validity of using this model of open system under dynamic equilibrium shall be verified by observed data.

UNIT STREAM POWER AND SEDIMENT TRANSPORT

Running water has the ability to carry and transport sediment. Geologists are interested in the study of sediment transport because the whole stream denudation system observed today is the cumulative result of sediment transport. Engineers are also interested in this study because the rate that sediment is trans-

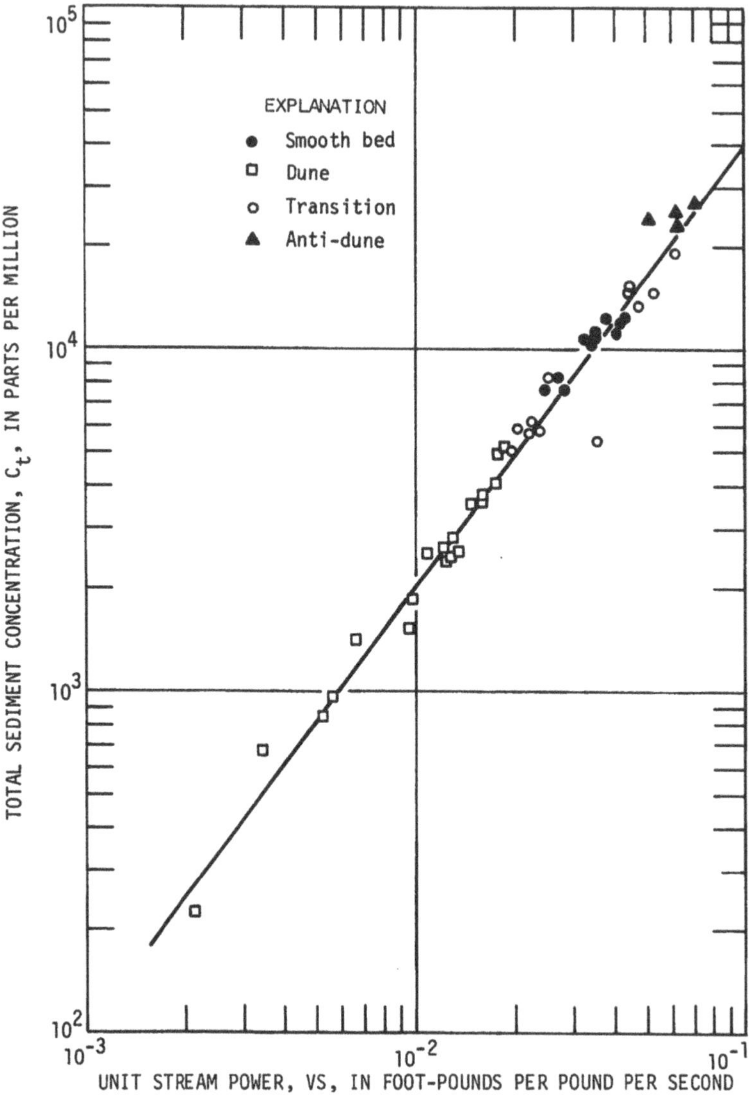

Figure 2 Relationship between total sediment concentration and unit stream power for Gilbert's test of 0.505 mm sand in 1.32-foot wide flume.

ported has a direct impact on the stability and longevity of engineering projects. The rate of sediment transport is related to many variables, such as water discharge, average velocity, energy slope, shear stress, etc. Most of the published equations are derived from the assumption that the rate of sediment transport can be determined from one or two of the mentioned variables. A review of existing data and literature (Yang, 1972) indicates that these assumptions hold true only under special conditions. Equations thus derived should not be considered as being general and their applicability is limited.

Another approach is to treat the process of sediment transport as a problem of dynamic equilibrium. In this case we shall take a unit mass of water into consideration. The rate of work done by this unit mass of water is proportional to its unit stream power which is defined as the time rate of potential energy expenditure per unit weight of water. The rate of load imposed on this unit mass of water is proportional to the total sediment concentration. According to our definition of dynamic equilibrium, there should be an approximate balance between total sediment concentration and unit stream power. Based on this concept, more than 1000 sets of laboratory data and field data were used to verify this relation (Yang, 1972, 1973). The equation

$$\log C_t = 5.435 - 0.286 \log \frac{\omega d}{\nu} - 0.457 \log \frac{U_*}{\omega}$$

$$+ (1.799 - 0.409 \log \frac{\omega d}{\nu} - 0.314 \log \frac{U_*}{\omega}) \log (\frac{VS}{\omega} - \frac{V_{cr}S}{\omega}) \quad (1)$$

was proposed by Yang (1973) for the prediction of total sediment concentration in alluvial channels, where C_t is the total sediment concentration in parts per million by weight, ω is the fall velocity of sediment particle, d is the average diameter of sediment, ν is the kinematic viscosity, U_* is the shear velocity, VS is the unit stream power, and $V_{cr}S$ is the unit stream power required at incipient motion. Equation 1 was shown to provide accurate predictions of total sediment concentrations for diversified flow and sediment conditions (Yang, 1973).

Because a stream in dynamic equilibrium can adjust its bed configuration and channel pattern, the approximate balance between the rate of work done and the rate of load imposed should be maintained through different stages of adjustment. Figure 2 shows the relationship between unit stream power and total sediment concentration for Gilbert's test of 0.505 mm sand in a 1.32-foot wide flume. This figure indicates that the approximate balance between unit stream power and sediment concentration is maintained through different stages of bed configuration development, i.e. smooth bed, dune, transition, and antidune. Schumm and Khan (1972) measured the total sediment concentration during the development of different channel patterns, from straight to meandering thalweg and, finally, to braided channels. Figure 3 indicates that the balance between total sediment concentration and unit stream power is maintained even if a freely developed channel changes its pattern.

It can be concluded that dynamic equilibrium is maintained through the process of sediment transport. As a result of sediment transport, channels are carved to form stream networks observed today. Data collected from natural streams will be used to show the general pattern of stream systems in the following sections. The concept of entropy shall be used to explain some of the phenomena observed in nature under the condition of dynamic equilibrium.

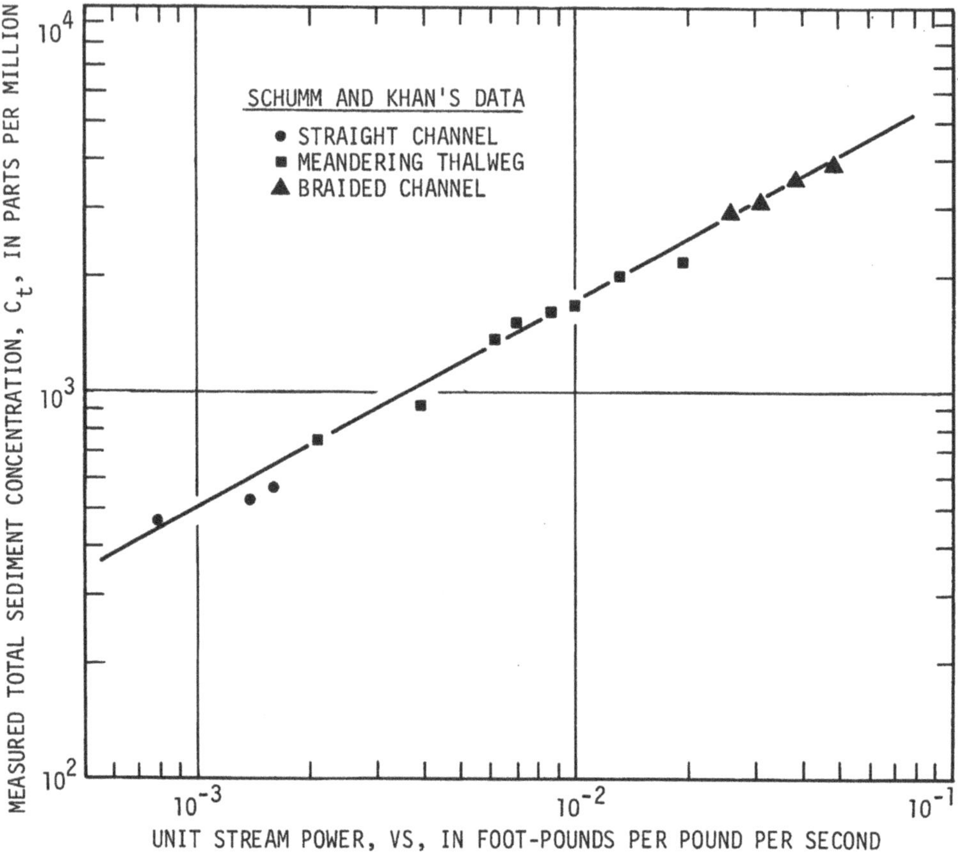

Figure 3 Relationship between total sediment concentration and unit stream power during the process of channel pattern development from straight to meandering to braided.

COMPOSITION OF STREAM SYSTEMS

About thirty years ago most geomorphologists studied stream systems on a descriptive basis which was concerned with the history of landscape evolution. This descriptive approach provides very little useful information for engineering practice. With the impetus given by Horton (1945), more geomorphologists abandoned the descriptive approach and adopted the quantitative approach for the study of landscape evolution. The first step in the quantification of river basin description is the designation of stream order. Many methods of ordering streams have been proposed. Each method has its advantages and disadvantages, depending on the purpose of the study. Any usefulness which the stream order system may have depends on the premise that on the average, if a sufficiently large sample is treated, the order number is directly proportional to the relative dimensions of certain parameters which describe some basic characteristics of a stream system. Some of these parameters are number of streams, stream length, stream slope, stream fall, drainage area, etc.

Strahler (1957) designated the first order stream as the smallest unbranched stream shown on a map. Where two streams of the same order join, the stream order is increased by one. Strahler's stream order is used in this study for two reasons. First, Strahler's stream order provides a systematic description of a river basin in accordance with Horton's law of stream numbers, law of stream lengths, law of stream slopes, and law of drainage areas. Secondly, observed data by using Strahler's stream order are consistent with the concepts of entropy and dynamic equilibrium.

By the use of Strahler's stream order, the number of streams, the average stream lengths, and average stream slopes for different order streams in the Rogue River Basin were measured from 15-minute (scale 1:62,500) and 2-degree (scale 1:250,000) topographical maps. There results are plotted on figure 4. Horton's law of stream numbers, law of average stream length, and law of average stream slope, can be described by the following equations:

$$\ln N_u = A - Bu \tag{2}$$

$$\ln L_u = C - Du \tag{3}$$

$$\ln S_u = E - Fu \tag{4}$$

where u is the stream order, N_u is the number of u-th order streams, L_u is the average length of u-th order streams, S_u is the average slope of u-th order streams, and A, B, C, D, E, and F are coefficients. Equations 2, 3, and 4 are shown as straight lines on figure 4. From equations 2, 3, and 4, the following parameters can be obtained for the Rogue River Basin:

$$\text{bifurcation ratio of stream number} = e^B = N_u/N_{u+1} = 3.72 \tag{5}$$

$$\text{stream length ratio} = e^D = L_u/L_{u+1} = 0.48 \tag{6}$$

$$\text{stream concavity} = e^F = S_u/S_{u+1} = \frac{Y_u/L_u}{Y_{u+1}/L_{u+1}} = 2.12 \tag{7}$$

where Y_u is the average stream fall of the u-th order streams. The three parameters shown in equations 5, 6, and 7 are independent of the scale of maps from which these parameters were obtained (Yang and Stall, 1971).

If stream systems observed today are the cumulative result of erosion and sediment transport under the condition of dynamic equilibrium, and if the concept of entropy can be applied to

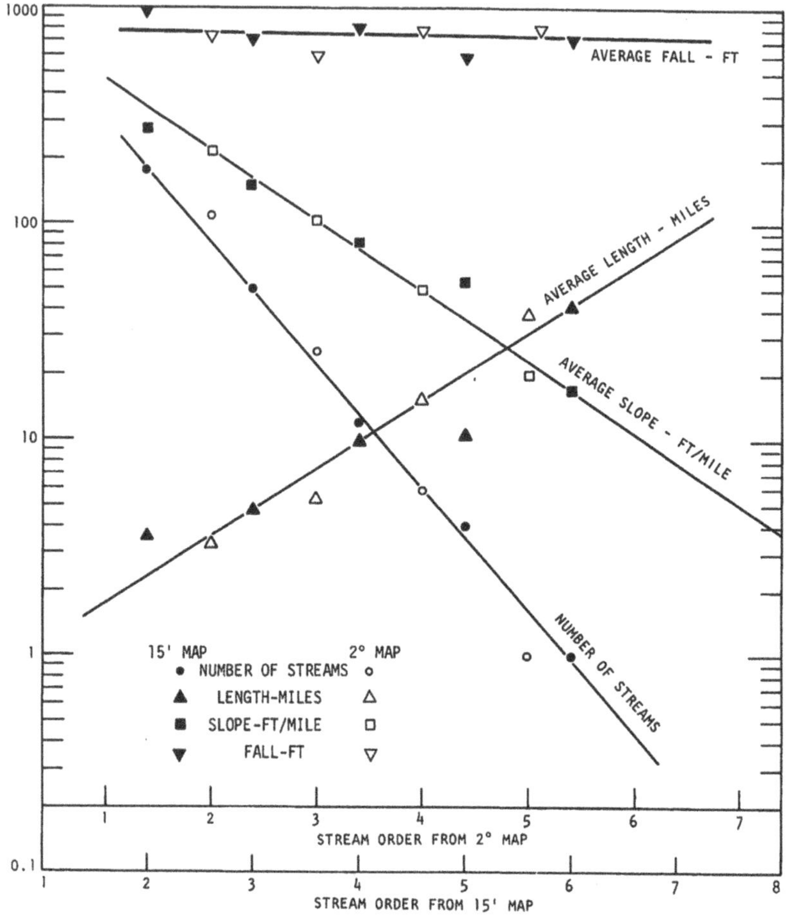

Figure 4 The relationships among number of streams, average length, average slope, average fall, and Strahler's stream order, for the Rogue River basin in Oregon.

stream systems, it can be shown theoretically that average stream fall ratio between any two different order streams in the same river basin should be unity (Yang, 1971). The average stream falls of different order streams are plotted on figure 4. The average stream fall ratio can be obtained from the product of stream length ratio and stream concavity, i.e.

$$\text{stream fall ratio} = Y_u/Y_{u+1} = e^D e^F = 1.02 \tag{8}$$

which is very close to unity. That the amount of average stream fall in a basin is approximately equal was first discovered by Broscoe (1959) in his study of six river basins in the United States. Data collected from 14 river basins in the United States (Yang, 1971) confirmed that the average stream fall for each order stream was equal within a basin. These results indicate that the concepts of entropy and dynamic equilibrium can provide a meaningful explanation of the structure of a stream system if an appropriate method of ordering streams is used.

HYDRAULIC GEOMETRY AND STREAMBED PROFILE

As a cumulative result of running water and sediment transport, the land surface has been carved into channels of particular cross-sectional shapes and longitudinal profiles. These channels are called streams or rivers. The close relationships among discharge, channel width, average water depth, velocity, and suspended load were noticed by Leopold and Maddock (1953). They found that width, depth, velocity, and suspended load can be expressed by power functions of water discharge. They used the term 'hydraulic geometry' to describe these relationships. Stall and Fok (1968) studied the hydraulic geometry of Illinois streams and found a consistent pattern among water discharge, flow frequency, and drainage area. They modified Leopold and Maddock's hydraulic geometry equations into the following general form.

$$\ln X = H + IF' + J \ln A_d \tag{9}$$

where X represents a variety of parameters including water discharge Q, channel cross-sectional area A, average flow velocity V, channel width W, or average depth D; F' is the flow frequency; A_d is the drainage area; and H, I, J are coefficients. Equation 9 enables us to study the variation of variable X with respect to time (flow frequency F') and location (drainage area A_d). Observed data indicates that equation 9 gives an adequate description of the general pattern of hydraulic geometry in rivers of Illinois as well as those in the United States (Stall and Yang, 1970). According to the concept of dynamic equilibrium close relationships should exist among unit stream power, total sediment concentration, channel geometry, and longitudinal streambed profile. The variations of unit stream power in stream systems will be shown here.

The hydraulic geometry equation for the average velocity is

$$\ln V = H + IF' + J \ln A_d \tag{10}$$

where V is the average velocity in feet per second, F' is flow frequency, and A_d is the drainage area in square miles. According to Horton's law of drainage area and law of average stream slope

$$\ln A_d = K + Mu \tag{11}$$

and

$$\ln S_u = E - Fu \tag{4}$$

where K, M, E, F are coefficients, and S_u has the dimension of feet per mile. By combining equations 4, 10, and 11, we have the

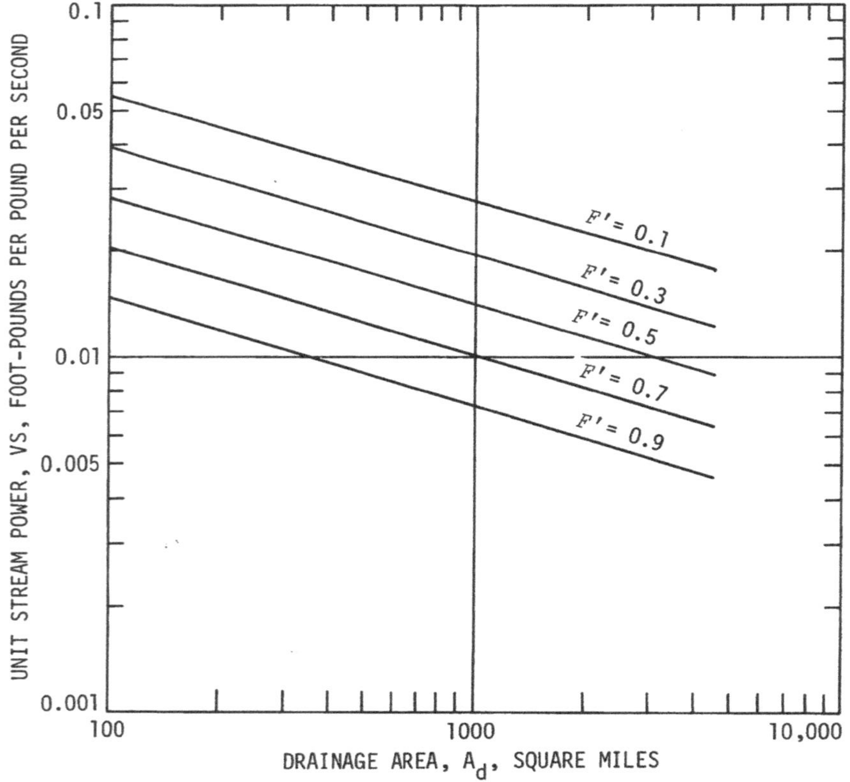

Figure 5 Relationships among unit stream power, drainage area, and flow frequency for the Upper Colorado River basin in Colorado.

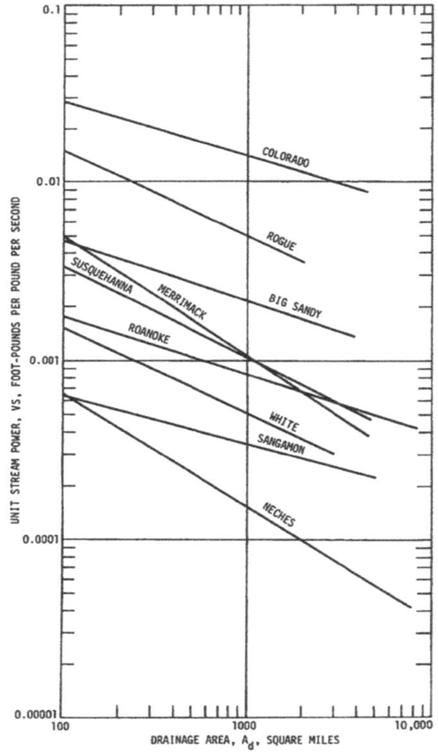

Figure 6 Unit stream power at median flows as related to drainage area for 9 stream systems of the United States.

hydraulic geometry-unit stream power equation.

$$VS = \frac{\exp\,(E - FK/M + H + IF')}{5280} A_d^{(J + F/M)} \qquad (12)$$

In equation 12, the unit stream power VS has the dimension of foot-pounds per pound per second, and the drainage area A_d has the dimension of square miles. The coefficients in equation 12 obtained for 9 river basins (Stall and Yang, 1972) which follow Horton's laws very well are tabulated in Table 1. Coefficients E, F, K, and M were obtained from 2-degree topographic maps. These are the only maps available with a common scale for all the river basins shown in Table 1. Coefficients, H, I, and J were obtained from the regression analyses of velocity data measured at U.S. Geological Survey gaging stations.

By using the coefficients shown in Table 1, the variation of unit stream power in a river basin can be computed by equation 12. Figure 5 is an example of the graphical presentation of equation 12 for the Upper Colorado River Basin. Figure 5 shows that unit stream power decreases with increasing drainage area and flow frequency. Figure 6 is plotted for the median flow condition in accordance with equation 12 by using the data shown in Table 1. These results indicate that on the average, unit stream power always decreases in the downstream direction. These actual data

Table 1 Coefficients for Hydraulic Geometry--Unit Stream Power Equations for 9 River Systems of the United States

River and state	E	F	H	I	J	K	M	J+E/M
Merrimack, NH	7.17	-0.96	0.59	-1.60	0.09	-0.85	1.27	-0.67
Susquehanna, NY	5.47	-0.93	0.37	-1.57	0.14	0.28	1.43	-0.51
Roanoke, NC	4.92	-0.74	-0.08	-0.88	0.13	-1.51	1.64	-0.32
Big Sandy, KY	6.48	-0.89	-0.12	-1.90	0.22	-1.20	1.59	-0.34
White, IN	4.42	-0.85	0.22	-1.60	0.11	0.78	1.45	-0.48
Sangamon, IL	3.23	-0.62	-1.01	-0.95	0.26	1.30	1.19	-0.26
Neches, TX	4.50	-0.90	0.03	-1.04	0.04	0.26	1.33	-0.64
Colorado, CO	7.49	-0.75	0.40	-1.66	0.16	-1.42	1.62	-0.30
Rogue, OR	6.90	-0.75	0.30	-1.21	0.16	0.0	1.18	-0.48

from river basins are considered as strong evidence of the existence of the law of least time rate of energy expenditure (Yang, 1971). This requires that a stream to minimize or reduce its unit stream power in the downstream direction whenever it is permitted to do so by its environment. Because the law of least time rate of energy expenditure was derived from concepts of entropy and dynamic equilibrium, and because equation 12 represents the result of the overall adjustments of hydraulic geometry and river profile in a river basin, it can be concluded that the concepts of entropy and dynamic equilibrium are indeed applicable to the study of stream morphology.

According to the concept of dynamic equilibrium a decrease of unit stream power should be accompanied by a decrease of total sediment concentration. This relationship is governed by equation 1. The decrease of unit stream power in the downstream direction explains the reason why sediment concentration should decrease in the downstream direction as was observed by Leopold and Maddock (1953).

The law of least time rate of energy expenditure requires a stream to minimize its unit stream power in the downstream direction. Observed data (Leopold and Maddock, 1953; Stall and Fok, 1968; and Stall and Yang, 1970) indicate that the average velocity increases slightly in the downstream direction. In order to reduce the velocity-slope product in the downstream direction, an increase of velocity must be accompanied by a reduction of slope. This decrease of slope in the downstream direction explains the reason why the general longitudinal streambed profile should be concave.

SUMMARY AND CONCLUSIONS

The morphological process in a stream system is very complex. Because of the large number of variables involved in this process, and because some of these variables are interrelated, it is extremely difficult for us to predict precisely the behavior of a natural stream at a particular time and location. Yet observed data indicate that there are some general patterns which are predictable in a river basin. The authors consider a stream system as an open system under a dynamic equilibrium condition. The concept of dynamic equilibrium between the time rate of work done and the time rate of load imposed enables us to establish a reliable relationship between unit stream power and total sediment concentration transported in natural streams. Observed data, when using Strahler's stream order, indicate that the average fall ratio between different order streams in a river basin is very close to unity. The net effect of all the adjustments made by a river is to minimize its unit stream power in the downstream direction. These phenomena suggest that the concept of entropy can be applied to the study of stream morphology in a stream system.

REFERENCES

Broscoe, A. J., 1959, Quantitative analysis of longitudinal stream profiles of small watersheds: Office of Naval Research Technical Report No. 18, Dept. of Geology, Columbia University, N. Y., 73p.

Chorley, R. J., 1962, Geomorphology and general systems theory: U.S. Geological Survey Professional Paper 500-B, 10p.

Davis, W. M., 1889, The rivers and valleys of Pennsylvania: Natl. Geog. Mag., v. 1, p. 183-253.

Gilbert, G. K., 1914, The transportation of debris by running water: U.S. Geol. Survey Prof. Paper 86, 263p.

Horton, R. E., 1945, Erosional development of streams and their drainage basins; Hydrophysical approach to quantitative morphology: Geol. Soc. of America Bull., v. 45, p. 273-370.

Hack, J. T., 1960, Interpretation of erosional topography in humid temperate regions: American Jour. of Sci., v. 258-A, p. 80-97.

Leopold, L. B., and Maddock, T. Jr., 1953, The hydraulic geometry of stream channels and some physiographic implications: U.S. Geol. Survey Prof. Paper 252, 57p.

Schumm, S. A., and Khan, H. R., 1972, Experimental study of channel patterns: Geol. Soc. of America Bull., v. 83, p. 1755-1770.

Stall, J. B., and Fok, Y. S., 1968, Hydraulic geometry of Illinois streams: University of Illinois Water Resources Center Research Report No. 15, Urbana, Ill. 47p.

Stall, J. B., and Yang, C. T., 1970, Hydraulic geometry of 12 selected stream systems of the United States: University of Illinois Water Resources Center Research Report No. 32, Urbana, Ill. 73p.

Stall, J. B., and Yang, C. T., 1972, Hydraulic geometry and low streamflow regimen: University of Illinois Water Resources Center Research Report No. 54, Urbana, Ill. 31p.

Strahler, A. N., 1957, Quantitative analysis of watershed geomorphology: Transactions, American Geophysical Union, v. 38, No. 6, p. 913-920.

Yang, C. T., 1971, Potential energy and stream morphology: Water Resources Research, v. 7, No. 2, p. 311-322.

Yang, C. T., and Stall, J. B., 1971, Note on the map scale effect in the study of stream morphology: Water Resources Research, v. 7, No. 3, p. 709-712.

Yang, C. T., 1972, Unit stream power and sediment transport: ASCE Jour. of the Hydraulics Division, v. 98, No. HY10, p. 1805-1826.

Yang, C. T., 1973, Incipient motion and sediment transport: ASCE Jour. of the Hydraulics Division, v. 99, HY10, p. 1679-1704.

13

GEOMORPHIC THRESHOLDS AND COMPLEX RESPONSE OF DRAINAGE SYSTEMS

S. A. Schumm

ABSTRACT

The alluvial and morphologic details of drainage systems are much too complex to be explained by progressive erosion alone. Within the constraints of the erosion cycle these complexities (terraces, alluvial deposits) must be explained by external variables such as climatic, tectonic, isostatic or land-use changes. However, field and experimental research into the details of fluvial landform development indicate that some abrupt modifications of such a system can be inherent in its erosional development and that two additional concepts are required for comprehension of drainage system evolution. These are 1) geomorphic thresholds and 2) complex response of drainage systems.

INTRODUCTION

Due to the complexity of Quaternary climatic and tectonic histories, topographic and stratigraphic discontinuities can be conveniently explained as a result of climatic and tectonic events. In this way the compulsion to fit geomorphic and stratigraphic details into a Quaternary chronology is satisfied, as is the basic scientific need to identify cause and effect. As the details of Holocene stratigraphic and terrace chronologies are studied, a bewildering array of changes are required to explain the behavior of a drainage system. In fact, it is now accepted that some major erosional adjustments can be induced by rather insignificant changes in the magnitude and frequency of storm events (Leopold, 1951).

The numerous deviations from an orderly progression of the erosion cycle has led many to discount the cycle concept completely. Current practice is to view the evolutionary development of the landscape within the conceptual framework of the erosion cycle, but to consider much of the modern landscape to be in dynamic equilibrium. There are obvious shortcomings in both concepts. For example, although the cycle involves continuous slow change, evidence shows that periods of relatively rapid system adjustment results from external causes. This, of course, is equally true of geomorphic systems in dynamic equilibrium. That is, for a change to occur in either the cycle or a system in dynamic equilibrium there must be an application of an external stimulus. Hence landscape changes and changes in rates of depositional or erosional processes are explained by the influence of man, by climatic change or fluctuations, by tectonics, or by isostatic adjustments.

One cannot doubt that major landscape changes and shifting patterns of erosion and deposition have been due to climatic change and tectonic influences and that man's influence is substantial. Nevertheless, it is the details of the landscape, the last inset fill, the low Holocene terrace, modern periods of arroyo cutting and gulleying, alluvial-fan-head trenching, channel aggradation and slope failure that for both scientific and practical reasons of land management require explanation and prediction. These geomorphic details are of real significance, but often they cannot be explained by traditional approaches.

Another aspect of the problem is that within a given region all landforms did not respond to the last external influence in the same way and, indeed, some have not responded at all. This is a major geomorphic puzzle that is commonly ignored. If land systems are in dynamic equilibrium conponents of the system should respond in a similar way to an external influence. Hence, the effects of hydrologic events of large magnitude should not be as variable as they appear to be.

The cyclic and dynamic equilibrium concepts are not of value in the location of incipiently unstable landforms because within these conceptual frameworks system change is always due to external forces. There is now both experimental and field evidence to indicate that this need not be true. The answer lies in the recognition of two additional concepts that are of importance for an understanding of landscape development. These may be termed 1) geomorphic thresholds and 2) complex response of geomorphic systems. Basically, they suggest that some geomorphic anomalies are, in fact, an inherent part of the erosional development of landforms and that the components of a geomorphic system need not be in phase.

These concepts are certainly not new, and one need not search long to find a geomorphic paper or book mentioning thresholds or the complexities of geomorphic systems (for example, Chorley and Kennedy 1971, Tricart 1965, Pitty 1971). However, these concepts have not been directed to the solution of specific problems nor has their significance been fully appreciated. The assumption that all major landform changes or changes in the rates and mechanics of geomorphic processes can be explained by climatic or tectonic changes has prevented the geologist from considering that landform instability may be inherent.

THRESHOLDS

Thresholds have been recognized in many fields and their importance in geography has been discussed in detail by Brunet (1968). Perhaps the best known to geologists are threshold velocities required to set in motion sediment particles of a given size. With a continuous increase in velocity, threshold velocities are encountered at which something begins and with a progressive decrease in velocity, threshold velocities are encountered at which something ceases, in this case, sediment movement. These are Brunet's (1968, pp. 14 and 15) "thresholds of manifestation" and "thresholds of extinction", and they are the most common types of thresholds encountered. When a third variable is involved, Brunet (1968, p. 19) identified "thresholds of reversal." An example of this is Hjulstrom's (1935) curve showing the velocity required for movement of sediment of a given size. The curve shows that velocity decreases with particle size until cohesive forces become significant, and then the critical velocity increases with decreasing grain size. Another example of this type of relationship is the Langbein-Schumm (1958) curve which shows sediment yield as directly related to annual precipitation and runoff until vegetative cover increases sufficiently to retard erosion. At this point there is a decrease in sediment yield with increased runoff and precipitation. Perhaps thresholds is not a good word to describe the critical zones within which these changes occur, but it is a simple and easily understood term.

The best known thresholds in hydraulics are described by the Froude and the Reynolds numbers, which define the conditions at which flow becomes supercritical or turbulent. Particularly spectacular are the changes in bed form characteristics at threshold values of stream power.

In the examples cited, an external variable changes progressively thereby triggering abrupt changes or failure within the affected system. Responses of a system to an external influence occur at what will be referred to as extrinsic thresholds. That is, the threshold exists within the system, but it will not be crossed and change will not occur without the influence of an external variable.

Thresholds can also be exceeded when input is relatively constant, that is, the external variables remain relatively constant, yet a progressive change of the system itself renders it unstable, and failure occurs. For example, it has been proposed that the progressive erosion of a region will cause short but dramatic periods of isostatic adjustment. With an essentially constant rate of denudation a condition is reached when isostatic compensation is necessary, and this probably takes place during a short period of relatively rapid uplift (Schumm 1963). Somewhat analagous to this is the long period of rock weathering and soil development required before a catalytic storm event precipitates mass movement. Following failure a long period of preparation ensues before failure can occur again (Tricart, 1965, p. 99). These intrinsic thresholds are probably common in geologic systems, but it is only geomorphic examples that will be considered here.

A geomorphic threshold is one that is inherent in the manner of landform change; it is a threshold that is developed within the geomorphic system by changes in the system itself through time. It is the change in the geomorphic system itself that is most important, because until the system has evolved to a critical situation, adjustment or failure will not occur. It may not always be clear whether the system is responding to geomorphic thresholds or to an external influence, but when a change of slope is involved, the control is geomorphic, and the changes whereby the threshold is achieved is intrinsic to the system.

EVIDENCE FOR GEOMORPHIC THRESHOLDS

Recent field and experimental work support the concept of geomorphic thresholds, and it has been used to explain the distribution of discontinuous gullies in the oil-shale region of western Colorado and to explain channel pattern variation along the Mississippi River.

Discontinuous gullies

Field studies in valleys of Wyoming, Colorado, New Mexico, and Arizona revealed that discontinuous gullies, short but troublesome gullied reaches of valley floors, can be related to the slope of the valley-floor surface (Schumm and Hadley, 1958). For example, the beginning of gully erosion in these valleys tends to be localized on steeper reaches of the valley floor. Carrying this one step farther, with the concept of geomorphic thresholds in mind, it seems that for a given region of uniform geology, land use and climate, a critical threshold valley slope will exist above which the valley floor is unstable. In order to test this hypothesis, measurements of valley-floor gradient were made in the Piceance Creek Basin of western Colorado. The area is underlain by oil shale, and the potential environmental problems that will be associated with the development of this resource are considerable.

Within this area, valleys were selected in which discontinuous gullies were present. The drainage area above each gully was measured on maps, and valley slopes were surveyed in the field. No records of runoff or flood events exist so drainage basin area was selected as a variable, reflecting runoff and flood discharge. When valley slope is plotted against drainage area, the relationship is inverse (Fig. 1), with gentler valley slopes being characteristic of large drainage areas. As a basis for comparison, similar measurements were made for valleys in which there were no gullies, and these data are also plotted on Fig. 1. The lower range of slopes of the unstable valleys coincide with the higher range slopes of the stable valleys. In other words, for a given drainage area it is possible to define a valley slope above which the valley floor is unstable.

Note that the relationship does not pertain to drainage basins smaller than about four square miles. In these small basins variations in vegetative cover, which are perhaps related to the aspect of the drainage basin or to variations in the properties of the alluvium, prevent recognition of a critical threshold slope. Above four-square miles there are only two cases of stable valley floors that plot above the threshold line, and one may conclude that these valleys are incipiently unstable and that a major flood will eventually cause erosion and trenching in these valleys.

Using Fig. 1, one may define the threshold slope above which trenching or valley instability will take place in the Piceance Creek area. This has obvious implications for land management for, if the slope at which valleys are incipiently unstable can be determined, corrective measures can be taken to artificially stabilize such critical reaches, as they are identified.

It seems possible that future work will demonstrate that similar relationships can be established for other alluvial deposits. For example, trenching of alluvial fans is common,and the usual explanation for fan-head trenches is renewed uplift of the mountains or climatic fluctuations. However, the concept of geomorphic thresholds should also be applicable to this situation. That is, as the fan grows it may steepen until it exceeds a threshold slope, when trenching occurs. Preliminary results from experimental studies of alluvial-fan growth reveal that periods of trenching do alternate with deposition at the fan head.

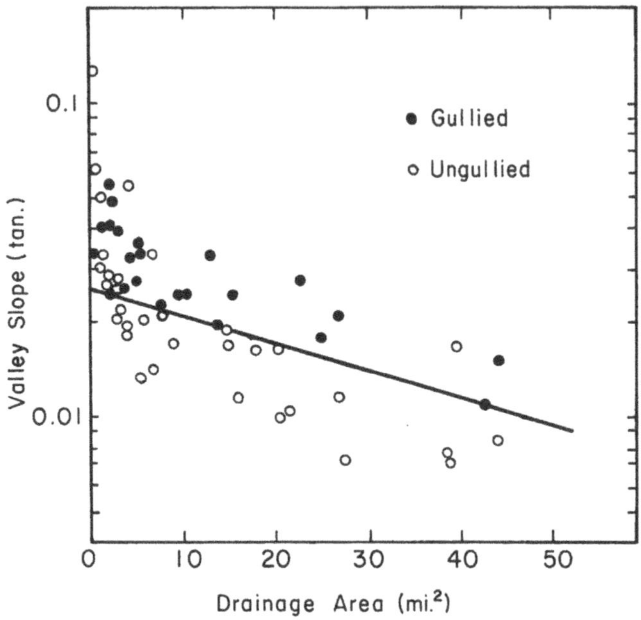

Figure 1 Relation between valley slope and drainage area, Piceance Creek basin, Colorado. Line defines threshold slope for this area (from P. C. Patton, 1973, unpublished M.S. thesis).

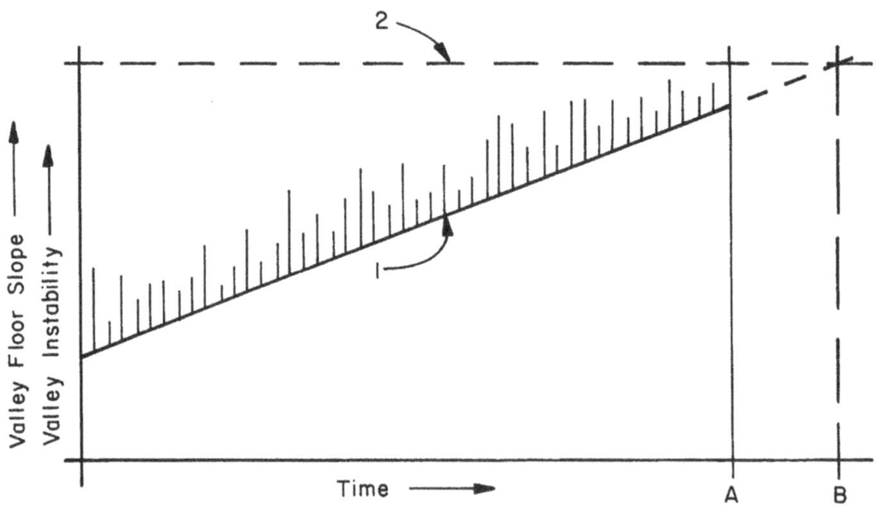

Figure 2 Hypothetical relation between valley-floor gradient and valley-floor instability with time. Superimposed on line 1, representing an increase of valley-floor slope, are vertical lines representing instability of the valley floor as related to flood events. When the ascending line of valley-floor slope intersects line 2 representing the maximum slope at which the valley is stable, failure or trenching of the valley alluvium will occur at time B. However, failure occurs at time A, as the apparent direct result of a major storm or flood event.

The concepts of thresholds as applied to alluvial deposits in the western U.S.A. is illustrated by Fig. 2, where the decreasing stability of an alluvial fill is represented by a line indicating increase of valley slope with time. Of course, a similar relation would pertain if, with constant slope, sediment loads decrease slowly with time. Superimposed on the ascending line of increasing slope are vertical lines showing the variations of valley floor stability caused by flood events of different magnitudes. The effect of even large events is minor until the stability of the deposit has been so reduced by steepening of the valley gradient that during one major storm, erosion begins at time A. It is important to note that the large event is only the most apparent cause of failure, as it would have occurred at time B in any case.

These studies of alluvial deposits in dry lands suggest that large infrequent storms can be significant but only when a geomorphic threshold has been exceeded. It is for this reason that high-magnitude, low-frequency events may have only minor and local effects on a landscape.

River patterns

Experimental studies of river meandering have been performed by both hydraulic engineers and geologists over many years. Such a study was designed to investigate the influence of slope and sediment loads on channel patterns (Schumm and Khan, 1972). It was found, during these experiments in which the water discharge was held constant, that if a straight channel was cut in alluvial material at a very low slope, the channel would remain straight. However, at steeper slopes the channel meandered. As the slope of the alluvial surface on which the model stream was flowing (valley slope) steepened, the velocity of flow increased, and shear forces acting on the bed and bank of the miniature channel increased. At some critical value of shear, bank erosion and shifting of sediments on the channel floor produced a sinuous course. The conversion from straight to sinuous channel at a given discharge occurred at a threshold slope (Fig. 3). As slope increased beyond this threshold, meandering increased until at another higher threshold the sinuous channel became a straight braided channel. The experiments revealed that there is not a continuous change in stream patterns with slope from straight through meandering to braided, but rather the changes occur at threshold slopes. Slope in this case is an index of sediment load and the hydraulic character of the flow in the channel; nevertheless, the relationship can be used to explain the variability of stream patterns.

If, in fact, the slope of the valley floor of a river varies due to the sediment contribution from tributaries or due to variations in deposition during the geologic past, then the river should reflect these changes of valley slope by changes in pattern. A comparison of the experimental results with Mississippi River patterns was made possible by data obtained from the Vicksburg District, Corps of Engineers. These data show that variations in channel pattern of the Mississippi River are related to changes of the slope of the valley floor (Fig. 4). Variations in valley slope reflect the geologic history of the river, and if today the valley slope exceeds a geomorphic threshold the river will show a dramatic change of pattern.

When the valley slope is near a threshold, major flood events will significantly alter the stream pattern. This conclusion has bearing on the work of Wolman and Miller (1960), concerning the geomorphic importance of events of high magnitude. They concluded that, although a major amount of work is done by events of moderate magnitude and relative frequent occurrence, nevertheless, the large storm or flood may have a major role in landscape modification. However, evidence on the influence of rare and large events on the landscape is equivocal. Major floods have destroyed the flood plain of Cimarron River (Schumm and Lichty, 1963), but equally large events have not significantly affected the Connecticut River (Wolman and Eiler, 1958).

These and other observations indicate that a major event may be of major or minor importance in landscape modification, and an explanation of the conflicting evidence requires further consideration of the threshold

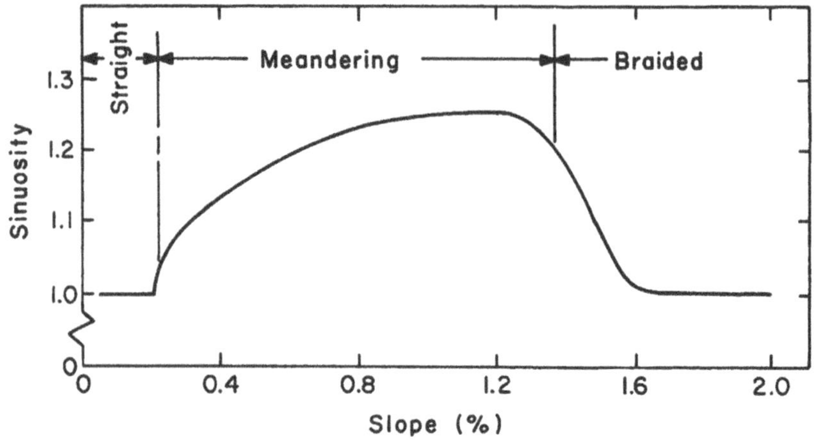

Figure 3 Relation between valley slope and sinuosity (ratio of channel length to length of flume or length of valley) during experiments. The change from a straight to a sinuous pattern, and from a sinuous to a braided pattern occurs at two threshold slopes. The absolute value of slope at which such changes occur will be influenced by discharge. Discharge was maintained at 0.15 cfs during the experiments. (After Schumm, S. A. and Khan, H. R., 1972.)

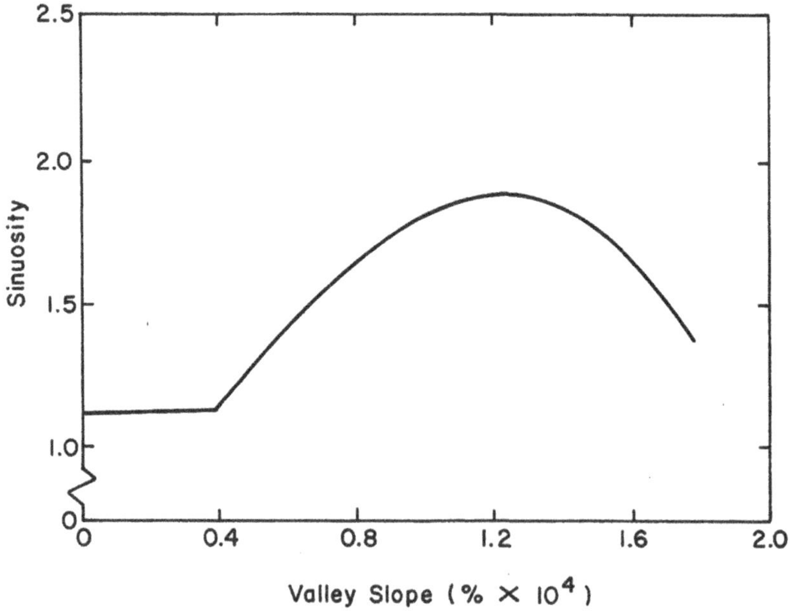

Figure 4 Relation between valley slope and sinuosity between Mississippi River between Cairo, Illinois and Head of Passes, Louisiana. Data from Potamology Section, U. S. Army Corps of Engineers, Vicksburg, Mississippi. Scatter about curve reflects natural variation of sinuosity. (After Schumm, et al., 1972)

concept. Some landscapes or components of a landscape have apparently evolved to a condition of geomorphic instability and these landforms fail; that is, depending on their development, they will be significantly modified by a large infrequent event whereas others will be unaffected. Therefore, there will be, even within the same region, different responses to the same conditions of stress.

When within a landscape some conponents fail by erosion whereas others do not, it is clear that erosional thresholds have been exceeded locally. One of the most significant problems of the geomorphologist or land manger is the location of incipiently unstable components of a landscape. The recognition of geomorphic thresholds within a given region will be a significant contribution to both an understanding of the details of regional morphology as well as providing criteria for identification of unstable land forms. At least one location, Piceance Creek basin, such a threshold has been identified. It remains to be seen how successful the concept can be applied elsewhere. It is very possible that major geomorphic thresholds associated with alluvial deposits will be identified most readily in subhumid, semiarid and arid regions, where the stabilizing influence of vegetation is least effective. However, slope stability and river pattern thresholds will exist in humid and tropical regions.

COMPLEX RESPONSE OF GEOMORPHIC SYSTEMS

Geomorphic histories tend to be complicated, and considering the climatic changes of the past few million years of earth history, one would expect them to be so. For example, throughout the world, geologists and archaeologists have studied the details of the most recent erosional and depositional history of valleys. This consists of identifying the sequence in which alluvial deposits were emplaced and then eroded. Because of worldwide climate changes during the Quaternary, it is reasonable to assume that alluvial chronologies applicable to large regions can be established. That is, a particular alluvial layer should be identifiable regionally and correlations of these deposits over large areas can then be made. There is no question that major climatic changes have affected erosional and depositional events, but when the alluvial chronologies of the last 10 to 15 thousand years are examined, it is not convincing that each event was in response to one simultaneous external change. In fact, investigations in southwestern United States reveal that during the last 10,000 years the number, magnitude and duration of erosional and depositional events in the valleys of this region not only varied from valley to valley but also varied within the same valley. For example, Kottlowski, Cooley, and Ruhe (1965) describe the situation as follows:

> "Late recent time is represented by epicycles of erosion and alluviation in the canyons and valleys of the southwest; however, the number, magnitude, and duration of the events differ from basin to basin and along reaches of the same stream."

This situation is so complex that correlations of terrace surfaces and alluvial fills over large areas seem impossible, but in the search for order, correlations are made. Haynes (1968) summarized the results of extensive radiocarbon dating of alluvial deposits in southwestern United States. His data demonstrate that, during the last 5000 years of record, there is significant temporal overlap among the three most recent alluvial deposits. This indicates that deposition was not in phase everywhere and that apparently deposition did not occur in response to a single event. Part of the complexity, at least, as related to the most recent events, may be explained by the threshold concept with erosion occurring, as described for the Piceance Creek area, when a geomorphic threshold is exceeded. However, within one region not all valleys will achieve the threshold at the same time.

There is, in addition, another explanation. We know very little about the response of a drainage system to rejuvenation. Climatic change, uplift, or lowering of baselevel can cause incision of a channel. This incision will convert the flood plain to a terrace, the geomorphic evidence of an erosional episode. However, a drainage system is composed

Figure 5   Drainage evolution research facility at Colorado State University. Simulated storms can be generated by a sprinkler system over the 9 x 15 m container, and the erosional development of small drainage systems can be studied.

of channels, hillsides, divides, flood plains, and terraces; it is complex. The response of this complex system to change will also be complex. That this is true was demonstrated during an experimental study of drainage system evolution at Colorado State University (Schumm and Parker, 1973).

During experimentation, a small drainage system (Fig. 5) was rejuvenated by a slight (10 cm) change of base level. As anticipated, base-level lowering caused incision of the main channel and development of a terrace (Fig. 6a,b). Incision occurred first at the mouth of the system, and then progressively upstream, successively rejuvenating tributaries and scouring the alluvium previously deposited in the valley (Fig. 6b). As erosion progressed upstream, the main channel became a conveyor of upstream sediment in increasing quantities, and the inevitable result was that aggradation occurred in the newly cut channel (Fig. 6c). However, as the tributaries eventually became adjusted to the new baselevels, sediment loads decreased, and a new phase of channel erosion occurred (Fig. 6d). Thus, initial channel incision and terrace formation was followed by deposition of an alluvial fill, channel braiding, and lateral erosion, and then, as the drainage system achieved stability, renewed incision formed a low alluvial terrace. This low surface formed as a result of the decreased sediment loads when the braided channel was converted into a better-defined channel of lower width-depth ratio. The low surface was not a flood plain because it was not flooded at maximum discharge.

Somewhat similar results were obtained by Lewis (1949) in a pioneering experiment performed in a small wooden trough four m. long and 50 cm wide. Lewis cut a simple drainage pattern in sediment (four parts sand, one part mud). He then introduced water at the head of the flume into both the main channel and into two tributaries. The main channel debouched onto a "flood plain" before entering the "sea" (tail box of flume).

During the experiment, the break in slope or knickpoint at the upstream edge of the "flood plain" eroded back, rejuvenating the upstream drainage system. Initially, erosion in the head waters was rapid as the channels adjusted, and deposition occurred on the "flood plain," thereby increasing its slope. As the upstream gradients were decreased by erosion and the stream courses stabilized, sediment supply to the "flood plain" decreased. Because of the reduction of the sediment load the stream cut into the alluvial deposits in the upper part of the flood plain to form a terrace. Lewis concludes that "perhaps the most significant fact revealed by the...experiment is that terraces were built in the lower reaches without any corresponding change of sea levels, tilt or discharge." From our results and those of Lewis, it seems that an event causing an erosional response within a drainage basin (tilting, changes of base level, climate and/or land use) automatically creates a negative feedback (high sediment production) which results in deposition; this is eventually followed by incision of alluvial deposits as sediment loads decrease.

A similar sequence of events may be under way in Rio Puerco, a major arroyo in New Mexico, as well as in other Southwestern channels. For example, the dry channel of Rio Puerco, although previously trenched to depths of 13 m, is now less than four m deep near its mouth. This is due to deposition caused by very high sediment loads produced by the rejuvenated drainage system.

Within a complex natural system, one event can trigger a complex reaction (morphologic and/or stratigraphic) as the components of the system respond progressively to change. This principle provides an explanation of the complexities of the alluvial chronologies, and it suggests that an infrequent event, although performing little of the total work within a drainage system, may, in fact be the catalyst that causes the crossing of a geomorphic threshold and the triggering of a complex sequence of events that will produce significant landscape modification.

CONCLUSIONS

Although we continually speak and write about the complexity of geomorphic systems, nevertheless, we constantly simplify in order to understand these systems. The experimental studies discussed in this

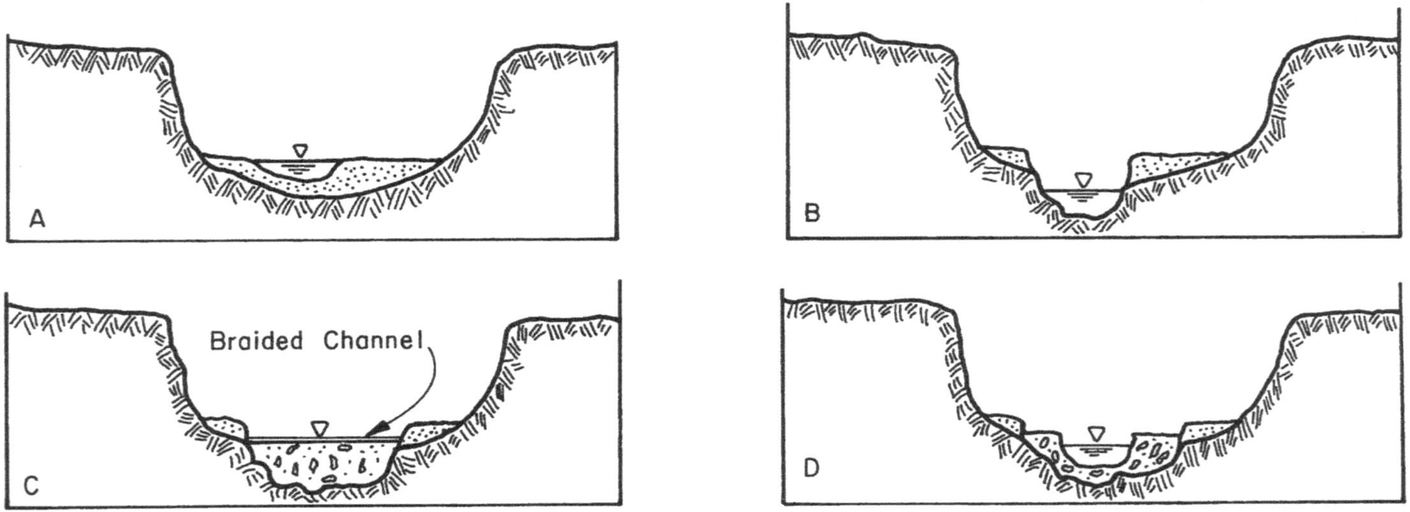

Figure 6

Diagrammatic cross sections of experimental channel 1.5 m from outlet of drainage system (base level) showing response of channel to one lowering of base level.

A Valley and alluvium, which was deposited during previous run, before base level lowering. The low width-depth channel flows on alluvium.

B After base level lowering of 10 cm, channel incises into alluvium and bedrock floor of valley to form a terrace. Following incision, bank erosion widens channel and partially destroys terrace (Figure 2C).

C An inset alluvial fill is deposited, as the sediment discharge from upstream increases. The high width-depth ratio channel is braided and unstable.

D A second terrace is formed as the channel incises slightly and assumes a low width-depth ratio in response to reduced sediment load. With time, in nature, channel migration will destroy part of the lower terrace, and a flood plain will form at a lower level.

paper are good examples of this approach. Simplification and the search for order in simplicity caused instrinsic thresholds to be overlooked in preference to explanations based on external controls. For example, the explanation of the details of the Holocene record and the distribution of modern erosion and depositional features in the western United States may well depend on our understanding of thresholds and complex response. These concepts do not conflict with the cycle of erosion or with the concept of dynamic equilibrium; rather, they supplement them.

It need not be true that landscape discontinuities or what appear to be abrupt changes in the erosional evolution of drainage systems must always be related to external influences. The evolution of land forms, at least in semiarid and arid regions, need not be progressive in the sense of constant and orderly development; in fact, change may occur both progressively and by saltation; that is, by jumps from one dynamic equilibrium to a new one. Obviously, hydrologic and meteorological events are discontinuous (Tricart, 1962) but even if they occur at a constant rate, changes in the geomorphic system itself (changing landform morphology and sediment loads) will cause abrupt adjustments of the system due to the existence of intrinsic thresholds.

It is very possible that, without the influence of external variables and over long time spans progressive erosion reduction of a landscape will be interrupted by periods of rapid readjustment, as geomorphic thresholds are exceeded. Readjustment of the system will be complex as morphology and sediment yields change with time. The timing of these changes unquestionably will be related to major flood or storm events, but such events, as emphasized earlier, may be only the catalyst that induces the change at a particular time. That is, it is the existence of geomorphic thresholds, and the complex feedback response of geomorphic systems, that permit high magnitude events to play a major role in landscape evolution.

Newtonian physical principles are utilized by engineers to control the landscape and by geomorphologists to attempt an explanation for the inception, evolution, and character of geomorphic systems. These physical laws apply to natural geomorphic situations, but their predictive power is reduced by the complexity of the field situation. For example, an increase in gravitational forces would probably not everywhere cause an equal acceleration in erosional rates. That is, increasing stress may not produce commensurate strain, but local failures will occur. Thus, the application of stress over time will not everywhere have the same result especially as the system to which stress is applied is itself changing through time. The logical consequence of the above situation as outlined in this paper is that high magnitude events will not everywhere produce dramatic erosional events; rather the result depends on the character of the geomorphic system.

The importance of this approach to the investigation of landforms is in its potential for application to prediction of landform response to both natural and man-induced change. The fact that, at least locally, geomorphic thresholds of instability can be defined quantitatively indicates that they can be identified elsewhere and then used as a basis for recognition of potentially unstable landforms in the field. This approach provides a basis for preventive erosion control. Using geomorphic principles the land manager can spend his limited funds in order to prevent erosion rather than spending it in a piecemeal fashion to attempt to restore seriously eroding areas to their natural conditions.

ACKNOWLEDGMENTS

The research reviewed herein has been supported by the National Science Foundation, Army Research Office and the Colorado Agricultural Experiment Station. William Weaver reviewed an early draft of this paper.

REFERENCES

Brunet, Roger, 1968, Les phénomenes de discontinuité en geographie: Centre de Recherches et Documentation Cartographic et Geographique. v. 7, 117 p. Editions du Centre National de la Recherche Scientifique.

Chorley, R. J. and Kennedy, B. A., 1971, Physical geography, a systems approach: London, Prentice-Hall International, 370 p.

Hayes, C. V., Jr., 1968, Geochronology of late-Quaternary alluvium: in- Means of correlation of Quaternary succesions (editors: H. E. Wright, Jr., and R. B. Morrison), Salt Lake City, University of Utah Press, p. 591-631.

Hjulstrom, Filip, 1935, Studies of the morphological activity of rivers as illustrated by the River Fyris: University Upsala Geological Institute, v. 25, p. 221-527.

Kottlowski, F. E., Cooley, M. E. and Ruhe, R. V., 1965, Quaternary geology of the Southwest: in Quaternary of the United States (editor: H. E. Wright, Jr.), Princeton, Princeton University Press, p. 287-298.

Langbein, W. B. and Schumm, S. A., 1958, Yield of sediment in relation to mean annual precipitation: Amer. Geophys. Union, Trans., v. 39, p. 1076-1984.

Leopold, L. B., 1951, Rainfall frequency: An aspect of climatic variation: Amer. Geophys. Union, Trans., v. 32, p. 347-357.

Lewis, W. V., 1944, Stream trough experiments and terrace formation: Geol. Mag. v. 81, p. 241-253.

Pitty, H. F., 1971, Introduction to geomorphology: London, Methuen and Company, 526 p.

Schumm, S. A., 1963, The disparity between present rates of denudation and orogeny: U. S. Geol. Survey Prof. Paper 454-H, 13 p.

______, and Hadley, R. F., 1957, Arroyos and the semarid cycle of erosion: Amer. Jour. Sci., v. 255, p. 161-174.

______, and Lichty, R. W., 1963, Channel widening and floodplain construction along Cimarron River in Southwestern Kansas: U. S. Geol. Survey Prof. Paper 352-D, p. 71-88.

______, and Khan, H. R., 1972, Experimental study of channel patterns: Geol. Soc. America Bull., v. 85, p. 1755-1770.

______, et al., 1972, Variability of river patterns: Nature (Physical Science) v. 237, p. 75-76.

______, and Parker, R. S., 1973, Implications of complex response of drainage systems for Quaternary alluvial stratigraphy: Nature (Physical Science) v. 243, p. 99-100.

Tricart, Jean, 1962, Les discontinuités dans les phénomenes d'erosion: International Assoc. Scientific Hydrology Pub. 59, Comm. d'erosion Continentale, p. 233-243.

Tricart, Jean, 1965, Principés et Méthods de Géomorphologie: Paris, Masson et Cie, 495 p.

Wolman, M. G., and Eiler, J. P., 1958, Reconaissance study of erosion and deposition produced by the flood of August 1953 in Connecticut: Amer. Geophys. Union, Trans., v. 39, p. 1-14.

______, and Miller, J. P., 1960, Magnitude and frequency of forces in geom